Text Book of

Design of Steel Structure

(Subject Code CIV 604)

For

Semester VI

Third Year Diploma in Civil Engineering Group

As Per New Syllabus of SBTE, Jharkhand

R. R. Gadpal

M.E. (Structures), M.I.E., MISTE
Applied Mechanics Deptt.,
Government Polytechnic, AMRAVATI

NIRALI PRAKASHAN
ADVANCEMENT OF KNOWLEDGE

N4731

Design of Steel Structure (SBTE - Sem. VI) ISBN 978-93-89944-26-6

First Edition	:	February 2020
©	:	Authors

Published By :
NIRALI PRAKASHAN

Abhyudaya Pragati, 1312, Shivaji Nagar

Off J.M. Road, PUNE – 411005

Tel - (020) 25512336/37/39, Fax - (020) 25511379

Email : niralipune@pragationline.com

➤ DISTRIBUTION CENTRES

PUNE

Nirali Prakashan : 119, Budhwar Peth, Jogeshwari Mandir Lane, Pune 411002, Maharashtra

(For orders within Pune) Tel : (020) 2445 2044, 66022708, Mobile : 9657703145

Email : niralilocal@pragationline.com

Nirali Prakashan : S. No. 28/27, Dhayari, Near Asian College Pune 411041

(For orders outside Pune) Tel : (020) 24690204; Mobile : 9657703143

Email : bookorder@pragationline.com

MUMBAI

Nirali Prakashan : 385, S.V.P. Road, Rasdhara Co-op. Hsg. Society Ltd.,

Girgaum, Mumbai 400004, Maharashtra; Mobile : 9320129587

Tel : (022) 2385 6339 / 2386 9976

Email : niralimumbai@pragationline.com

➤ DISTRIBUTION BRANCHES

JALGAON

Nirali Prakashan : 34, V. V. Golani Market, Navi Peth, Jalgaon 425001, Maharashtra,

Tel : (0257) 222 0395, Mob : 94234 91860; Email : niralijalgaon@pragationline.com

KOLHAPUR

Nirali Prakashan : New Mahadvar Road, Kedar Plaza, 1st Floor Opp. IDBI Bank, Kolhapur 416 012

Maharashtra. Mob : 9850046155; Email : niralikolhapur@pragationline.com

NAGPUR

Nirali Prakashan : Above Maratha Mandir, Shop No. 3, First Floor,

Rani Jhanshi Square, Sitabuldi, Nagpur 440012, Maharashtra

Tel : (0712) 254 7129; Email : pratibhabookdistributors@gmail.com

DELHI

Nirali Prakashan : 4593/15, Basement, Agarwal Lane, Ansari Road, Daryaganj

Near Times of India Building, New Delhi 110002 Mob : 08505972553

Email : niralidelhi@pragationline.com

BENGALURU

Nirali Prakashan : Maitri Ground Floor, Jaya Apartments, No. 99, 6th Cross, 6th Main,

Malleswaram, Bengaluru 560003, Karnataka; Mob : 9449043034

Email: niralibangalore@pragationline.com

Other Branches : Hyderabad, Chennai

niralipune@pragationline.com | www.pragationline.com

Also find us on www.facebook.com/niralibooks

Preface ...

This book is written as per New syllabus pattern prescribed by S.B.T.E. for 6th Semester (3rd Year) Diploma in Civil Engineering Students.

Topics are on Introduction, Limit State Design and Design of Connections and Detailing. Design of Tension Member by L.S.M., Design of Compression Members and Column Bases by L.S.M., Slab base and Gusseted base, Design of Flexural Members for BM and SF by L.S.M. and Steel Roof Truss and Plastic Analysis. The various topics dealt in this book are concise and self-contained with maximum possible pictorial illustrations for easy understanding and clear conception.

While writing this book efforts have been made to make this book student oriented. The subject matter has been presented in simple language and illustrated by incorporating a good number of solved, unsolved and well-graded examples which have been asked in earlier S.B.T.E. Examinations.

All the chapters are enlarged and huge number of problems which were asked in MSBTE examinations are solved and at the end of each chapter an exercise is given which will quench the thirst of knowledge of students preparing for the examination as well as for the completion of term work by their own efforts. We feel profound pleasure that the key for solution for almost each and every problem in exercise is given at the end for which lot of pains have been taken.

The authors take this opportunity to thank their colleagues and friends for their constant inspiration, valuable guidance and suggestions. The authors are very much thankful to publisher to bring out this book within a short time. Special thanks to Mr. Akbar Shaikh (D.T.P.), Mrs. Anagha Kaware (Proof Reading), Miss Chaitali Takle (Fig. Drawing)

Any suggestions for the improvement of this book are welcome and highly appreciated and will be incorporated in the next edition.

Author

■■■

Syllabus ...

1. Introduction   [03 Hours]

- Advantages and Disadvantages of Steel as Construction Material. Types of Sections, Grades of Steel (IS 2062) and Strength Characteristics; Use of Steel Table (SP6-Part1); Types of Loads on Steel Structure and Its I. S. Code Specification. Methods of Design and Comparison between them.

2. Limit State Design   [04 Hours]

- Basis for Design - Classification of Limit States - Characteristic and Design Actions - Ultimate and Design Strengths - Partial Safety Factors for Loads and Materials - Factors Governing the Ultimate Strength: Stability, Fatigue and Plastic Collapse - Serviceability.

3. Design of Connections and Detailing   [12 Hours]

- General - Types of Connections - Bolted, Riveted and Welded connections - Rigid and Flexible Connections - Components of Connections - Basic Requirements of Connections - Clearance for Holes - Minimum and Maximum Spacing of Fasteners - Minimum Edge/End Distances - Requirements of Tacking Fasteners. Bolted Connection - Types of Bolts - Bearing Type Bolts - Nominal and Design Shear Strengths of Bolts - Reduction Factors for Long Joints, Large Grip Lengths, Thick Packing Plates - Nominal and Design Bearing Strengths of Bolts - Reduction Factors for Over Sized and Slotted Holes - Nominal and Design Tensile Strengths (Tension Capacity) of Bolts-Simple Problems. Welded Connection- Types of Welds - Fillet Welds - Minimum and Maximum Sizes - Effective Length of Weld - Fillet Welds on inclined Faces - Design Strengths of Shop/Site Welds - Butt Welds - Effective Throat Thickness and Effective Length of Butt Weld - Simple Problems. Design Problems related to Eccentric Riveted/Bolted and Welded Connections.

4. Design of Tension Member by L.S.M.   [06 Hours]

- Tension Members - Effective Length and Effective Sectional Area of Tension Members - Design Strength of Tension Members against Yielding of Gross Section Requirements: Deflection Limits, Vibration, Durability and Fire Resistance, against Rupture of Critical Section and due to Block Shear. Problems on Determination of Design Strength of given Members and Designing Tension Members using Rolled Steel Sections for given Loads - Design of Bolted/Riveted and Welded Connections for Tension Members -Problems.

5. Design of Compression Members and Column Bases by L.S.M.   [12 Hours]

- Compression Members - Effective Length and Effective Sectional Area of Compression Members - Design Stress and Design Strength - Buckling Class of Cross-Sections - Imperfection Factor - Stress Reduction Factor - Thickness of Elements - Analysis and Design of Axially Loaded Column. Introduction to Lacing and Battening (No Numerical Problem on Lacing and Battening)
- Slab Base and Gusseted Base - Code Provisions (IS:800-2007) - Minimum Thickness and Effective Area of Base Plate - Design of Slab Base for Axially Loaded Columns using Bolts/Riveted/Welds. Introduction to Gusseted Base (No Numerical Problems on Gusseted Base).

6. Design of Flexural Members for BM and SF by L.S.M.   [07 Hours]

- General - Effective Span of Beams, Design Strength of Bending, (Flexure), Limiting Deflection of Beams - Design of Laterally supported Simple Beams for Bending Moment and Shear Force using Single / Double Rolled Steel Sections (Symmetrical Cross-sections only) - Problems. Names of various Components of Plate Girder and their Functions with Usual IS Recommendations.

7. Steel Roof Truss   [08 Hours]

- Types of Steel Roof Truss and its Selection Criteria. Calculation of Panel Point Load for Dead Load; Live Load and Wind Load as per I.S. 875-1987 Analysis and Design of Steel Roof Truss. Design of Angle Purlin as per I. S. Arrangement of Members at Supports.

8. Plastic Analysis   [12 Hours]

- Plastic Analysis: Analysis of Steel Structures - Methods - Elastic, Plastic and Advanced method of Analysis based on IS: 800-2007 - Idealized Stress Vs Strain curve - Problems. For Structural Steel - Requirements and Assumptions of Plastic Method of Analysis - Formation of Plastic Hinges in Flexural Members - Plastic Moment of Resistance and Plastic Modulus of Sections - Shape Factors of Rectangular / Circular/ I / T-Sections - Collapse Load.

■■■

Contents ...

■■■

Chapter **1**

INTRODUCTION

Syllabus

- Advantages and Disadvantages of Steel as Construction Material. Types of Sections, Grades of Steel (IS 2062) and Strength Characteristics; Use of Steel Table (SP6-Part1); Types of Loads on Steel Structure and Its I. S. Code Specification. Methods of Design and Comparison between them.

About this Chapter

After reading this chapter students can understand:

- Uses of Steel
- Steel Structure and
- Types of Sections

1.1 INTRODUCTION

- Structural steel has been used in the construction of structures for well over a century.

- It is perhaps the most versatile of structural materials and has been used extensively in the construction of multi-storeyed buildings, railways, bridges, industrial structures, transmission towers, overhead tanks, chimneys, bunkers, silos etc.

1.2 ADVANTAGES AND DISADVANTAGES OF STEEL AS A CONSTRUCTION MATERIAL

Advantages:

1. Extensively useful for large span industrial structures, bridges, towers and communication networks, steel overhead tanks.
2. Steel has many good mechanical properties like malleability, ductility, elasto-plasticity (i.e. more ultimate strength and too large strains).
3. It is most appropriate material to construct earthquake resistant structures due to more ductile nature.
4. It is easy to fabricate by riveting or welding to any desired shape.
5. Can sustain tension, compression, shear bending and torsional forces.
6. Pre-fabricated parts of steel can be transported and erected at site which results in speedy construction, saving in time and expenses.
7. The steel structures can be dissembled and reused and recycled.
8. Repairs and retrofitting and strengthening of steel structures is much simpler.
9. Steel is gas resistant.

Disadvantages:

1. Steel is a very costly material.
2. It is susceptible to corrosion and hence requires corrosion treatment periodically.
3. It loses their strength rapidly during fire.
4. It requires skill labour for erection.
5. Creates noise and requires electricity during connection of members.

1.3 TYPES OF SECTIONS USED

- The following types of sections are standardized by the Indian Standards Institution.

1. I – Sections:

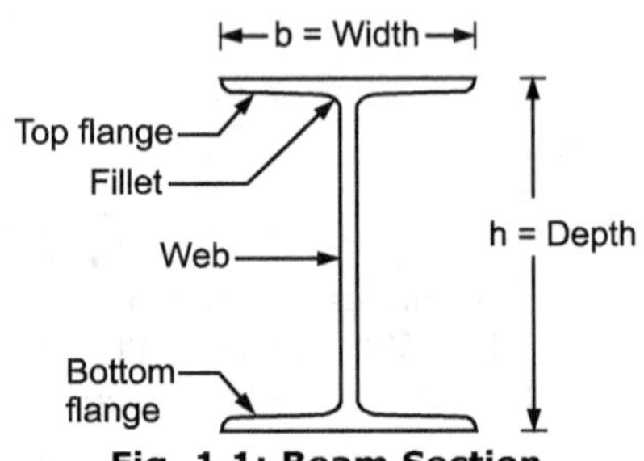

Fig. 1.1: Beam Section

  (a) Indian Standard Junior Beams (ISJB).

  (b) Indian Standard Light Beams (ISLB).

  (c) Indian Standard Medium Weight Beams (ISMB).

  (d) Indian Standard Wide Flange Beams (ISWB).

  (e) Indian Standard Column Sections (SC) or (ISSC).

  (f) Indian Standard Heavy Weight Beams (ISHB).

- All above I-sections are designated with the depth of the respective section in mm.

- For example, ISLB 200 where 200 indicates the depth in milimetres. These sections are also known as Rolled Steel Joists (R.S.J.).

2. Channel Sections

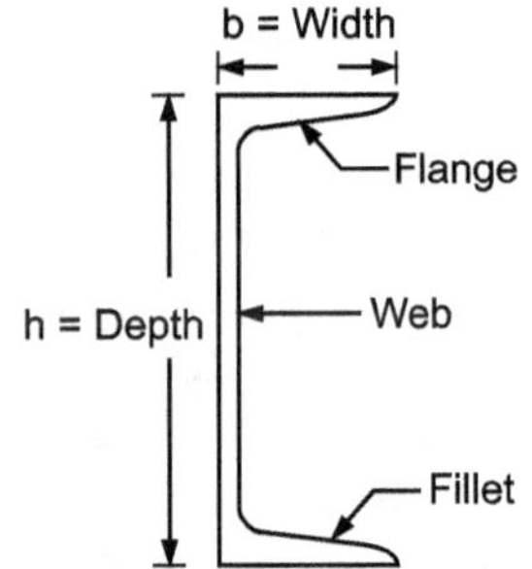

Fig. 1.2: Channel Section

  (a) Indian Standard Junior Channel (ISJC).

  (b) Indian Standard Light Channel (ISLC).

  (c) Indian Standard Medium Weight Channel (ISMC).

  (d) Indian Standard Medium Weight Channel with parallel flange (MCP).

3. Angle Sections

  (a) Indian Standard Equal Angles.

  (b) Indian Standard Unequal Angles.

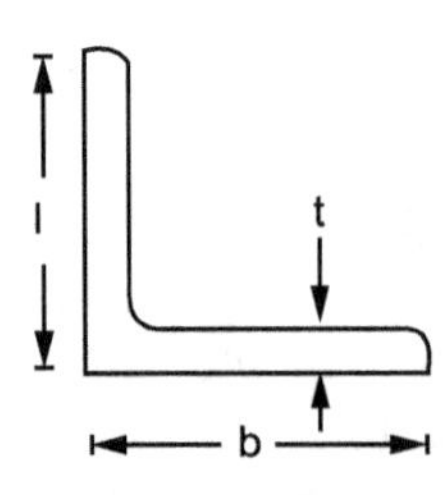

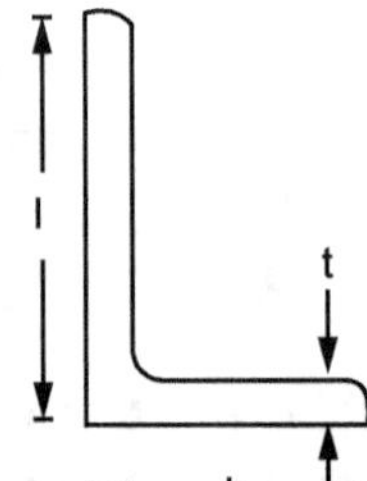

l = Length of longer leg   b = Width of shorter leg

(a) Equal Angle Section (l = b)   **(b) Unequal Angle Section ($l \neq$ b)**

Fig. 1.3

- Angle sections are designated by abbreviation ISA alongwith the lengths of both legs and their thickness.

- For example, ISA 50 × 50 × 6 mm means an equal-angle section of 6 mm thickness and with both legs 50 mm long.

4. Tee Sections

  (a) Indian Standard Normal Tee bars – (ISNT).

  (b) Indian Standard Wide Flange Tee bars – (ISHT).

  (c) Indian Standard Long Legged Tee bars – (ISST).

(d) Indian Standard Light Tee bars – (ISLT).

(e) Indian Standard Junior Tee bars – (ISJT).

(f) Indian Standard Deep Legged Tee bars – (ISDT).

(g) Indian Standard Slit Medium Weight Tee bars – (ISMT).

(h) Indian Standard Slit Tee bars from I sections – (ISHT).

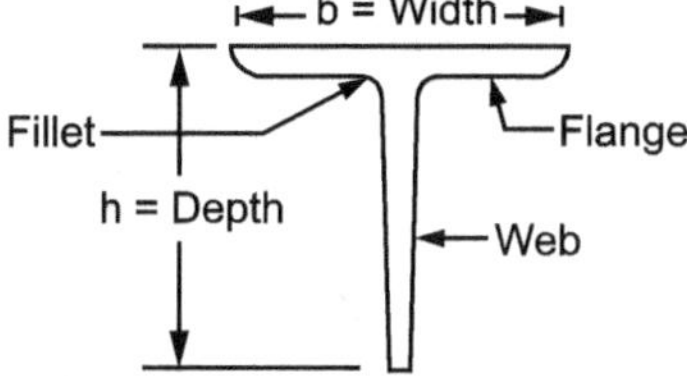

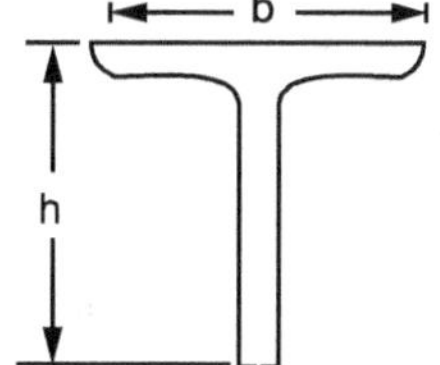

(a) Rolled Normal Tee Bar   **(b) Slit Tee Bar and Deep Legged Tee Bar**

Fig. 1.4

1.4 GRADES OF STEEL AND STRENGTH CHARACTERISTICS

Table 1.1 gives the various types of structural steels and their strengths.

Table 1.1: Types of Structural Steel

Type of steel	Class of product	Nominal thickness (mm)	Tensile strength (N/mm^2)	Yield stress (N/mm^2)
IS: 226/75				
(Standard	Plates, sections,	Up to 20	410 to 530	250
quality)	angles, tees, beams,	> 20 to 40	410 to 530	240
	channels, etc. flats.	Over 40	410 to 530	230
	Bars – round,			
	square and	Up to 20	410 to 530	250
	hexagonal	Over 20	410 to 530	240
IS: 961/75				
(High tensile)	Plates, sections,	Up to 28	570	350
St 58 HT	angles, beams,	> 28 to 45	570	340
	channels etc.	> 45 to 63	570	320
	bars, flats.	Over 63	540	290
St 55 HTW	Plates, sections,	Up to 16	540	350
	bars, flats	> 16 to 32	540	340
		> 32 to 63	510	330
		Over 63	490	280
IS: 2062/84				
(Fusion welding	Plates, sections -	Up to 20	410	250
quality)	angles, beams, tees,	> 20 to 40	410	240
	etc. flats	Over 40	410	230

contd. ...

IS: 1977/75				
(Ordinary quality)	Plates, sections -	Up to 20	410 – 530	250
Fe 410-0	angles, beams etc.	> 20 to 40	410 – 530	240
		Over 40	410 – 530	230
	Flats, bars	Up to 20	410 – 530	250
		Over 20	410 – 530	240
IS: 8500/77				
Fe 440 HTI	Plates, sections -	< 6	440 to 560	300
and	angles, beams,	> 6 to 20	440 to 560	300
Fe 440 HT2	channels etc.	> 20 to 40	440 to 560	290
	bars, flats	> 40 to 63	440 to 560	280
Fe 540 HT,	Plates, sections –	< 6	540 to 660	410
Fe 540 HTA	angles, beams,	> 6 to 20	540 to 650	400
and	channels, etc.	> 20 to 40	540 to 660	390
Fe 540 HTB	bars, flats	> 40 to 63	540 to 660	380
Fe 570 HT	– do –	< 6	570 to 720	450
		> 6 to 20	570 to 720	440
		> 20 to 40	570 to 720	430
		> 40 to 63	570 to 720	420
Fe 590 HT	– do –	< 6	590 to 740	490
		> 6 to 20	590 to 740	480
Fe 640 HT	– do –	< 6	640 to 790	540
		> 6 to 20	640 to 790	530

1.5 USE OF STEEL TABLE AND RELEVANT I.S. CODES

- Steel table gives the general properties of the standard steel sections. It is published by special publication of Bureau of Indian standards and numbered as SP6 - Part 1.

- The properties are designation, weight per metre, sectional area, depth of section, width and thickness of flange, thickness of web, moments of inertia, radii of gyration, moduli of section etc.

- Referring to steel tables one can easily choose the section according to the design requirements.

- The properties of some of the sections are given at the end of this book in Appendix C for ready reference.

- Bureau of Indian Standards (BIS) has evolved a rational, efficient and economical series of Indian Standards.

1. For Steel:

 (A) Structural Steel as per

  IS : 226, IS : 2062, IS : 3502, IS : 1977, IS : 961, IS : 8500.

 (B) Steel for reinforced concrete

  IS : 432, IS : 1139, IS : 1786, IS : 2090.

(C) Steel for bars, rivets etc.

     IS : 1148, 1149, 1570, 2073, 7383, 4431, 4432 and 5517.

(D) Steel for tubes and pipes

     IS : 1239, 1914, 1978.

2. For Code of Practice for Design of Steel Structures:

     IS : 800 – 1984.

3. For size of weld and stresses in weld :

     IS : 816 – 1969.

4. For code of practice for design loads:

     IS : 875 – 1987.

     Part 1 : Dead loads – Unit weights of building materials and stored materials.

     Part 2 : Imposed loads.

     Part 3 : Wind loads.

     Part 4 : Snow loads.

     Part 5 : Special loads and load combinations.

1.6 TYPES OF LOADS ON STEEL STRUCTURES

- The various loads which are likely to act on the steel structure are as given below:

    1. Dead load.

    2. Live load (Imposed load).

    3. Wind load.

    4. Snow load.

    5. Seismic load.

1. Dead Load:

- The dead loads (viz. the self weight of the structural members), superimposed dead loads and loads due to filling materials are referred as permanent loads.

- Dead loads are loads which are constant in magnitude and fixed in position throughout the lifetime of the structure.

- Dead loads in a building comprise the self weight of the structure and all other superimposed dead loads (viz. all permanent constructions and installations including weight of walls, partitions, floors and roofs).

- IS: 875 (Part 1) – 1987 gives the unit weight of building materials, building parts and components.

2. Live Load:

- Live loads are the loads which vary in magnitude and/or in positions.

- Live loads are also known as imposed or transient loads.

- Live loads include any external loads imposed upon the structure when it is serving its normal purpose.

- Live loads are assumed to be produced by intended use of occupancy in buildings including distributed, concentrated, impact and vibration and snow loads.

- Live loads are expressed as uniformly distributed static loads (u.d./.).

- Live loads include the weight of materials stored, furniture and movable equipments.

- Code IS: 875 (PART B) – 1987 defines the principal occupancy for which a building or part of a building is used or intended to be used. The buildings are classified according to occupancy as under as per IS: 875 (part 2) 1987.

    1. Assembly buildings.

    2. Business buildings.

    3. Educational buildings.

    4. Industrial buildings.

    5. Institutional buildings.

    6. Mercantile buildings.

    7. Residential buildings.

    8. Storage buildings.

- Imposed floor loads for different occupancies are given in IS: 875 (part 2) – 1987.

3. Wind load:

- The wind exerts pressure on the structures.

- The pressure exerted may be both external and internal.

- The external pressure will depend upon the geographical location of the structure, its height and its proximity to other structures which cause a hindrance to the air flow.

- The internal wind pressure will depend upon the permeability of the structure.

- A structure with large openings in its claddings will be subjected to more internal pressure.

- In the design of the structures, the combined effect of external and internal pressure of wind will be taken. This is done as per recommendations of IS: 875 - 1987 (Part 3).

4. Snow loads:

- In the areas of snow fall, an allowance for snow loads will be taken as per IS: 875.

- Actual loads caused by snow depends upon the shape of the roof and its capacity to retain the snow.

- It is a common practice of not combining the wind loads with the snow loads.

- However, when they are assumed to act vertically over an entire area, due allowance for this combined effect shall be made.

5. Seismic loads:

- Structures situated in the regions subjected to earthquakes shall be designed to resist the horizontal force produced in the structures due to the earthquake tremors.

- These loads shall be assumed as per IS: 1893–2002. (Part I) "Criteria for Earthquake Resistant Design of Structures - Part 1: General Provisions and Buildings". Reference may be made in this regard to IS: 875 for full details.

1.7 METHODS OF DESIGN

- Steel structures may be designed based on the following three methods:

    (i)   Elastic design method

    (ii) Plastic design method

    (iii) Limit state design method

- Elastic design method is the oldest and traditional method commonly in use. We know steel is considered to be absolutely elastic up to the yield point and elastic theory is an acceptable theory for analysis for steel component stressed with the yield point. Accordingly a structural steel component can be designed so that the stresses induced do not exceed a permissible stress. Accordingly permissible stressed are specified in direct, bending and shearing stresses.

$$\text{Permissible stress} = \frac{\text{Yield stress}}{\text{Factor of safety}}$$

- Plastic design method has been developed to take into account the behaviour of structural member stressed beyond the yield point and to determine the load at the stage of collapse of a member. In this approach the working load is equal to the collapse load divided by a load factor.

$$\text{Working load} = \frac{\text{Ultimate load}}{\text{Load factor}}$$

- Limit state design method is a further development over the plastic design method. In this method the basis in dependent entirely on the actual behaviour of materials in structures and the performance of structures determined by tests as well as long-term observations.

- This method has been developed to take into account every possible condition that can be lead a structure to become unfit for use. The philosophy behind this theory is, that the object of a structural design is to make a safe and economical structure which should satisfactorily fulfill its required purpose.

- This method aims to achieve not only strength but also stability and serviceability. Besides having the desired strength, a structure should not overturn or sway objectionably and should remain stable.

- Deflections, vibrations, fatigue should not render the structure unfit for use. That means the structure should also be designed for serviceability. Thus, this method specifies a number of limits states.

- Compared with the elastic method of design, the limit state method is more general.

- It is possible to apply different safety factors to different limit states. This is more justified than applying just one load factor as in plastic design method.

- It is also possible to implement any new limit state based on the development in the knowledge base regarding loading or structural behaviour or materials.

Important Points

- Types of Sections used in steel structures are I-sections, Channel sections, Angle sections and Tee sections.

- Steel is useful for large span industrial structures, bridges, communication towers and overhead tanks, gantry and crane girders, steel columns and chimneys, building frame etc.

- Steel table gives the general properties of like weight per metre, sectional area, dimensions, moment of inertia, modulus of section and radii of gyration etc. SP6 - Part 1 gives the steel table.

- IS: 800 - 2007 is the code of practice for Design of Steel structures.

- Types of loads on steel structures are: Dead load, Live load, Wind load, Snow load and Seismic load.

- The three methods of design are: Elastic design, Plastic design and Limit state design.

Practice Questions

1. What are the types of loads to be considered while designing the structures?

2. State four types of loads to which structures are subjected.

3. State any two/four types of structural steel sections.

4. Draw neat sketch of unequal angle section showing all components and giving geometrical properties.

5. Draw a neat sketch of I.S. channel section, I section, T section.

6. Draw neat sketches of rolled steel channel sections and equal angle section showing their all components and giving geometrical properties.

7. State I.S. code useful for getting load on the structures.

8. State any one advantage and any one disadvantage of use of steel as construction material.

9. Enumerate any four types of forces to be considered in designing steel structures.

10. State the full form of ISWB and ISMC.

11. State any four rolled steel sections used and draw the sketch of any one.

12. Enlist types of loads to be considered while designing steel structures.

13. State the meaning of ISLB and ISHB.

14. State any two grades of steel and characteristic strength in steel structures.

15. What are the loads on steel structures and its I.S. Code provision? (Any two)

16. Draw neat labelled sketch of I-section of steel.

17. State any two advantages of steel as construction material.

18. State use of I.S. 875.

19. State and explain the methods of design.

■■■

Chapter **2**

LIMIT STATE DESIGN

Syllabus

- Basis for Design - Classification of Limit States - Characteristic and Design Actions - Ultimate and Design Strengths - Partial Safety Factors for Loads and Materials - Factors Governing the Ultimate Strength: Stability, Fatigue and Plastic Collapse - Serviceability.

About this Chapter

After reading this chapter students can understand:

- Limit state and its types.
- Characteristic values, partial safety factors.
- Factors governing the ultimate strength.

2.1 INTRODUCTION

- Limit state method is an entirely new concept, based on statistical probability.
- In elastic design (working stress method) the factor of safety was about 3 for concrete, and 1.8 for steel.
- The elastic method fails to give the maximum load on the structure, serviceability, exact economy in design and idea of reserve strength of member. The actual stresses developed in the structure at the working loads often differed considerably from theoretical values.
- The reason is that the assumption, that concrete is a perfectly elastic material is not correct; on the contrary it is a more complex material.

2.1.1 Basis for Design

- Limit state method gives a new approach of statistical probability.
- A structure of failure may reach a limit state, due to coincidental occurrence of both overload and excessive weakening of materials at a critical section.
- The strength of materials and the load are the two factors which are subjected to considerable variation.
- Statistical probability of strength and load can be estimated, which will be a limiting state of material strength and load.
- Statistical base is covered by characteristic strength and characteristic load.
- Partial factors of safety are introduced to reduce the probability of failure to zero.
- *Limit state* may be defined as "the acceptable limit for the safety and serviceability of the structure before failure occurs".
- Thus the concept of design with limit state is to achieve acceptable probabilities, so that the structure will not become unfit for use and will not reach a limit state.

2.2 CLASSIFICATION OF LIMIT STATES

- The two limit states are as follows:
  1. Limit state of collapse.
  2. Limit state of serviceability.

2.2.1 Limit State of Collapse

- The limit state of collapse is reached when the structure as a whole or part of the structure collapses.
- Collapse may occur due to the rupture of one or more members or on account of formation of mechanism or due to elastic or inelastic stability or from loss of equilibrium etc.
- This limit state is taken care of by providing resistance greater than the forces to which it is subjected and keeping a margin of safety through safety factors.
- This limit state corresponds to: (a) flexure, (b) shear and (c) torsion.

2.2.2 Limit State of Serviceability

- Limit state of serviceability is related to the satisfactory performance of the structure at working load. There are four major types of serviceability limit states applicable to steel structures. They are: (i) Deflection, (ii) Durability, (iii) Vibration, (iv) Fire resistance
  - **(i) Deflection:** Excessive deflection poses number of problems viz. feeling lack of safety, impairing strength of structure or its components and damage to finishing. The maximum deflection limits have been specified by the Code depending on the type of building. They are given in Table 2.1.

Table 2.1: Deflection Limits

Time of Building	Deflection	Design Load	Member	Supporting	Maximum Deflection
(1)	(2)	(3)	(4)	(5)	(6)
Industrial Buildings	Vertical	Live Load / Wind load	Purlins and Girts	Elastic cladding Brittle cladding	Span/150 Span/150
		Live load	Simple Span	Elastic cladding Brittle cladding	Span/240 Span/300
		Live load	Cantilever Span	Elastic cladding Brittle cladding	Span/120 Span/150
		Live load/Wind load	Rafter supporting	Profiled Metal Sheeting Plastered Sheeting	Span/180 Span/240
		Crane load (Manual operation)	Gantry	Crane	Span/500
		Crane load (Electric operation upto 50 t)	Gantry	Crane	Span/750
		Crane load (Electric operation upto 50 t)	Gantry	Crane	Span/1000
	Lateral	No Cranes	Column	Elastic cladding Masonry/Brittle cladding Crane (absolute)	Height/150 Height/240 Span/400
		Crane + Wind	Gantry (Lateral)	Relative displacement between rails supporting crane	10 mm
		Crane + Wind	Column/Frame	Gantry (Elastic cladding; pendent operated) Gantry (Brittle cladding; cab operated)	Height/200 Height/400

contd. ...

Other Buildings	Vertical	Live load	Floor and Roof	Elements not susceptible to cracking	Span/300
				Elements susceptible to craking	Span/360
		Live load	Cantilever	Elements not susceptible to cracking	Span/150
				Elements susceptible to craking	Span/180
	Lateral	Wind	Building	Elastic cladding	Height/300
				Brittle cladding	Height/500
		Wind	Inter storey drift	–	Storey height/300

(ii) Durability: *Durability* is defined as ability of the structure to maintain its level of reliability and performing the desired function in the working environment under anticipated exposure conditions, without deterioration of cross-sectional area and loss of strength due to corrosion during its intended life span.

(iii) Vibration: Suitable provision shall be made for vibration set in due to machinery operating loads and impact loads.

(iv) Fire Resistance: Temperature causes variation of mechanical properties of steel such as variation in yield stress, variation of modulus of elasticity etc. Design provision shall be made to resist fire.

2.3 CHARACTERISTIC AND DESIGN ACTIONS

- To achieve the design objectives, the design is based on characteristics values for material strengths and applied loads. The design values are derived from the characteristic values through the use of partial safety factors. The reliability of design is ensured by satisfying the requirement:

    Design action $\leq$ Design strength.

2.3.1 Characteristic Values

Characteristic strength:
- Characteristic strength of a material is the value of material below which not more than 5% of the test results are expected to fall.
- It means that characteristic strength has 95% realiability or there is only 5% probability of actual strength being less than the characteristic strength.
- Characteristic strength of general structural steel (f_y) is 250 MPa.

Characteristic loads:
- Characteristic loads means those loads which have a 95% probability of not being exceeded during the life time of the structure. In absence of statistical data characteristic loads shall be as per IS: 875 – 1987 having following parts:

    Part 1   :   Dead loads
    Part 2   :   Imposed loads
    Part 3   :   Wind loads
    Part 4   :   Snow loads
    Part 5   :   Special loads and Load combinations.

- In addition to this IS: 1893 – 2002 (Part 1 to 5) Criteria for Earthquake Resistant Design of Structures shall also be read to account for seismic loads.

2.4 ULTIMATE AND DESIGN STRENGTHS

- The design strength S_d is given by:

$$S_d = \frac{S_u}{\gamma_m}$$

where, S_u = Ultimate strength

γ_m = Partial safety factor for material strength

2.4.1 Partial Safety Factor

- The concept of limit states acknowledges that there can be variations in both loads and material strength.

- Thus, the safety of the structure depends on each of the two principal design factors (i.e. load and material strength) which are not dependent on each other.

- Hence, two different safety factors one for load and the other for material strength are used. Because each of the two safety factors contribute partially to safety, they are known as *partial safety factors*.

2.4.2 Partial Safety Factors for Material Strength

- The partial safety factors for material strength allows for uncertainty of element behaviour and probability of strength reduction due to fabrication and tolerances, variation of member sizes, uncertainty in calculation of strength and imperfection in materials.

- The design strength (f_d) is obtained by dividing the yield strength (f_y) by partial safety factors of material strength (γ_m) i.e. $f_d = \frac{f_y}{\gamma_m}$.

Table 2.2: Partial Safety Factors for Material Strength γ_m

Sr. No.	Definition		Partial Safety factor
(i)	Resistance governed by yielding, γ_{mo}		1.1
(ii)	Resistance of member to buckling, γ_{mo}		1.1
(iii)	Resistance governed by ultimate stress, γ_{m1}		1.25
(iv)	Resistance of connections	Shop fabrication	Field fabrication
	(a) Bolts friction type γ_{mf}	1.25	1.25
	(b) Bolts bearing type γ_{mb}	1.25	1.25
	(c) Rivets γ_{mr}	1.25	1.25
	(d) Welds γ_{mw}	1.25	1.50

2.4.3 Partial Safety Factor for Loads

- The partial safety factor for loads allows for possible deviation of loads, reduced possibility of all loads acting together, inaccurate assessment of load, and uncertainty in assessment of effects of loads. The partial safety factor for loads is a load factor which multiplied to characteristic load gives the **design load**.

$$\text{Design load} = \gamma_f \times \text{characteristic load}$$

The various load factors recommended by IS : 800 are given in Table 2.3.

Table 2.3: Safety Factors for Loads

Combination	Limit state of strength					Limit State of Serviceability			
	DL	LL[1]		WL/EL	AL	DL	LL		WL/EL
		Leading	Accompanying				Leading	Accompanying	
DL + LL + CL	1.5	1.5	1.05	–	–	1.0	1.0	1.0	–
DL + LL + CL	1.2	1.2	1.05	0.6	–	1.0	0.8	0.8	0.8
+ WL/EL	1.2	1.2	0.53	1.2					
DL + WL / EL	1.5 $(0.9)^2$	–	–	1.5	–	1.0	–	–	1.0
DL + ER	1.2 $(0.9)^2$	1.2	–	–	–	–	–	–	–
DL + LL + AL	1.0	0.35	0.35	–	1.0	–	–	–	–

Notes:

1. *When action of various live loads is simultaneously considered, the leading live load shall be considered to be the one causing the higher load effects in the member/section.*

2. *This value is to be considered when the dead load contributes to stability against overturning is critical or dead load causes reduction in stress due to other loads.*

   DL = Dead load, LL = Imposed Load (Live load), WL = Wind load,

   CL = Crane load, (Vertical/Horizontal), AL = Accidental load,

   ER = Erection load, EL = Earthquake load

- In industrial structures dead load and imposed loads are normally small, crane load in combination of wind load are critical in most of the cases.

- Therefore, crane load is a major load which should be multiplied by the factor 1.5 in combination with dead load and imposed load.

- Even crane load combined with wind load along with dead load and imposed load are critical in most of the cases.

- When stability against overturning or sliding or reversal of stresses is under consideration lesser value of safety factor for dead load equal to 0.9 is considered for greater safety.

2.5 FACTORS GOVERNING THE ULTIMATE STRENGTH

(i) Stability: The structure as a whole and also each of its component should be stable. This should include the overall frame stability against overturning and sway.

(ii) Fatigue: Except in cases when a structure or any of its components is subjected to stress fluctuations, this aspect may not be relevant. In cases where fatigue needs consideration, the partial safety for factor for load γ_f equal to unity shall be used for the load causing the stress fluctuation and stress range.

(iii) Plastic collapse: Step-by-step loading of flexural member changes the bending stress variation due to yield spread in cross-section. When the yield spreads completely over the section, then it is no longer resists further increase in load causes collapse.

Important Points

- Limit state may be defined as "the acceptable limit for the safety and serviceability of the structure before failure occurs.

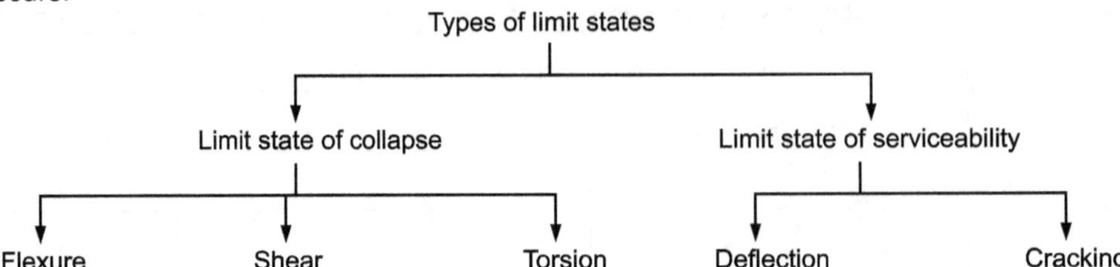

- **Characteristics strength** of the material is the value of material below which not more than 5% of the test results are expected to fall.
- Similarly steel grades are also designated by their characteristic strength f_y.
- **Characteristic load** means those loads which have a 95% probability of not being exceeded during the life time of the structure.
- Partial safety factor for material strength

$$f_d = \frac{f}{\gamma_m}$$

- Partial safety factor for load

$$f_d = \gamma_f \cdot F$$

- Characteristic load: It means that the value of load which has a 95% probability of not being exceeded during the life of the structure.

Practice Questions

1. Define limit state and state its types.
2. What safety factors are used for material strength, load and moments?
3. Explain the concept of limit state of flexure.
4. Define 'Characteristic strength' and 'Characteristic load'.

■■■

Chapter **3**

DESIGN OF CONNECTIONS AND DETAILING

Syllabus

- General - Types of Connections - Bolted, Riveted and Welded connections - Rigid and Flexible Connections - Components of Connections - Basic Requirements of Connections - Clearance for Holes - Minimum and Maximum Spacing of Fasteners - Minimum Edge/End Distances - Requirements of Tacking Fasteners. Bolted Connection - Types of Bolts - Bearing Type Bolts - Nominal and Design Shear Strengths of Bolts - Reduction Factors for Long Joints, Large Grip Lengths, Thick Packing Plates - Nominal and Design Bearing Strengths of Bolts - Reduction Factors for Over Sized and Slotted Holes - Nominal and Design Tensile Strengths (Tension Capacity) of Bolts-Simple Problems. Welded Connection- Types of Welds - Fillet Welds - Minimum and Maximum Sizes - Effective Length of Weld - Fillet Welds on inclined Faces - Design Strengths of Shop/Site Welds - Butt Welds - Effective Throat Thickness and Effective Length of Butt Weld - Simple Problems. Design Problems related to Eccentric Riveted/Bolted and Welded Connections.

About this Chapter

After reading this chapter students can understand:
- Types of Connections
- Basic requirements of Connections
- Bolted connections, Types of bolts, Design shear, Bearing and Tension strength of bolts
- Welded Connections, Types of Welds, Requirements and Design of Strength of Shop/Site Welds

3.1 INTRODUCTION

- Every steel structure is made-up of various components which are connected together either by means of rivets, bolts, pins or by welds.

3.2 TYPES OF CONNECTIONS

- Broadly the types of connections are based on:
- (i) Type of fastener used.
- (ii) Transfer of axial an shear forces and moments.

3.2.1 Type of Connection based on type of Fastner

- If the joint between the members is secured by the use of rivets, it is called *riveted joint*.
- If the joint between the members is secured by the use of bolts, it is called *bolted joint*.
- If the joint between the members is secured by the use of welds, it is called *welded* joint.

3.2.2 Types of Connection based on Transfer of Forces

(a) Rigid connections
(b) Flexible connections.
(c) Semi-rigid connections.

3.2.2.1 Rigid Connections

- A rigid joint is a joint which transfers not only the axial and shear forces but also moments. Rigid connection transfers significant moment to the supporting structure and undergoes negligible deformation at the joint. Fig. 3.1 (c) shows a t-stub connection transferring shear and moment from beam end to column.

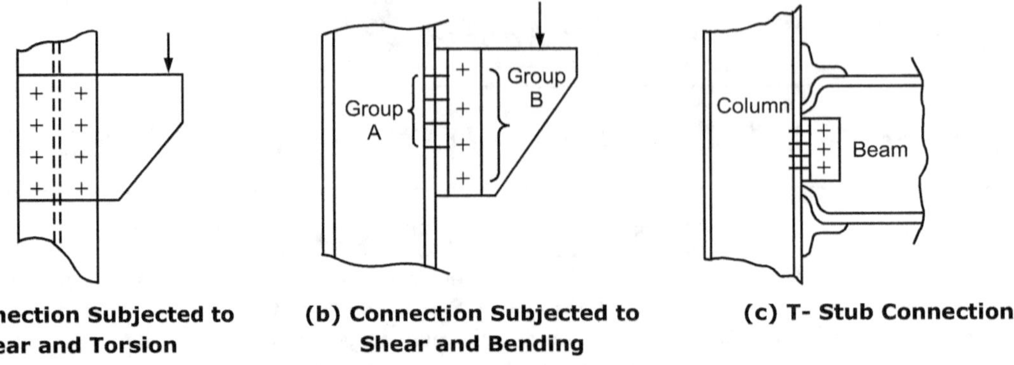

(a) Connection Subjected to Shear and Torsion **(b) Connection Subjected to Shear and Bending** **(c) T- Stub Connection**

Fig. 3.1: Rigid Connection

- In a rigid joint, the ends of the member framing into it undergo the same displacement with the result the angle between them remains the same even on loading Le. it does not allow relative rotation between the members as in the case of portal frame, even though the joint may undergo rotation after loading Fig. 3.1 (a). Brackets attached to columns are also examples of rigid connections which transfer bending torsion and shear. Fig. 3.1 (a) a shows the connection which transfers eccentric load by torsion and shear. In Fig. 3.1 (b) eccentric load is transferred by bending tension and shear by bolts group A and bolts group B transfer load by torsion and shear.

3.2.2.2 Flexible Connection

- It is also known as Hinged, pinned or simple shear connection. It is a connection at a joint which offers no restraint to rotation. This means that original angle between the connected members changes during loading causing relative rotation between connected members. It transfers shear and/or axial forces and no movements. Ideally in a hinged connection two members are connected by a single rivet/bolt or pin. Connection which transfers only shear can be classified as simple connection.

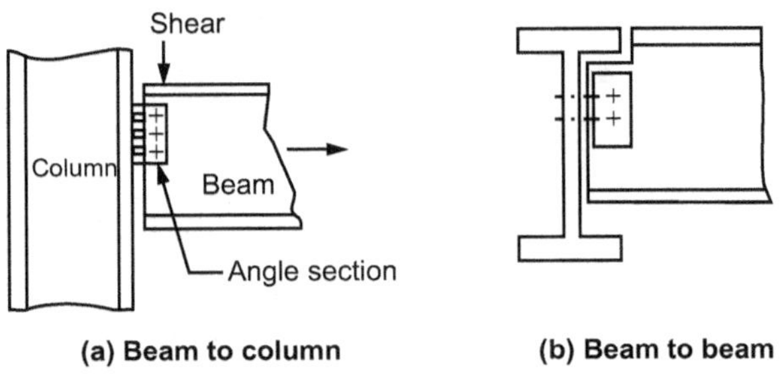

(a) Beam to column   **(b) Beam to beam**

Fig. 3.2: Simple/Pinned Connection

- For example, beam to column connection or beam to beam connection are simple connections are shown in Fig. 3.2. Actually a small amount of moment will be developed but that is normally neglected.

3.2.2.3 Semirigid Connection

- It is a connection which transfers the axial force and shear force fully but the moments only partially Fig. 3.3. In reality, all the connections are semirigid. However, we assume some of them as rigid and some as hinged depending upon the method of connection.

- The design of semirigid connection is based on rotation capacity of the joint and shall be determined based on experiments or to idealize the connection as equivalent rotational spring with the bilinear or non-linear moment capacity characteristics. In semirigid joint the ends of the members framing into it under small rotation so that the angle between them gets slightly reduced.

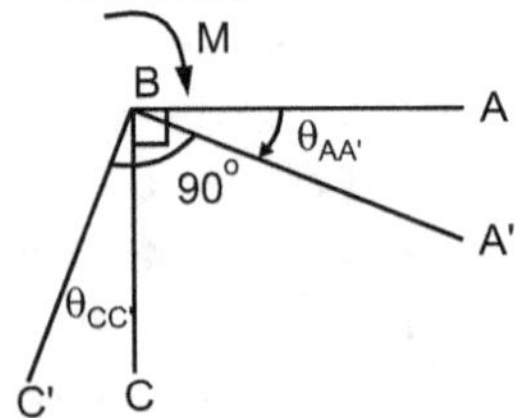

Fig. 3.3: Rigid and Semirigid Joints

3.3 COMPONENTS OF CONNECTIONS

- The various components of connection are cleats, gusset plate, brackets, connecting plates, connections etc. The capacities components other than connectors shall be assessed as per the provisions of IS : 800-200%. The connectors are nothing but the fasteners such as rivets, bolts, pins and welds.

3.4 TYEPS OF FASTENERS

- Structural members are connected to one another through bolts/rivets or weld. Use of rivets is becoming obsolete. This is because riveting requires preheating, more labour, supervision and is noise pollution. Hence, more stress is given on the design of bolted connections. However, it may be mentioned that the design of riveted connection is similar to the design of bolted connection. Welded connections. which are predominantly used in practice are discussed in the next topics.

3.5 BASIC REQUIREMENTS OF CONNECTIONS

- The connections shall satisfy the following basic requirements:
  - Clearances for Holes for fasteners.
  - Minimum and maximum spacing of fasteners.
  - Minimum edge/end distances.
  - Tacking fasteners.

3.5.1 Clearances for Holes for Fasteners

- Bolts may be located in standard size, over size, short or long slotted hole.
  - **(a) Standard clearance hole:** The diameter of standard clearance hole for fasteners shall be given in Table 3.1.
  - **(b) Oversize hole:** Holes of size larger than the standard clearance holes as given in Table 3.1 may be used in slip resistant connections and hold down bolted connections.
  - **(c) Short and long slots:** Slotted holes of size larger than the standard clearance hole as given in Table 3.1 may be used in slip resistant connections and hold down bolted connections only where specified.

Table 3.1: Clearances for Fastener Holes

Sr. No.	Nominal size of fastener 'd' mm	Size of Hole = Nominal diameter of Fastener + Clearances mm			
		Standard clearance in diameter and width of slot	Oversize clearance in diameter	Clearance in length of slot	
				Short slot	Long slot
1.	12 - 14	1.0	3.0	4.0	2.5 d
2.	16 - 22	2.0	4.0	6.0	2.5 d
3.	24	2.0	6.0	8.0	2.5 d
4.	Length than 24	3.0	8.0	10.0	2.5 d

3.5.2 Pitch 'P'

- Pitch is the distance between centres of any two adjacent holes in the direction of load/stress.

3.5.3 Minimum Spacing

- The distance between centre of fasteners shall not be less than 2.5 times the nominal diameter of the fasteners.
- Minimum pitch = 2.5 d

  where, d = Nominal diameter of fastener.

3.5.4 Maximum Spacing

- In tension members the pitch shall not exceed 16t or 200 mm whichever is less.
- In compression members pitch shall not exceed 12t or 200 mm whichever is less, where, t is the thickness of the thinner plate.
- **Maximum Pitch:** The maximum distance between the centres of any two adjacent fasteners shall not exceed 32t or 300 mm whichever is less, where t is the thickness of the thinner plate.
- Distance between two consecutive fasteners: The distance between the centres of any two consecutive fasteners in a line adjacent and parallel to an edge of an outside plate shall not exceed (100 mm + 4t) or 200 mm whichever is less both in tension and compression members, where t is the thickness of the thinner outside plate

3.5.5 Minimum Edge and End Distance

- **Edge distance:** The edge distance i.e. the distance from the centre of the fastener hole to the nearest edge of an element measured perpendicular to the direction of load.
- **End distance:** Distance from the centre of the fastener hole to the edge of an element measured parallel to the direction of load.
- The minimum edge and end distance from the centre of any hole to the nearest edge of the plate shall not be less than 1.7 times the hole diameter In the sheared or hand-flame out edges, and 1.5 times the hole diameter In case of rolled, machine flame cut, sawn and planed edges.
- The maximum edge distance to the nearest line of fasteners from an edge of any unstiffened part $< 12\ t \times \sqrt{(250/f_y)}$, where, t is the thickness of the thinner outer plate.

3.5.6 Gauge 'g'

- It is the distance between adjacent parallel lines of fasteners, perpendicular to the direction of load/stress.

3.5.7 Requirements of Tacking Fasteners

- Tacking fasteners: Tacking fasteners are not subjected to any stress. They are required to be provided when the maximum distance specified above are exceeded. They stitch two sections to act as one unit. Tacking fasteners shall have a spacing in line not exceeding 32 t or 300 mm whichever is less where, t is the thickness of the thinner outside plate.
- In tension members composed of two flats, angles, channels or ties, tacking fasteners with solid distance pieces shall be provided at a spacing not exceeding 1000 mm.
- For compression members tacking fasteners in a line shall be spaced at a distance not exceeding 600 mm.

(A) BOLTED CONNECTIONS

3.6 TYPES OF BOLTS

1. **Bearing Types Bolts:** In bearing type connection, the fasteners bear against the sides of the holes in connection.
2. **Friction Type Bolts:** In friction connections, the fasteners are tightened to the clamp and the connected parts are under high pressure.
- The bolts connecting structural members are broadly classified into two types:
1. Black bolt (Bearing type).
2. High strength bolt (Friction type).

3.6.1 Black Bolt

- Black bolts are made from mild steel rods with a square or hexagonal head and nuts as shown in Fig. 3.4 conforming to IS: 1363. Black bolts are ordinary, rough unfinished and commonly used bolts. They are least expensive bolts.

- They are mainly used for light structures and are not recommended for connections subjected to impact, fatigue or dynamic loads.

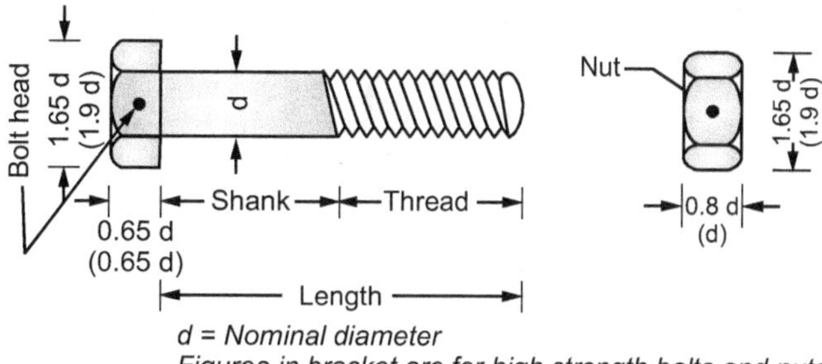

d = Nominal diameter
Figures in bracket are for high strength bolts and nuts

Fig. 3.4: Hexagonal Head Black Bolt and Nut

3.6.2 Advantages of Black Bolts over Riveted or Welded Connections

1. It requires simple tools and unskilled labour.
2. No specialized equipment is required.
3. Minor discrepancies in dimensions get eliminated.
4. It is noiseless.
5. As soon as the bolts are tightened the connection starts supporting loads.
6. Lapse of time for cooling (as in case of riveting or welding) is not required.
7. Progress of work is fast.
8. It is easy to dismantle and reuse the materials.

3.6.3 High Strength Bolts

- High strength bolts are made from the medium carbon steel. The bolts of property class 8.8 and 10.9 are commonly used in steel construction.
- They are identified by class identification symbol 8.8S, 10.9S etc. which is embossed on the head of these bolts. The suffix 'S' denotes a high strength bolts with a hexagon head.
- The bolts with induced initial tension are called *High Strength Bolts*. In high strength bolt initial pretension in bolts develops clamping force at the interfaces of elements being joined. *High Strength Friction Grip* (HSFG) bolts are commonly used in practice.
- They slip into bearing at ultimate load only. The frictional resistance to slip between the plate surfaces subjected to clamping force opposes slip due to externally applied shear.
- High strength friction grip bolts (HSFG) and nut shall conform to IS: 3757. HSFG bolts are tightened by part-turn method, direct tension indicator tightening or wrench tightening (torque control method).

3.6.4 Advantages of HSFG Bolts

1. It does not allow slip between the connected members.
2. Loads are transferred by friction only.
3. Due to high strength less number of bolts are required, with the result that size of the gusset plate gets reduced.
4. There is no noise pollution.
5. Deformation is minimized.

- Where slip in the serviceability limits state is to be avoided in the connection, high strength bolts in a friction type joint, fitted bolts or welds shall be used.

3.7 TYPES OF JOINTS

- **Lap joint:** When one member is placed above the other and both are connected by means of bolts/weld the joint is known as lap joint. The minimum lap shall not be less than four times thickness of the thinner part joined or 40 mm, whichever is less. The lap joints have single row chain bolting or staggered bolting Fig. 3.5 (a).

- **Butt joint:** In butt joint the plates to be connected shall butt against each other and connected by providing a cover plate on one side or on both sides of the joint.
- If cover plate is provided on one side it is called as *single cover butt joint* and where cover plates are provided on both sides it is called as *double cover butt joint*. (Fig. 3.5 (b)).

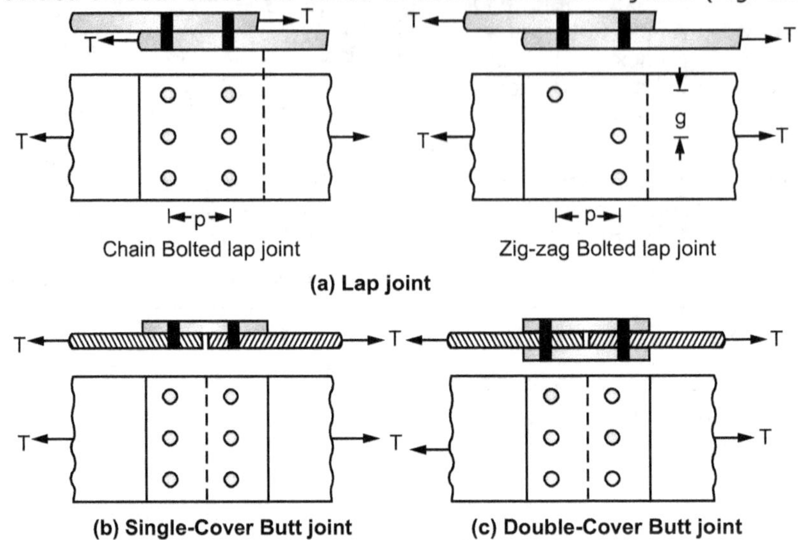

Fig. 3.5: Types of Joints

3.8 FAILURE OF BOLTED JOINT

- The bolted joints may fail in one of the following modes:
  1. Shear failure mode - Shear failure or bolt or tearing failure of plate.
  2. Tensile failure mode - Tensile failure of bolt or tensile failure of plate.
  3. Bearing failure mode - Bearing failure of bolt or bearing failure of plate.

Shear failure mode:
- Plates bolted together and subjected to tensile load may result in the shearing of bolts.
- In case of lap joint when the shearing of bolt occurs at one cross-section, the same is referred to as single shear failure (Fig. 3.6 (a)).
- If it occurs at two cross-sections, as in the case of butt joint with two cover plates, it is called as double shear failure.
- When the strength of the plate is less than the shearing strength of bolt, the tearing failure of plate may occur. To avoid this type of failure minimum edge distance shall be provided.

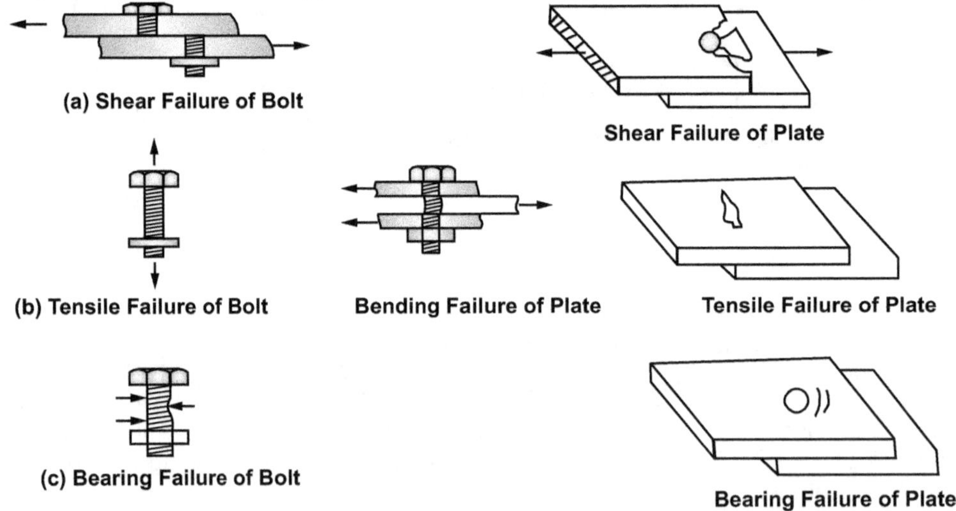

Fig. 3.6

Tension failure:
- The bolt subjected to tensile force fails if factored tensile force is greater than the tensile capacity of the bolt. The tensile capacity depends upon tensile strength of the bolt and minimum cross-sectional area of the threaded length of the bolt. (Fig. 3.6 (b)).

Bearing failure:
- Normally the bolt material is of much higher strength than that of steel plate through which the bolts passes. As a result bearing failure takes place in the plate material. The bolt may deform due to high local bearing stresses between the bolt and the plate (Fig. 3.6 (c)).

3.9 NET TENSILE STRESS AREA

- The ratio of net tensile area at threads, to nominal plain shank area of bolt is taken as 0.78.
- But IS : 800-2007 specifies that where the net tensile stress area is not defined, it shall be taken as the area at the root of the threads.
- Where it can be shown that threads do not occur in the shear plane the area may be taken as the cross-section area A_s at the shank.
- In steel construction, in general, bolts of property class may be expressed symbolically as **a.b.** In property class number 'a' represents $\frac{1}{100}^{th}$ of the nominal tensile strength in N/mm^2 and number 'b' indicates the ratio of yield stress to ultimate stress.
- For example: class 4.6 in which number 4 is $\frac{1}{100}^{th}$ of nominal ultimate stress of bolt i.e. $f_{ub} = 4 \times 100 = 400\ N/mm^2$ and yield stress, f_{yb} is $0.6 \times 400 = 240\ N/mm^2$. Similarly, for bolt property class 8.8 has ultimate stress of $f_{ub} = 8 \times 100 = 800\ N/mm^2$ and yield strength $f_{yb} = 0.8 \times 800 = 640\ N/mm^2$.

3.10 DEFINITIONS AND DETAILING OF BOLTS

- **Gross cross-sectional area A_g:** It is the area of the unthreaded portion of the bolt or shank.
- **Net effective cross-sectional area A_n:** It is the cross-sectional area at the root of the threads. It is taken equal to $0.78 \times A_g$ (see Section 3.9).
- **Bolt Holes:** Bolts where possible, shall be drilled. Punching of holes reduces ductility and toughness and it may lead to brittle failure. Hence punching of holes shall be permitted only in material whose yield stress is less than 360MPa and its thickness does not exceed 5600/fy mm.
- Bolt holes are made larger than the bolt diameter to facilitate erection and to allow for inaccuracies in fabrication.
- For the bolts which are normally used in practice, the diameter of hole shall be 2 mm more than the diameter of bolt.
- The diameter of bolt, pitch and edge distances are summarized in Table 3.2.

Table 3.2: Diameter of bolt, Pitch and Edge distances

Nominal bolt Dia. mm	12	14	16	18	20	22	24	27	30	Above 36
Dia. of hole mm	13	15	18	20	22	24	26	30	33	Bolt dia. + 3 mm
Minimum edge distance in mm										
(a)for sheared or rough edge	20	26	30	34	37	40	44	51	56	1.7 × Hole dia.
(b)for rolled, sawn	19	23	27	30	33	36	39	45	50	1.5 × Hole dia.

Max., edge distance = $12\ t \times \sqrt{(250/f_y)}$

Pitch (minimum)	2.5 × Nominal diameter of bolt
Pitch (maximum)	32 t or 300 mm
(i) Parts in tension	16 t or 200 mm, whichever is less
(ii) Parts in compression	12 t or 200 mm, whichever is less
(iii) tacking fasteners	32 t or 300 mm, whichever is less
	16 t or 200 mm, whichever is less for plates exposed to weather

where, t is the thickness of the thinner outside plate or angle.

3.11 DESIGN STRENGTH OF BOLT

3.11.1 Bolts in Shear

- The **Nominal shear strength** of the Bolt is given by:

$$V_{nsb} = \left(\frac{f_u}{\sqrt{3}}\right)(n_n A_{nb} + n_s A_{sb}) \qquad \ldots (3.1)$$

where, A_{sb} = Nominal plane shank area of the bolt

A_{nb} = Net shear area of bolt at threads, may be taken as the area Corresponding to root diameter at the thread.

f_u = Ultimate tensile strength of a bolt.

n_n = Number of shear planes with threads intercepting the shear plane

n_s = Number of shear planes without threads intercepting shear plane

- The **Design Shear Strength** of Bolt is

$$V_{dsb} = \frac{V_{nsb}}{\gamma_{mb}} = \frac{\left(\frac{f_u}{\sqrt{3}}\right)(n_n A_{nb} + n_s A_{sb})}{1.25} \qquad \ldots (3.2)$$

3.11.2 Reduction Factors

- The shear capacity of bolts as determined by using equation (3.2) shall be reduced by the factors when long joints, large grip lengths and packing plates are used.

3.11.2.1 Long Joints

- When the length of the joint, l_j of a splice or end connection in a compression or tension member containing more than two bolts. (i.e. the distance between the first and last rows of bolts in the joint, measured in the direction of the load transfer) excess 15d in the direction of the load, the nominal shear capacity, V_{dsb} shall be reduced by the factor β_{lj}, given by:

$$\beta_{lj} = 1.075 - \frac{l_j}{200\,d} \text{ but } 0.75 \le \beta_{ij} \le 1.0$$

$$\beta_{lj} = 1.075 - 0.005\,\frac{l_j}{d} \qquad \ldots (3.3)$$

where, d = Nominal diameter of the fastener.

3.11.2.2 Large Grip Lengths

- When the grip length, lg (equal to the total thickness of the connected plates) exceeds 5 times the diameter, d of the bolts, the shear capacity shall be reduced by factor β_{lg} given by:

$$\beta_{lj} = \frac{8d}{3d + lg} = \frac{8}{3 + \frac{lg}{d}} \qquad \ldots (3.4)$$

β_{lj} shall not be more than β_{lj} given in equation (3.3). The grip length, lg shall in no case be grater than 8d.

3.11.2.3 Packing Plates

- The design shear capacity of bolts carrying shear through a packing plate in excess of 6 mm shall be decreased by a factor, β_{pk} given by:

$$\beta_{pk} = (1 - 0.0125\,t_{pk}) \qquad \ldots (3.5)$$

where, t_{pk} = thickness of thicker packing, in mm

3.11.2 Bolts in Bearing

- Design bearing strength of bolts on any plate is given by:

$$V_{dpb} = \frac{V_{nob}}{\gamma_{mb}} \qquad \ldots (3.6)$$

- Nominal bearing strength of Bolts on any Plate is given by,

$$V_{dpb} = 2.5\, k_b\, d\, t_p\, f_u \qquad \qquad \dots (3.7)$$

where, k_b is smaller for $\left[\dfrac{e}{(3d_o)};\ \left(\dfrac{p}{3d_o}\right) - 0.25;\ \dfrac{f_{ub}}{f_u};\ 1.0\right]$

$\quad$ e, p = End and pitch distances of fastener along bearing direction.

$\quad$ d_o = Diameter of the bolt hole.

$\quad$ d = Nominal diameter of the bolt.

$\quad$ f_{ub} = Ultimate tensile stress of bolt.

$\quad$ f_u = Ultimate tensile stress of plate.

$\quad$ t_p = Σ thickness of connected plates experiencing bearing stress in the same direction.

Thus, Design bearing strength of Bolt on any Plate

$$V_{dpb} = \frac{V_{npb}}{\gamma_{mb}} = \frac{2.5\, k_b\, d\, t_p\, f_u}{1.25} \qquad \qquad \dots (3.8)$$

3.11.2.1 Reduction Factors for Oversized and Slotted Holes

- The bearing resistance (in the direction normal to the slots in slotted holes) of bolts in holes other than standard clearance holes may be reduced by multiplying the bearing resistance obtained by equation 2.4 (b) v_{npb}, by the factors given below:

(i) Over size and short slotted holes = 0.7 and

(ii) Long slotted holes = 0.5

3.11.3 Bolts in Tension

- Nominal tensile strength of Bolts is given by:

$$T_{ntb} = 0.9\, f_{ub}\, A_n < f_{yb}\, A_{sb}\left(\frac{\gamma_{mb}}{\gamma_{mo}}\right) \qquad \qquad \dots (3.9)$$

where, $\qquad A_n$ = Net tensile area as specified by IS: 1367

$\qquad\qquad\qquad$ (for bolts where tensile stress is not defined. A_n shall be as the area of the root of threads).

taken $\qquad A_{sb}$ = Shank area of the bolt.

$\qquad\qquad f_{ub}$ = Ultimate tensile stress of the bolt.

$\qquad\qquad f_{yb}$ = Yield stress of bolts.

Factored tensile force

$$T_b \leq \frac{T_{ntb}}{\gamma_{mb}} \qquad \qquad \dots (3.10)$$

Standard values of strength of bolts:

- As the bolt material offers bearing resistance twice the ultimate tensile stress, it is generally not necessary to consider bolt bearing in design.

- The shear strength, tensile strength of a bolt are often required in design. They have been given in Table 3.3.

Table 3.3: Design strength of Black bolts in kN based on Net Tensile Area

Bolt size mm	Tensile stress area mm²	Tension strength kN	Single Shear strength kN
(12)	84.3	24.2	15.6
16	157	43.8	29.0
20	245	68.5	45.3
(22)	303	82.9	56.0
24	353	98.7	65.2
(27)	459	124.9	84.8
30	561	154.2	103.6
36	817	222.0	150.9

Notes: For tension strength values obtained using Eq. 3.10.

For single shear strength values obtained using Eq. 3.2.

Sizes in brackets are not preferred.

Table 3.4: Strength of Bolts in N/mm² in Clearance Holes

	Bolt grade 4.6 N/mm²	Bolt grade 8.8 N/mm²	Connecting Equation
Shear strength v_{nsb}	185	370	$\dfrac{f_u}{(\sqrt{3} \times 1.25)}$
Bearing strength v_{npb}	800	1600	$\dfrac{2.5\,f_u}{1.25} = 2\,f_u$
Tension strength T_{ntb}	272	576	$\dfrac{0.9\,f_{ub}}{1.25} = 0.72\,f_{ub}$ but not greater than $\dfrac{f_{yb} \times 1.25}{1.1} = 1.136\,f_{yb}$

Table 3.5: Bearing strength in N/mm² of connecting parts of black bolts

Grade 410	Grade 540	Grade 570	Connection equation
820	1080	1140	$\dfrac{2.5\,f_u}{1.25} = 2\,f_u$

3.11.4 Tension Strength of Plate

- Due to presence of holes the plate may fail through the weakest section.
- The tensile strength of plate governed by rupture of net cross-sectional area, A_n, at bolts is given by

$$T_{dn} = \frac{0.9\,f_u\,A_n}{\gamma_{m1}} \qquad \qquad \dots (3.11)$$

where,

γ_{m1} = Partial safety factor for failure at ultimate stress.

A_n = Net effective area of plate.

$A_n = (b - n\,d_h) \times t$ for chain bolting.   ... (3.11 (a))

$A_n = \left[b - n\,d_h + \left(\dfrac{\Sigma p^2}{4g} \right) \right] \times t$ for zig-zag bolting.   ... (3.11 (b))

b, t = Width and thickness of plate.

d_h = Diameter of bolt hole (2 mm in addition to the diameter of bolt for directly punched holes).

g = Gauge distance i.e. the distance between the two consecutive bolts in a chain measured at right angles to the direction of tension (stress) in the member.

p = Pitch length parallel to the tensile force.

n = Number of bolt holes.

3.12 EFFICIENCY OF A JOINT

- It is obvious that if the reduction of cross-sectional area of any member is minimum the joint will be more efficient. For better efficiency the section should have the least number of holes at critical section.
- The efficiency of the joint is the ratio of actual strength of connection to the gross strength of connected member, expressed in % as

$$\text{Efficiency} = \frac{\text{Minimum actual strength of joint}}{\text{Gross strength of solid plate}} \times 100$$

Strength of solid plate = $0.9 \times f_u \times$ Cross-sectional area of plate

3.13 ANALYSIS AND DESIGN AXIALLY LOADED BOLTED JOINT

- Generally the truss members are axially loaded and they are joined through gusset plate by bolting.
- The centroid of connecting members should meet at a joint to avoid eccentricity.
- The thickness of gusset plate should be more than the thickness of members meeting at the joint.
- The joint of the truss is designed, based on number of bolts required to transfer the force (equations given in article 2.6 shall be used to find bolt value).

SOLVED EXAMPLES

Ex. 3.1: *Determine the bolt value of 20 mm diameter bolt connecting 10 mm plate in (i) single shear, (ii) double shear. Bolts used are 4.6 grade, plate of 410 grade. Take area of bolt = 245 mm².*

Sol.: For 4.6 grade bolts,

$f_{ub} = 4 \times 100 = 400$ N/mm² and $f_{yb} = 0.6 \, f_{ub} = 0.6 \times 400 = 240$ N/mm².

Area of bolt $\rightarrow A_{nb} = 0.78 \, A_g = 0.78 \times \dfrac{\pi}{4} \times 20^2 = 245$ mm²

The strength of bolt in single shear,

$$V_{dsb} = \frac{V_{nsb}}{\gamma_{mb}} = \frac{f_{ub}}{\sqrt{3}} \frac{(n_n A_{nb} + \eta_s \cdot A_{sb})}{\gamma_{mb}} \qquad \ldots (\gamma_{mb} = 1.25)$$

$$= \frac{400}{\sqrt{3}} \frac{(1 \times 245 + 0)}{1.25} = 45260 \text{ N} = 45.26 \text{ kN}$$

Similarly, double shear strength

$$= 2 \times 45.26 = 90.52 \text{ kN}$$

Nominal bearing strength of bolt

$$V_{dpb} = \frac{V_{npb}}{\gamma_{mb}} = \frac{2.5 \, k_b \cdot d \cdot t_p \cdot f_{ub}}{\gamma_{m.b}}$$

From table 2.1, for a 20 mm diameter bolt, $d_o = 22$ mm. Assuming rolled edge, e = 33 mm and pitch = 50 mm.

k_b will be smaller of $\left[\dfrac{e}{3 d_o}; \quad \dfrac{p}{3 \, d_o} - 0.25; \quad \dfrac{f_{ub}}{f_u}; \quad 1.0 \right]$.

$\therefore \quad k_b = \left[\left(\dfrac{33}{3 \times 22} \right) = 0.5, \dfrac{50}{3 \times 22} - 0.25 = 0.5, \dfrac{400}{400} = 1.0 \text{ and } 1.0 \right] = 0.5$ (Smaller values)

$$V_{dpb} = \frac{2.5 \times 0.5 \times 20 \times 10 \times 410}{1.25} = 82000 \text{ N} = 82 \text{ kN}$$

$\therefore$ Bolt value for single shear

$\qquad$ = (Smaller of single shear strength V_{dsb} and bearing strength V_{dpb}) = **45.26 kN**

Bolt value for double shear

$\qquad$ = (Smaller of double shear strength and V_{dpb}) = **82 kN**

Ex. 3.2: *Two flats (Fe 410 Grade Steel), each 210 mm × 8 mm are to be joined using 20 mm diameter, 4.6 grade bolts, to form a lap joint. The joint is supposed to transfer a factored load of 250 kN. Design the joint and determine suitable pitch for the bolts.*

Sol.: For Fe410 grade steel, $f_u = 410$ MPa.

$\qquad$ For bolts of grade 4.6, $f_{ub} = 400$ MPa

$\qquad$ For 20 mm dia. bolt, $A_{nb} = 245$ mm² (Table 2.2)

$\qquad$ Diameter of hole $d_o = d_h = 20 + 2 = 22$ mm (Table 2.1)

Single shear strength of bolt,

$$V_{dsb} = \frac{V_{nsb}}{\gamma_{mb}} = \frac{f_{ub}}{\sqrt{3}} \frac{(n_n A_{nb} + n_s A_{sb})}{1.25}$$

$$= \frac{400}{\sqrt{3}} \frac{(1 \times 245 + 0)}{1.25} = 45264 \text{ N}$$

$$\left[\begin{array}{l} n_s = 0, \text{ becaue no shear} \\ \text{plane without threads is} \\ \text{expected to intercept in} \\ \text{this type of failure} \end{array}\right]$$

$$= 45.26 \text{ kN}$$

Number of bolts required $= \dfrac{\text{Factored load}}{V_{dsb}} = \dfrac{250}{45.26} = 5.52$ say 6

Arranging the bolts in three lines as shown in Fig. 2.5.

Strength of joint per pitch length on the basis of shear on bolts $= 2 \times 45.26 = 90.52$ kN.

Equating this to tensile strength of flat per pitch length

$$T_{dn} = 0.9 \frac{f_u}{\gamma_{m1}} \cdot A_n \quad \text{where, } A_n = (p - d_h) \cdot t \text{ where } d_h = \text{diameter of the hole}$$

$$= 0.9 \frac{f_u}{\gamma_{m1}} (p - d_h) \cdot t$$

or $\quad 0.9 \times \dfrac{410}{1.25} (p - 22) \times 8 = 90.52 \times 10^3$

$\therefore \qquad\qquad p = 60.33 \text{ mm} > (2.5 \text{ d} = 2.5 \times 20 = 50 \text{ mm})$

$\therefore \qquad\qquad$ Adopt $p \simeq$ **65 mm**

Edge distance available $= \dfrac{210 - 2 \times 65}{2} = 40$ mm,

Check for bearing strength

k_b is smaller of $\left[\dfrac{e}{3d_o}; \quad \dfrac{p}{3d_o} - 0.25; \quad \dfrac{f_{ub}}{f_u}; \quad 1.0\right]$

i.e. $\left[\dfrac{40}{3 \times 22} = 0.60; \quad \dfrac{65}{3 \times 22} - 0.25 = 0.73; \quad \dfrac{400}{410} = 0.975; \quad 1.0\right]$.

Hence, $\qquad\qquad k_b = 0.60$

Bearing strength of bolt

$$V_{dpb} = \frac{V_{npb}}{\gamma_{mb}}$$

$$= \frac{2.5 \, k_b \, d \, t_p \, f_u}{\gamma_{mb}}$$

$$= \frac{2.5 \times 0.6 \times 20 \times 8 \times 410}{1.25}$$

$$= 78.72 \text{ kN} > 45.26 \text{ kN}$$

Hence design needs no revision.

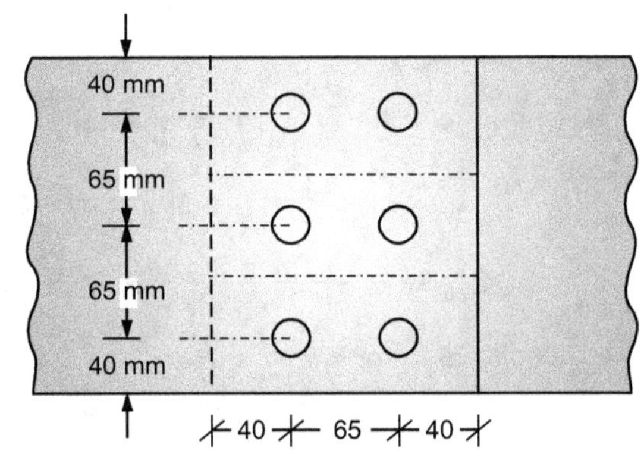

Fig. 3.7

Ex. 3.3: *Design the lap joint between the plates of sizes 100 × 16 mm thick and 100 × 10 mm thick so as to transmit a factored load of 110 kN using single row of 16 mm bolts of grade 4.6 and grade 410 plate.*

Sol.: For Fe 410 grade steel, $f_u = 410$ MPa.

For bolts of grade 4.6, $f_{ub} = 400$ MPa.

For 16 mm diameter bolts, $A_{nb} = 157 \text{ mm}^2$ (Table 2.2)

Diameter of hole, $d_h = 16 + 2 = 18$ mm. (Table 2.1)

Single shear strength of bolt,

$$V_{dsb} = \frac{V_{nsb}}{\gamma_{mb}} = \frac{f_{ub}}{\sqrt{3}} \frac{(n_n \cdot A_{nb} + n_s A_{sb})}{1.25}$$

$$= \frac{400}{\sqrt{3}} \frac{(1 \times 157 + 0)}{1.25} = 29006 \text{ N} = 29 \text{ kN}$$

Number of bolts required $= \dfrac{\text{Factored laod}}{V_{dsb}} = \dfrac{110}{29} = 3.79$ say 4

Arranging the bolts in single line as shown in Fig. 3.8.

Equating this to tensile strength of plate per pitch length

$$T_{dn} = 0.9 \frac{f_u}{\gamma_{m1}} (p - d_h) \cdot t$$

$$0.9 \times \frac{410}{1.25}(p - 18) \times 10 = 29000$$

$\therefore \qquad p = 27.82 < (2.5\ d = 2.5 \times 16 = 40)$

$\therefore$ Provide pitch, $p = 40$ mm and edge distance $e = 30$ mm.

Check for bearing strength

k_b is smaller of $\left[\dfrac{e}{3d_o}; \ \dfrac{p}{3d_o} - 0.25; \ \dfrac{f_{ub}}{f_u}; \ 1.0\right]$

i.e. $\left[\dfrac{30}{3 \times 18} = 0.56; \ \dfrac{40}{3 \times 18} - 0.25 = 0.49; \ \dfrac{400}{410} = 0.975; \ 1.0\right]$

Hence $k_b = 0.49$.

Bearing strength of bolt

$$V_{dpb} = \frac{V_{npb}}{\gamma_{mb}} = \frac{2.5\ k_b \cdot d \cdot t_p \cdot f_u}{\gamma_{mb}} = \frac{2.5 \times 0.49 \times 16 \times 10 \times 410}{1.25}$$

$$= 64288 \text{ N} = 64.29 \text{ kN} > 29 \text{ kN} \qquad \therefore \text{ O.K.}$$

$\therefore$ **Provide 4 – 16 mm ϕ bolts at 40 mm pitch with edge distance of 30 mm.**

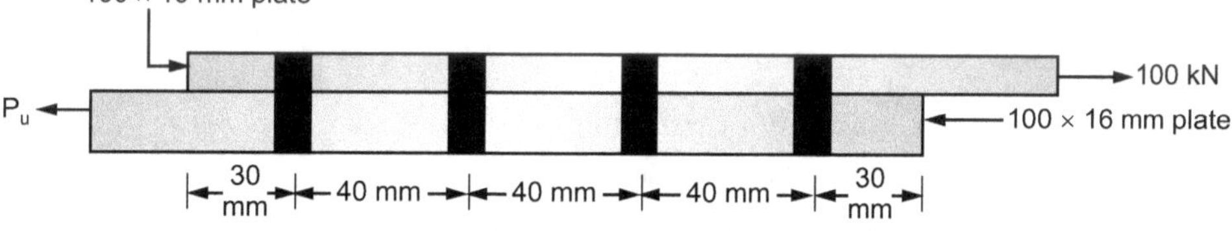

Fig. 3.8

Ex. 3.4: *A lap joint consists of two plates 200 × 12 mm connected by means of 20 mm diameter bolts of grade 4.6. All bolts are in one line. Calculate strength of single bolt and number of bolts to be provided in the joint.*

Sol.:

Nominal diameter of bolt $= 20$ mm

$\therefore$ Net area of bolt at thread $(A_{nb}) = 0.78 \times \dfrac{\pi}{4} \times d^2$

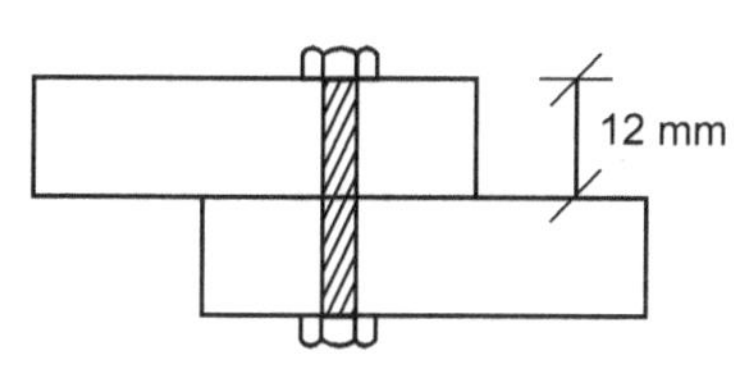

$$= 0.78 \times \frac{\pi}{4} \times 20^2$$

$$A_{nb} = 245.04 \text{ mm}^2$$

For Fe410 grade steel plate (assumed)

Ultimate stress for plate $\qquad\qquad f_y = 410$ N/mm^2

For 4.6 grade of bolt

Ultimate stress for bolt $(f_{ub}) = 4 \times 100 = 400$ N/mm^2

Yield stress for bolt $(f_{yb}) = 400 \times 0.6 = 240$ N/mm^2

Now find design shearing strength of bolt ($V_{ds}\,b$)

∴ We know that

∴
$$V_{d\,sb} = \frac{f_{ub}}{\sqrt{3} \times \gamma_{mb}}\,[nn \times A_{nb} + ns + A_{ns}]$$

Here number of shear plate with threat intercepting the shear plate $n_n = 1$.

Number of shear plate with out thread intercepting the shear plane $n_s = 0$.

∴
$$V_{d\,sb} = \frac{400}{\sqrt{3} \times 1.25} \times [1 \times 245.04 + 0]$$

γ_{mb} = Partial factor of safety for bolt material = 1.25

$V_{d\,sb} = 45.27 \times 10^3$ N

Now find design bearing strength of bolt ($V_{d\,sb}$)

$$V_{d\,pb} = 2.5 \times kb \times (d \times t) \times \frac{f_y}{\gamma_{mb}}$$

Here coefficient K_b is minimum of

1. $\left[\dfrac{\rho}{3dh},\ \dfrac{P}{3dh} - 0.25,\ \dfrac{f_{ub}}{f_u},\ 1\right]$

    (a)   Diameter of hole (dh) = Nominal diameter + 2 = 20 + 2 = 22 mm

    (b)   End distance (ρ) = 2d = 2 × 20 = 40 mm

    (c)   Pitch (ρ) = 2.5 d = 2.5 × 20 = 50 mm

        (i) $\dfrac{\rho}{3dh} = \dfrac{40}{3 \times 22} = 0.606$

        (ii) $\dfrac{\rho}{3dh} - 0.25 = \dfrac{50}{3 \times 22} - 0.25 = 0.507$

        (iii) $\dfrac{f_{ub}}{f_y} = \dfrac{400}{410} = 0.975$

and   (iv) 1

Hence K_b = 0.507 mm take minimum value

Now find design bearing strength of bolt (V_{dpb})

$$= 2.5 \times K_b \times (d \times t) \times \frac{f_u}{\gamma_{mb}} = 2.5 \times 0.507 \times (20 \times 12) \times \frac{410}{1.25}$$

$V_{dpb} = 99.77 \times 10^3$ N

Now find bolt value i.e. strength of bolt

∴   Bolt value = Minimum strength between shearing and bearing strength of bolt i.e. Minimum between V_{dsb} and V_{dpb}

$$= 45.27 \times 10^3 \text{ N}$$

Full strength of member = $0.9 \times \dfrac{f_u}{\gamma_m} \times$ Area of plate

$$= \frac{0.9 \times 410}{1.25}\,(250 - 1 \times 22) \times 12 = 630.54 \times 10^3 \text{ N}$$

Full strength of plate

∴ Number of bolts = $\dfrac{\text{Full strength of plate}}{\text{Bolt value}}$

$$= \frac{630.54 \times 10^3}{45.27 \times 10^3} = 13.92 \text{ say 14 Nos.}$$

Ex. 3.5: *A lap joint consists of two plates of 100 mm $\times$ 10 mm connected by 20 mm diameter bolts of grade 4.6. All bolts are in one line. Calculate strength of single bolt and number of bolts to be provided in the joint.*

Sol.:

$$\text{Nominal diameter of bolt} = 20 \text{ mm}$$

$$\text{Net area of bolt at thread} = 0.78 \times \frac{\pi}{4} \times d^2 = 0.78 \times \frac{\pi}{4} \times 20^2$$

$$A_{nb} = 245.04 \text{ mm}^2$$

For Fe 410 grade steel plate (assumed)

$$\text{Ultimate stress for plate, } f_u = 410 \text{ N/mm}^2$$

For 4.6 grade of bolt

$$\text{Ultimate stress of bolt, } f_{ub} = 4 \times 100 = 400 \text{ N/mm}^2$$

$$\text{Yield stress for bolt, } f_{yb} = 400 \times 0.6 = 240 \text{ N/mm}^2$$

Design shearing strength of bolt

$$V_{dsb} = \frac{f_{ub}}{\sqrt{3} \ \gamma_{mb}} \ [n_n - A_{nb} + n_s \ A_{ns}]$$

Here $n_n = 1$, $n_s = 0$ and $\gamma_{mb} = 1.25$

$$\therefore \qquad V_{dsb} = \frac{400}{\sqrt{3} \times 1.25} [1 \times 245.04 + 0] = 45.27 \times 10^3 \text{N} = 45.27 \text{ kN}$$

Design bearing strength of bolt

$$V_{dpb} = 2.5 \ k_b \times (d \times t) \times \frac{f_u}{\gamma_{mb}}$$

where coefficient k_b is minimum of $\left[\dfrac{e}{3d_h} \ ; \ \dfrac{p}{3d_h} - 0.25; \ \dfrac{f_{ub}}{f_u}; \ 1\right]$

(i) Diameter of hole, $d_h = d + 2 = 20 + 2 = 22$ mm

(ii) End distance, $e = 2d = 2 \times 20 = 40$ mm

(iii) Pitch, $p = 2.5 \ d = 2.5 \times 20 = 50$ mm

$$\therefore \qquad k_b = \left[\frac{40}{2 \times 22}; \ \frac{50}{3 \times 22} - 0.25; \ \frac{400}{410}; \ 1.0\right] \text{ whichever is minimum}$$

$$= [0.606; \ 0.507; \ 0.975; \ 1.0] \text{ whichever is minimum} = 0.507$$

$$\therefore \qquad V_{dpb} = 2.5 \times 0.507 \times (20 \times 10) \times \frac{410}{1.25} = 83.148 \times 10^3 \text{ N}$$

$$V_{dpb} = 83.15 \text{ kN}$$

Strength of single bolt

i.e. Bolt value, $B_v = $ Minimum of V_{dsb} and $V_{dpb} = 45.27$ kN

$$\text{Full strength of plate} = 0.9 \times \frac{f_u}{\gamma_m} \times \text{Area of plate}$$

$$= 0.9 \times \frac{410}{1.25} \times (100 - 1 \times 22) \times 10 = 230256 \text{ N}$$

$$\therefore \qquad \text{Number of bolts} = \frac{\text{Full of strength of plate}}{\text{Bolt value}} = \frac{230256}{45.27 \times 10^3} = 5.08 \text{ say 6 No.}$$

Ex. 3.6: *Design the Lap Joint between two plates of sizes 100 $\times$ 12 mm thick each, to transmit a factored load of 80 kN using single row of bolts of grade 4.6 and 410 grade of plate. (Assume data if required).*

Sol.: Data : P = 80 kN, f_u = 410 mPa, f_{ub} = 400 Mpa, Assume 15 mm dia. bolt.

Step 1: $d_n = d_o = 16 + 2 = 18$ mm

$$A_{nb} = 157 \text{ mm}^2$$

Step 2: Single shear strength of bolt - V_{dsb}

$$V_{dsb} = \frac{f_{ub}}{\sqrt{3}} \left[\frac{n_h\, A_{nb} + n_s\, A_{sb}}{V_{mb}} \right] = \frac{400}{\sqrt{3}} \left[\frac{157}{1.25} \right]$$

$$= 29000 \text{ N} = 29 \text{ kN in single shear}$$

Step 3: Bearing strength of bolt

where K_b is smaller

$$p = 50 \text{ mm}$$
$$\rho = 40 \text{ mm}$$
$$\frac{\rho}{3d_o} = \frac{40}{3 \times 18} = 0.74$$
$$\frac{P}{3d_o} - 0.25 = 0.67$$
$$\frac{400}{410} = 0.975 = 1.0$$

$\therefore$
$$K_b = 0.67$$
$$V_{dpb} = \frac{2.5\, K_b \cdot dt \cdot f_u}{1.25} = 105.48 > 29 \text{ kN} = 105.48 \text{ kN}$$

Step 4: Tensile strength of plate per pitch length

$$P = 50 \text{ mm}$$
$$\rho = 40 \text{ mm}$$
$$T_{dn} = 0.9 \frac{f_u}{\gamma_{m1}} \ (P - d_n)\, t = 0.9 \times \frac{410}{1.25} \ (50 - 18) \times 12$$

$$\boxed{T_{dn} = 113.35 \text{ kN}}$$

Least Bolt value, minimum of three

$$B_v = 29 \text{ kN} = \frac{80 \text{ kN}}{29} = 2.875 \text{ say 3 No. bolts required}$$

Ex. 3.7: *An inclined truss member consists of 2 angles 100 × 75 × 10 mm connected back-to-back with longer leg to gusset plate 12 mm thick. Design the bolted joint to transfer a design force of 750 kN. Steel Fe410 and bolts are of grade 4.6*

Ans. $G = 100 \times 75 \times 100$ mm, $P = 750$ kN, $f_u = 410$ mpa, $f_{ub} = 400$ mpa

For 16 mm dia., bolt $d_n = d_o = 16 + 2 = 18$ mm

$A_{nb} = 157$ mm^2

Single shear strength of bolt = ?

$$V_{dsb} = \frac{f_{ub}}{\sqrt{3}} \left(\frac{h_n \times A_{nb}}{\gamma_{mb}} \right) = \frac{400}{\sqrt{3}} \left(\frac{1 \times 157}{1.25} \right)$$

$$\boxed{V_{dsb} = 29 \text{ kN}}$$

Double shear strength of bolt = $2 \times 29 = 58$ kN

Bearing strength of thinner plate

$$V_{dph} = \frac{2.5\, kb \cdot d \cdot t \cdot f_{ub}}{\gamma_{mb}}$$

Assume
$$P = 3d = 3 \times 16 = 48 = 50 \text{ mm};$$
$$e = 2d = 2 \times 16 = 32 \approx 40 \text{ mm}$$
$$p = 50 \text{ mm} \qquad e = 40 \text{ mm}$$

k_b is least of,

(i)
$$\frac{\rho}{3d_o} = \frac{40}{3 \times 18} = 0.74$$

(ii)
$$\frac{P}{3d_o} - 0.25 = \frac{50}{3 \times 18} - 0.25 = 0.67$$

(iii)
$$\frac{f_{ub}}{f_u} = \frac{400}{410} = 0.975; \ 1.0$$

$\therefore$
$$\boxed{K_b = 0.67}$$

$\because$
$$V_{dpb} = 2.5 \times 0.67 \times 16 \times 12 \times \frac{400}{1.25} = 102912 \ N$$

$$\boxed{V_{dpb} = 102.91 \ kN}$$

$\therefore$ Least Bolt value

$$B_r = \frac{P_u}{No. \ of \ Bolt}$$

$$No. \ of \ Bolts \ reqd. \ = \frac{P_u}{B_v} = \frac{750 \times 10^3}{58}$$

$$\boxed{No. \ of \ Bolt = 12.93 \approx 14 \ Bolt}$$

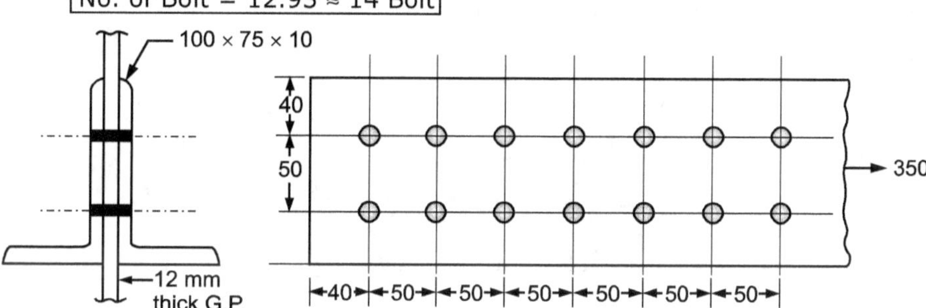

Fig. 3.10

Ex. 3.8: *Design the lap joint for the plates of sizes 100 × 12 mm and 100 × 8 mm thick connected, so as to transmit a factored load of 80 kN using single row of 16 mm diameter bolts of grade 4.6 and plates of 410 grade.*

Sol.: Given, Minimum thickness of plate = 8 mm,

Factored load, P_u = 80 kN, d = 16 mm

f_{ub} = 400 N/mm^2 and f_u = 410 N/mm^2

Dia. of hole, d_o = 16 + 2 = 18 mm

$A_{nb} = 0.78 \times \frac{\pi}{4} d^2 = 0.78 \times \frac{\pi}{4} \times 16^2 = 156.82$ mm^2

$\therefore$ Single shear strength of bolt

$$V_{dsb} = \frac{V_{nsb}}{\gamma_{mb}} = \frac{f_{ub}}{\sqrt{3}} \times \frac{(n_n \cdot A_{nb} + n_s \cdot A_{sb})}{1.25}$$

$$= \frac{400}{\sqrt{3}} \times \frac{(1 \times 156.82 + 0)}{1.25} = 28.97 \ kN$$

$$Number \ of \ bolts \ required \ = \frac{P_u}{V_{dsb}} = \frac{80}{28.97} = 2.76 \ say \ 3 \ Nos.$$

Arranging the bolts in single line

To calculate pitch, equate V_{dsb} to T_{dn}.

i.e. $V_{dsb} = T_{dn}$

$$28.97 \times 10^3 = 0.9 \frac{f_u}{\gamma_{m_1}} (p - d_o) \cdot t$$

$$28790 = 0.9 \times \frac{410}{1.25} (p - 18) \times 8$$

$\therefore$ $p = 30.26$ mm $< (2.5 \ d = 2.5 \times 16 = 40$ mm$)$

$\therefore$ Provide pitch, p = 40 mm and edge distance, e = 30 mm.

$$K_b = \left[\frac{e}{3d_o} \; ; \; \frac{p}{3d_o} - 0.25 \; ; \; \frac{f_{ub}}{f_u} \; ; \; 1.0\right] \text{ whichever is smaller}$$

$$= \left[\frac{30}{3 \times 18} = 0.55 \; ; \; \frac{40}{3 \times 18} - 0.25 = 0.49 \; ; \; \frac{400}{410} = 0.975; \; 1.0\right]$$

$$= 0.49$$

$\therefore$ Bearing strength $V_{dpb} = \dfrac{2.5 \, k_b \cdot d \cdot t_p \cdot f_u}{\gamma_{mb}} = \dfrac{2.5 \times 0.49 \times 16 \times 8 \times 410}{1.25}$

$$= 51430 \text{ N} = 51.43 \text{ kN} > V_{dsb} \quad \therefore \text{ O.K.}$$

Ex. 3.9: *A lap joint consists of two places 180 × 10 mm connected by means of 20 mm diameter bolts of grade 4.6. All bolts are in one line. Calculate strength of single bolt and number of bolts to be provided.*

Ans. Given : t = 10 mm, d = 20 mm, $\therefore$ d_o = 22 mm, f_{ub} = 400 N/mm^2

f_u = 410 N/mm^2

$$A_{nb} = 0.78 \times Ag = 0.78 \times \frac{\pi}{4} \times 20^2 = 245 \text{ mm}^2$$

(i) Design shear strength of bolt (single shear)

$$V_{dsb} = \frac{f_{ub}}{\sqrt{3}} \times \frac{(n_n \cdot A_{nb} + n_s \cdot A_{sb})}{1.25}$$

$$= \frac{400}{\sqrt{3}} \times \frac{(1 \times 245 \times 0)}{1.25} = 45260 \text{ N} = 45.26 \text{ kN}$$

(ii) Assume

$$p = 2.5 \, d = 2.5 \times 20 = 50 \text{ mm}$$

$$e = 2.0 \, d = 2.0 \times 20 = 40 \text{ mm}$$

$\therefore$

$$K_b = \left[\frac{e}{3d_o} \; ; \; \frac{p}{3d_o} - 0.25 \; ; \; \frac{f_{ub}}{f_u} \; ; \; 1.0\right] \text{ whichever is smaller}$$

$$= \left[\frac{40}{3 \times 22} = 0.60 \; ; \; \frac{50}{3 \times 22} - 0.25 = 0.507 \; ; \; \frac{400}{410} = 0.975 \; ; \; 1.0\right] = 0.507$$

$\therefore$ $V_{dpb} = \dfrac{2.5 \, k_b \cdot d \cdot t \cdot f_u}{1.25} = \dfrac{2.5 \times 0.507 \times 20 \times 10 \times 410}{1.25}$

$$= 83148 \text{ N} = 83.14 \text{ kN}$$

$\therefore$ Bolt value = Minimum of V_{dsb} and V_{dpb}

$$B_v = 45.26 \text{ kN}$$

$\therefore$ Number of bolts $= \dfrac{\text{Full strength of plate}}{B_v} = \dfrac{0.9 \, f_u \, (b - d_o) \cdot t/\gamma_{m_0}}{B_v}$

$$= \frac{0.9 \times 410 \, (180 - 22) \cdot 10/1.25}{45260} = 10.3 \text{ n say 11 No}$$

Ex. 3.10: *Determine the bolt value 16 mm diameter bolts of 4.6 grade used to connect two angles ISA 80 × 50 × 6 mm placed back to back on both sides of 8 mm thick gusset plate. Also determine the number of bolts required for the joint when it carries a direct factored load of 106.5 kN. Draw neat sketch of designed connection.*

Sol.: The angles are connected on both sides of gusset plate. Hence the bolts will be in double shear and will bear against 8 mm thick (least of 8 and 2 × 6 mm) gusset plate.

Assume Fe410 grade of angle; f_u = 410 MPa.

For bolts of grade 4.6, f_{ub} = 400 MPa.

For 16 mm diameter bolt, A_{nb} = 157 mm^2,

d_h and d_o = 18 mm.

$\gamma_{mb} = \gamma_{m1}$ = partial factor of safety for bolt and angles = 1.25.

Double shear strength of bolts

$$V_{dsb} = \frac{2\,V_{nsb}}{\gamma_{mb}} = 2\,\frac{f_{ub}}{\sqrt{3}}\,\frac{(n_n\,A_{nb} + n_s\,A_{sb})}{1.25}$$

$$= 2 \times \frac{400}{\sqrt{3}} \cdot \frac{(1 \times 157 + 0)}{1.25} = 58012 \text{ N} = 58.01 \text{ kN}$$

Bearing strength of thinner plate

$$V_{dpb} = \frac{V_{npb}}{\gamma_{nb}} = 2.5\,k_b \cdot d \cdot t \cdot \frac{f_{ub}}{\gamma_{mb}}$$

Assume $p = 3d = 3 \times 16 = 48 \approx 50$ mm and

$e = 2d = 2 \times 16 = 32 \approx 40$ mm.

k_b is least of $\left[\dfrac{e}{3d_o};\ \dfrac{p}{3d_o} - 0.25;\ \dfrac{f_{ub}}{f_u};\ 1.0\right]$

i.e. $\left[\dfrac{40}{3 \times 18} = 0.74;\ \dfrac{50}{3 \times 18} - 0.25 = 0.67;\ \dfrac{400}{410} = 0.975;\ 1.0\right]$

Hence $k_b = 0.67$.

$\therefore\ V_{dpb} = 2.5 \times 0.67 \times 16 \times 8 \times \dfrac{410}{1.25} = 70323$ N $= 70.32$ kN

$\therefore$ Bolt Value B_v = Least of V_{dsb} or V_{dpb} = **58.01 kN**

No. of bolts required $= \dfrac{P_u}{B_v} = \dfrac{106.5}{58.01} = 1.83$ **say 2 No.**

Min. Pitch $= 2.5\,d = 2.5 \times 16 =$ **40 mm**

Edge distance = 27 mm say **30 mm.**   (See Table 2.1 (b)).

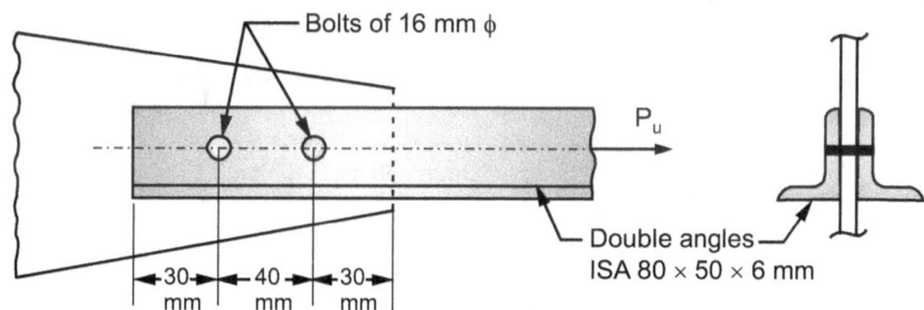

Fig. 3.11

Ex. 3.11: *Design a lap joint to carry a service load of 350 kN. The bolts are 20 mm diameter 4.6 grade and placed in double row, plates are made of two flats 10 mm thick (Fe 410 grade steel).*

Sol.: $f_u = 410$ MPa, $f_{ub} = 400$ MPa.

For 20 mm dia. bolts of grade 4.6, $A_{nb} = 245$ mm^2.

Diameter of hole, $d_h = d_o = 20 + 2 = 22$ mm.

Single shear capacity of bolt

$$V_{dsb} = \frac{f_{ub}}{\sqrt{3}}\,\frac{(n_n\,A_{nb} + n_s\,A_{sb})}{\gamma_{mb}} = \frac{400}{\sqrt{3}}\,\frac{(1 \times 245 + 0)}{1.25}$$

$$= 45264 \text{ N} = 45.26 \text{ kN}$$

No. of bolts required $= \dfrac{P_u}{V_{dsb}} = \dfrac{P \cdot \gamma_f}{V_{dsb}} = \dfrac{350 \times 1.5}{45.26} = 11.59$ say **12 No.**

Arranging the bolts in double lines as shown in Fig. 3.12. Strength of joint per pitch length on the basis of shear on bolts. $2 \times 45.26 = 90.52$ kN.

Equating this to tensile capacity of flat per pitch length

$$T_{dn} = 0.9 \frac{f_u}{\gamma_{m1}} (p - d_h) \times t = 90.52 \times 10^3$$

$$\therefore \quad 0.9 \times \frac{410}{1.25} (p - 22) \times 10 = 90.52 \times 10^3$$

$$\therefore \qquad\qquad p = 52.66 > (2.5\, d = 2.5 \times 2.5 \times 20 = 50 \text{ mm})$$

$$\therefore \qquad\qquad \text{Adopt } p \simeq \textbf{60 mm}$$

Providing edge distance = 33 mm say **40 mm**.

Check for bearing strength

k_b is smaller of $\left[\dfrac{e}{3d_o}; \quad \dfrac{p}{3d_o} - 0.25; \quad \dfrac{f_{ub}}{f_u}; \quad 1.0 \right]$

i.e. $\left[\dfrac{40}{3 \times 22} = 0.60; \quad \dfrac{60}{3 \times 22} - 0.25 = 0.65; \quad \dfrac{400}{410} = 0.975; \quad 1.0 \right]$

Hence $k_b = 0.60$

Bearing strength of bolt

$$V_{dp} = \frac{2.5\, k_b \cdot d \cdot t_p \cdot f_u}{\gamma_{mb}} = \frac{2.5 \times 0.60 \times 20 \times 10 \times 410}{1.25}$$

$$= 98400 \text{ N} = 98.40 \text{ kN} > 45.26 \text{ kN}$$

Hence design is O.K.

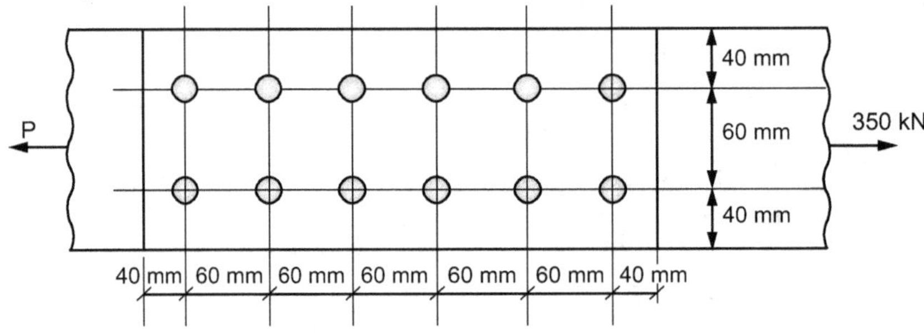

Fig. 3.12

Ex. 3.12: *A flat 250 × 10 mm is connected to a gusset plate using 20 mm φ bolts of 4.6 grade. Design the connection using lap joint. Also draw the sketch showing design connection.*

Sol.: d = 20 mm, ∴ $d_o = d_h = 20 + 2 = 22$ mm.

As the load is not given, we must calculate full strength of members

$$= \frac{0.9 \times f_u \times \text{c/s Area of plate}}{\gamma_m} = \frac{0.9 \times 410 \times (250 \times 10)}{1.25}$$

$$P_u = 738000 \text{ N} = 738 \text{ kN}$$

Single shear strength of 20 mm bolt.

$$V_{dsb} = \frac{f_{ub}}{\sqrt{3}} \frac{(n_n \times A_{nb})}{\gamma_{mb}} = \frac{400}{\sqrt{3}} \frac{(1 \times 245)}{1.25} = 45264 \text{ N} = 45.26 \text{ kN}$$

No. of bolts required $= \dfrac{P_u}{V_{dsb}} = \dfrac{738}{45.26} = 16.30$ say **18 No.** (even number of bolts)

Arranging the bolts in the three lines as shown in Fig. 3.13.

Strength of joint per pitch length on the basis of shear on bolts = 3 × 45.26 = 135.78 kN.

Equating this to tensile strength of flat per pitch length.

$$T_{dn} = 0.9 \frac{f_u}{\gamma_{m1}} (p - d_h) \cdot t$$

or $0.9 \times \dfrac{410}{1.25} (p - 22) \times 10 = 135.78 \times 10^3$

∴ $p = 68$ mm $> (2.5\ d = 2.5 \times 20 = 50$ mm) say **70 mm**.

Providing edge distance, e = 33 mm say **40 mm**.

Check for bearing strength

k_o is smaller of $\left[\dfrac{e}{3d_o};\ \dfrac{p}{3d_o} - 0.25;\ \dfrac{f_{ub}}{f_u};\ 1.0\right]$

i.e. $\left[\dfrac{40}{3 \times 22} = 0.60;\ \dfrac{70}{3 \times 22} - 0.25 = 0.81;\ \dfrac{400}{410} = 0.975;\ 1.0\right]$

Hence $k_b = 0.60$

Bearing strength of bolt.

$$V_{dpb} = \frac{2.5\ k_b \cdot d \cdot t_p \cdot f_u}{\gamma_{mb}} = \frac{2.5 \times 0.6 \times 20 \times 10 \times 410}{1.25}$$

$$= 98400\ N = \textbf{98.40 kN} > \textbf{45.26 kN} \qquad \therefore\ \text{Design is O.K.}$$

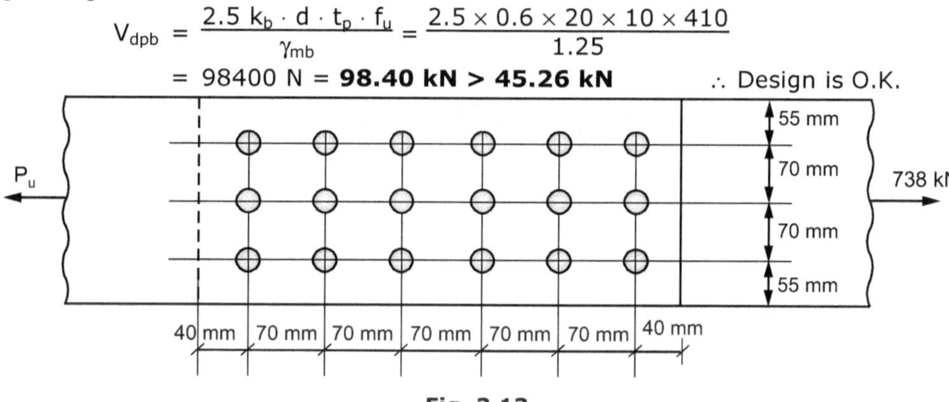

Fig. 3.13

Ex. 3.13: *A single jointed lap joint is used to connect two plates 10 mm thick. If bolts of 20 mm is provided at 50 mm pitch, calculate the efficiency of the joint. Use Fe 410 plate and 4.6 grade bolt.*

Sol.: d = 20 mm, ∴ $d_o = d_h = 20 + 2 = 22$ mm, $A_{nb} = 245$ mm^2.

Single shear strength of bolt.

$$V_{dsb} = \frac{f_{ub}}{\sqrt{3}} \frac{(n_n \cdot A_{nb})}{\gamma_{mb}} = \frac{400}{\sqrt{3}} \frac{(1 \times 245)}{1.25} = 45264\ N = 45.26\ kN$$

Assuming edge distance, e = 40 mm

$k_b = 0.5$

k_b is smaller of $\left[\dfrac{e}{3d_o};\ \dfrac{p}{3d_o} - 0.25;\ \dfrac{f_{ub}}{f_u};\ 1.0\right]$

i.e. $\left[\dfrac{40}{3 \times 22} = 0.60;\ \dfrac{50}{3 \times 22} - 0.25 = 0.50;\ \dfrac{400}{410} = 0.975;\ 1.0\right]$

Bearing strength of bolt

$$V_{dpb} = \frac{2.5\ k_b \cdot d \cdot t_p \cdot f_u}{\gamma_{mb}} = \frac{2.5 \times 0.5 \times 20 \times 10 \times 410}{1.25} = 82000\ N = 82\ kN$$

Strength of net section per pitch length

$$T_{dn} = \frac{0.9\ f_u\ (p - d_h) \cdot t}{\gamma_{m1}} = \frac{0.9 \times 410\ (50 - 22) \times 10}{1.25} = 82656\ N = 82.65\ kN$$

Strength of joint per pitch length = 45.26 kN (least of V_{dsb}, V_{dpb}, T_{dn})

Strength of plate per pitch length

$$= 0.9 \times f_u \times p \times \frac{t}{\gamma_{m1}} = 0.9 \times 410 \times 50 \times \frac{10}{1.25} = 147600\ N = 147.60\ kN$$

∴ Efficiency of joint $= \dfrac{45.26}{147.60} \times 100 = \textbf{30.66\%}$

Ex. 3.14: *A single bolted double cover butt joint is used to connect two plates 10 mm thick as shown in Fig. 3.15. Assuming 20 mm bolts of 4.6 grade at 50 mm pitch, calculate the efficiency of the joint. Use Fe 410 grade plate.*

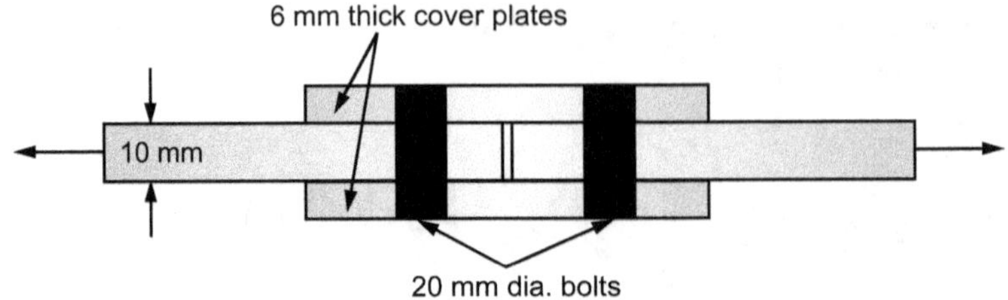

Fig. 3.14

Sol.: $d = 20$ mm, $\therefore d_o = d_h = 20 + 2 = 22$ mm, $A_{nb} = 245$ mm^2.

Double shear strength of bolt.

$$V_{dsb} = \frac{2f_{ub}}{\sqrt{3}} \frac{(n_n \cdot A_{nb})}{\gamma_{mb}} = \frac{2 \times 400}{\sqrt{3}} \frac{(1 \times 245)}{1.25} = 90528 \text{ N} = 90.53 \text{ kN}$$

Assuming edge distance $e = 40$ mm.

k_o is smaller of $\left[\dfrac{e}{3d_o}; \quad \dfrac{p}{3d_o} - 0.25; \quad \dfrac{f_{ub}}{f_u}; \quad 1.0\right]$

i.e. $\left[\dfrac{40}{3 \times 22} = 0.60; \quad \dfrac{50}{3 \times 22} - 0.25 = 0.50; \quad \dfrac{400}{410} = 0.975; \quad 1.0\right]$

i.e. $k_b = 0.5$

Bearing strength of bolt

$$V_{dpb} = \frac{2.5\, k_b \cdot d \cdot t_p \cdot f_u}{\gamma_{mb}} = \frac{2.5 \times 0.5 \times 20 \times 10 \times 410}{1.25} = 82000 \text{ N} = 82 \text{ kN}$$

Strength of plate in tearing

$$= \frac{0.9\, f_u\, (p - d_h) \cdot t_p}{\gamma_{m1}} = \frac{0.9 \times 410\, (50 - 22) \times 10}{1.25} = 82000 \text{ N} = 82 \text{ kN}$$

$\therefore$ Strength of solid plate per pitch length

$$= \frac{0.9\, f_u\, pt}{\gamma_{m1}} = \frac{0.9 \times 410 \times 50 \times 10}{1.25} = 147600 \text{ N} = 147.60 \text{ kN}$$

Efficiency of joint $= \dfrac{82}{147.60} \times 100 = \textbf{55.55\%}$

Ex. 3.15: *Design joint B of a roof truss as shown in Fig. 3.16. The members are connected by 16 mm diameter bolts of 4.6 grade to the gusset plate 12 mm thick. The members are subjected to factored forces indicated against arrow head.*

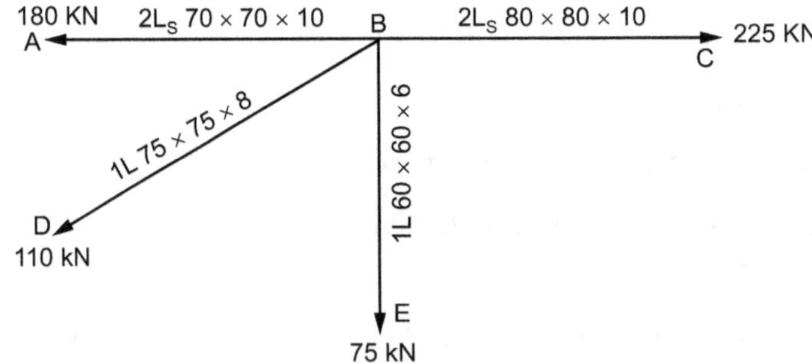

Fig. 3.15

Sol.: Here $f_u = 410$ MPa and $f_{ub} = 400$ MPa.

For 16 mm diameter bolt, $d_h = d_o = 16 + 2 = 18$ mm

$$A_{nb} = 157 \text{ mm}^2$$

Single shear strength of bolt

$$V_{dsb} = \frac{f_{ub}}{\sqrt{3}} \frac{(n_n \times A_{nb})}{\gamma_{mb}} = \frac{400}{\sqrt{3}} \frac{(1 \times 157)}{1.25} = 29006 \text{ N} = 29 \text{ kN}$$

∴ Double shear strength of bolt = $2 \times 29 = 58$ kN

Bearing strength of bolt

$$V_{dpb} = \frac{2.5 \, k_b \cdot d \cdot t \cdot f_{ub}}{\gamma_{mb}}$$

k_o is smaller of $\left[\dfrac{e}{3d_o}; \quad \dfrac{p}{3d_o} - 0.25; \quad \dfrac{f_{ub}}{f_u}; \quad 1.0\right]$

Assume $p = 2.5 \, d = 2.5 \times 16 = 40$ say 50 mm

$e = 2 \, d = 16 \times 2 = 32$ say 40 mm

∴ k_b is least of $\left[\dfrac{40}{3 \times 18} = 0.74; \quad \dfrac{50}{3 \times 18} - 0.25 = 0.67; \quad \dfrac{400}{410} = 0.975; \quad 1.0\right]$

Hence $k_b = 0.67$.

∴ Bearing strength of bolt on

(i) $t = 6$ mm; $V_{dpb} = \dfrac{2.5 \times 0.67 \times 16 \times 6 \times 410}{1.25} = 52742 \text{ N} \approx 52.74 \text{ kN}$

(ii) $t = 8$ mm; $V_{dpb} = \dfrac{2.5 \times 0.67 \times 16 \times 8 \times 410}{1.25} = 70323 \text{ N} = 70.32 \text{ kN}$

(iii) $t = 12$ mm*; $V_{dpb} = \dfrac{2.5 \times 0.67 \times 16 \times 12 \times 410}{1.25} = 105485 \text{ N} = 105.48 \text{ kN}$

[For double angles thickness of thinner plate will be minimum of 2 × 10 mm (thickness of angles) and 12 mm (thickness of gusset plate) i.e. 12 mm]

Design of Member AB: (P_u = 180 kN)

The member is made of double angles ISA 70 mm × 70 mm × 10 mm with 12 mm gusset plate in between. Hence the bolts will be in double shear and bearing strength will be against 12 mm thick gusset plate. Therefore strength of bolt will be smaller of 58 kN and 105.48 kN.

∴ Number of bolts required $= \dfrac{180}{58} = 3.10$ say **4**.

Design of Member BC: (P_u = 225 kN)

The member is made of double angles ISA 80 mm × 80 mm × 10 mm with 12 mm gusset plate in between. Hence the strength of bolt will be same as that for members AB i.e. 58 kN.

∴ Number of bolts required $= \dfrac{225}{58} = 3.88$ say **4**.

Design of Member BD: (P_u = 110 kN)

The member is single angle ISA 75 mm × 75 mm × 8 mm connected to 12 mm thick gusset plate. Hence the bolts will be in single shear and bearing capacity will be against 8 mm thick angle leg.

Hence the strength of bolt will be 29.0 kN (smaller of 29.0 kN and 70.32 kN)

∴ Number of bolts required $= \dfrac{110}{29} = 3.79$ say **4**.

Design of Member BE: (P_u = 75 kN)

The member is made of single angle ISA 60 mm × 60 mm × 6 mm and connected to 12 mm thick gusset plate. Hence the bolts will be in single shear and bearing against 6 mm thick angle leg.

Hence strength of bolt will be 29.0 kN.
(Smaller of 29.0 kN and 52.74 kN)

$$\therefore \text{ Number of bolts required } = \frac{75}{29} = 2.58$$

say **3**. The details of connections of the members at joint B are shown in Fig. 3.16.

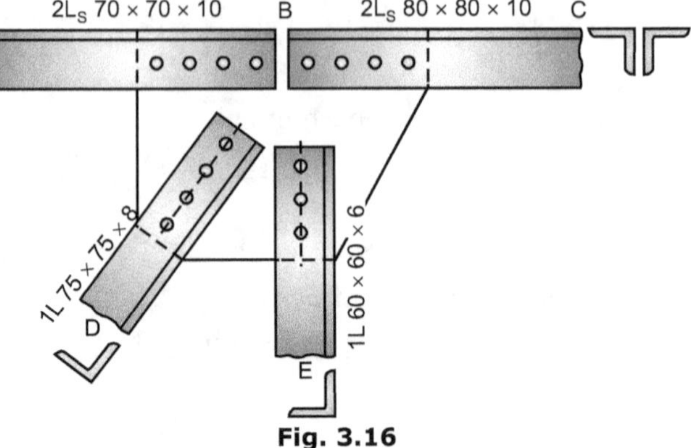

Fig. 3.16

(B) WELDED CONNECTIONS

3.14 INTRODUCTION

- When the various members of a structure are connected together by means of welds, the connection is called a welded connection.
- Welding is a bonding between two pieces of metals at faces rendered plastic or liquid by heat and pressure.
- The reference codes for welding are IS: 816, IS: 9595 and IS: 800 § 10.5.

3.15 TYPES OF WELDS

- Welds are of two types:
  1. Butt welds or Groove welds
  2. Fillet welds.

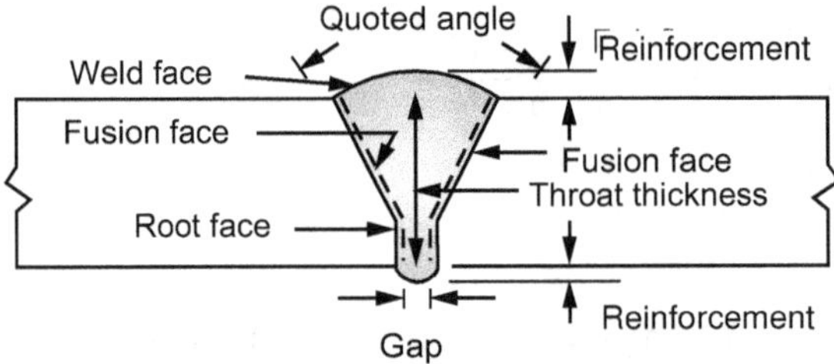

Fig. 3.17: Butt Weld

- When members are allowed to lap over each other, fillet welds are used.
- Butt welds are used when the members to be connected are in the same plane.

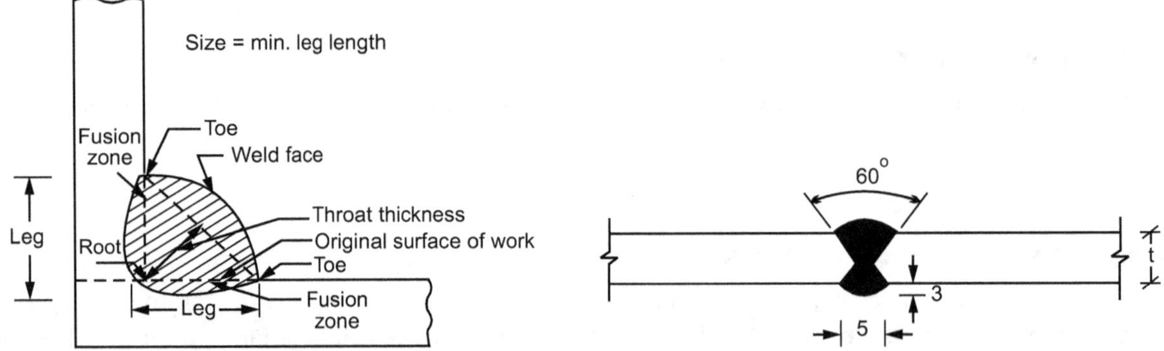

Fig. 3.18: Fillet Weld   **Fig. 3.19: A Typical Double V Butt Joint**

3.16 ADVANTAGES OF WELDED CONNECTIONS

1. As no holes are required for welding, the gross-sectional area of member is effective, hence more effective in taking loads.

2. Welded joints provide rigidity and strength.
3. Welded structures are comparatively lighter than same type of bolted structures.
4. A welded joint has a better finish and appearance.
5. No noise is produced in the welding process as in case of bolting process.
6. Welded joints are often economical as less labour and material are required for a joint.
7. For a welded structure maintenance and painting are easy.
8. The welding process offers an airtight and watertight joint and hence used in fluid retaining structures.
9. Welding process is quick and saves time of construction and requires less working space.
10. Any shape of joint can be made with ease.

3.17 DISADVANTAGES OF WELDED CONNECTIONS

1. Skilled labour and electricity are required for welding.
2. Internal stresses and warping are produced due to uneven heating and cooling.
3. Inaccuracies are caused by the deposition of weld metal in excess.
4. It is difficult to inspect a welded joint as compared to inspect a bolted joint.
5. Welded joints are more brittle and therefore their fatigue strength is less than the bolted joints.

3.18 SYMBOLS USED FOR WELDS

- A knowledge of welding symbols is essential for a site engineer so as to enable him to read the drawings.
- Symbols save lot of space in the drawing and as such descriptive notes can be omitted.

Table 3.6: Basic Types of Welds and Their Symbols

Form of weld	Section	Symbol
Fillet	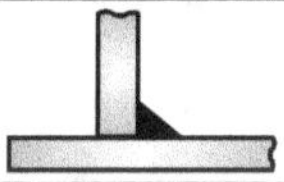	
Square butt		
Single-V-butt	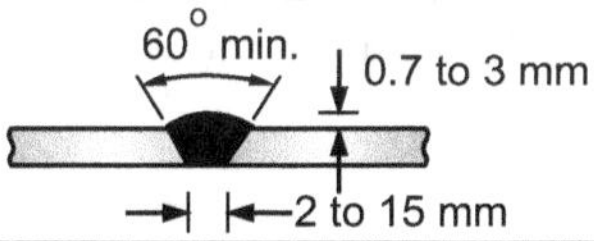 $60°$ min., 0.7 to 3 mm, 2 to 15 mm	
Double-V-butt	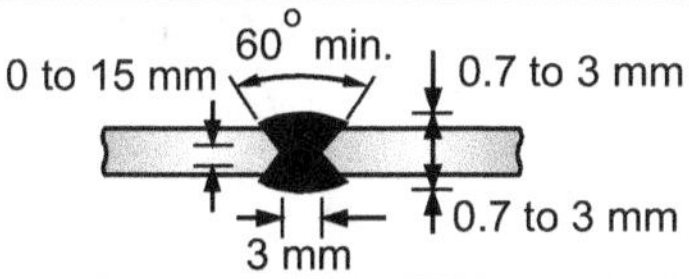 0 to 15 mm, $60°$ min., 0.7 to 3 mm, 3 mm, 0.7 to 3 mm	
Single-U-butt		
Double-U-butt	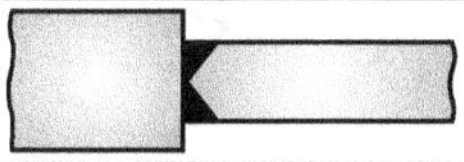	
Single-bevel-butt		
Double-bevel-butt	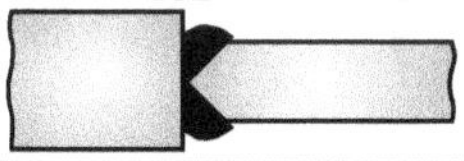	
Single-J-butt		
Double-J-butt	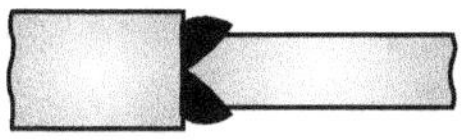	

3.19 SPECIFICATIONS OF FILLET WELD (IS: 816)

1. Size of fillet weld (s):

- The size of the normal fillet weld is taken equal to its minimum leg length as shown in Fig. 3.18. The size of fillet weld shall not be less than 3 mm.

2. Throat thickness of fillet weld (t):

- It is the perpendicular distance from the root of the fillet weld to the line joining its toes.
- It is taken equal to k × size of the fillet.
- The value of k for different angles between fusion faces is given below in Table 3.7.
- Almost in all cases right angle fillet welds are used for which k = 0.70.

Table 3.7: Values of k for Different Angles Between Fusion Faces

Angles between fusion faces	$60^\circ - 90^\circ$	$91^\circ - 100^\circ$	$101^\circ - 106^\circ$	$107^\circ - 113^\circ$	$114^\circ - 120^\circ$
Constant k	0.70	0.65	0.60	0.55	0.50

$\therefore$ $t = k \times$ size of fillet weld

        $t = 0.70 \times s$ ... (3.12)

where, $k = 0.70$ for angle between fusion faces 60° to 90°

(A most general case)

3. Effective length of fillet weld:

- The effective length of a fillet weld is taken equal to its actual length minus twice the weld size.

4. Minimum Length:

- The effective length of fillet weld shall not be less than 4 × size of the weld.

5. Intermittent weld:

- Fillet welds can be provided continuously or intermittently.
- Effective length of intermittent weld > larger of (4 × size of weld or 40 mm).

6. Clear spacing between effective length:

- Clear spacing between effective length of intermittent fillet weld < 12 t and 16 t for compression and tension joint respectively provided it shall not be more than 200 mm in any case, where t = thickness of thinner plate/part.

7. Minimum Size of fillet weld:

- The minimum size of the first run or of a single run fillet weld shall be as given in table 3.8.

Table 3.8: Minimum Size of Fillet Weld

Thickness of thicker part	Minimum size
Upto and including 10 mm	3 mm
Over 10 mm upto and including 20 mm	5 mm
Over 20 mm upto and including 32 mm	6 mm
Over 32 mm upto and including 50 mm	10 mm

8. Maximum size:

- Maximum size of fillet weld applied to a square edge of a plate of thickness above 6 mm should not exceed thickness of the edge minus 1.5 mm. (See Fig. 3.20 (a)).
- Maximum size of fillet weld applied to rounded toe of rolled steel section should not exceed 3/4 × the thickness of rounded toe. (See Fig. 2.16 (b)).

9. Overlap:

- The overlap in a lap joint should not be less than 5 times the thickness of thinner plate as shown in Fig. 3.20.

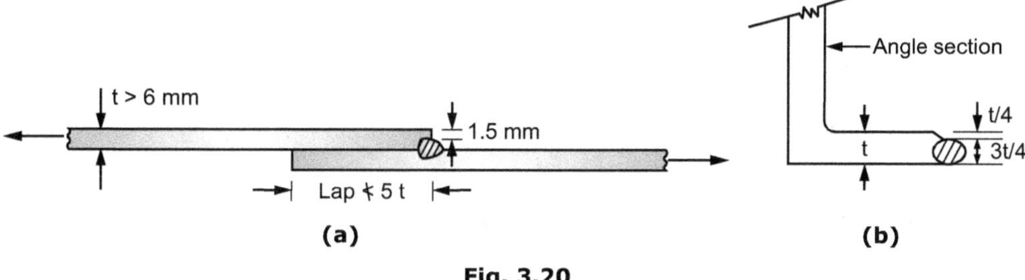

Fig. 3.20

10. Side fillet:

- In a lap joint made by a side or longitudinal fillet weld, the length of each fillet weld should not be less than the perpendicular distance between them.

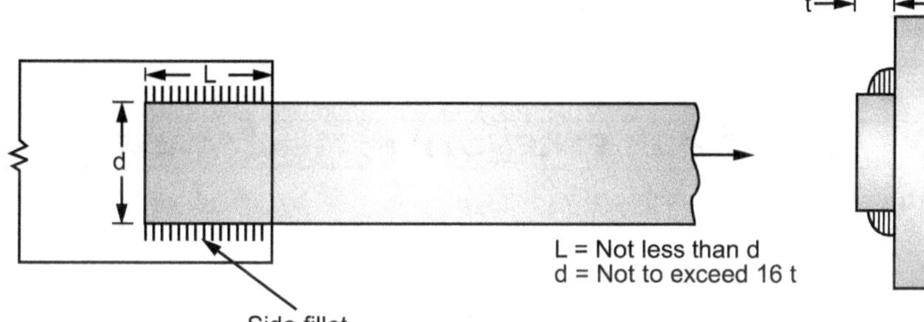

Fig. 3.21

11. End Return:

- Fillet welds terminating at ends or sides of parts or members should preferably be returned continuously around the corners for a distance not less than twice the weld size as shown in Fig. 3.22.

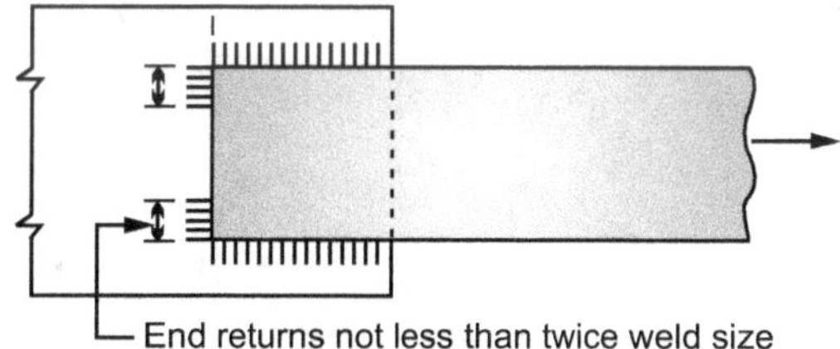

Fig. 3.22

12. Single fillet weld:

- A single line of fillet weld, when used, should not be subjected to bending about its longitudinal axis.

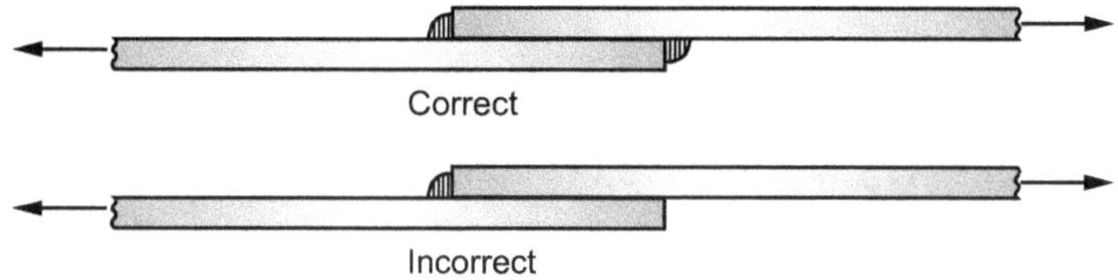

Fig. 3.23

13. Fillet welds on inclined faces:

- When faces of two members are inclined at an angle fillet weld is used. Fillet welds usually have a triangular cross-section. In order to do fillet welding, the angle between joining surfaces shall be within a range of 60° to 120°.

3.20 DESIGN STRESSES IN FILLET WELDS

(i) Shop welds

- Design stress of a fillet weld f_{wd} shall be based on throat area and is given by

$$f_{wd} = \frac{f_{wn}}{\gamma_{mw}} = \frac{\frac{f_u}{\sqrt{3}}}{\gamma_{mw}}$$

$$f_{wd} = \frac{f_u}{\sqrt{3}\,\gamma_{mw}} \qquad \text{... (3.13)}$$

where, f_u = smaller of ultimate stress of the weld or of the parent metal

γ_{mw} = Partial safety factor

= 1.25 for shop welds

(ii) Site Welds

In this case γ_{mw} is taken as 1.5.

∴ Design stress for site weld becomes

$$f_{wd} = \frac{f_u}{\sqrt{3}\cdot\gamma_{mw}} \quad \text{where} \quad \gamma_{mw} = 1.50$$

3.21 DESIGN STRENGTH IN FILLET WELD

- Design strength of a fillet weld of length L (P_{dw})

$$P_{dw} = f_{wd}\cdot L \cdot t_t \qquad \text{... (3.14)}$$

where,   L = Effective length of weld

t_t = Throat thickness of weld

For example, taking f_u = 410 N/mm^2

γ_{mw} = 1.25 (for shop weld) and 1.5 (for site weld)

$$f_{wd} = \frac{f_u}{\sqrt{3}\,\gamma_{mw}} = \frac{410}{\sqrt{3}\times 1.25} = 189.37 \text{ N/mm}^2$$

Design strength of shop weld for a length of 1 mm

$$= f_{wd}\times 1\times t_t = f_{wd}\times 0.7s = 189.37\times 0.7s = 132.56\,s$$

where,   s = Size of the weld in mm.

3.22 BUTT WELDS OR GROOVE WELDS

- These welds are made between the edges of two parts to be connected. These welds are used to connect two plates in the same plane. Single V, double V, Single U, double U butt welds are used. Butt welds are treated as parent metal with a thickness equal to the throat thickness and the stresses are not to exceed those permitted in the parent metal.

3.22.1 Effective Throat Thickness of Butt Weld

- For single grooved welds the throat thickness is taken equal to $\frac{5}{8}$ the thickness of the tinner plate.

- For double grooved welds the throat thickness is taken equal to the thickness of the tinner plate.

(a) Single V-groove weld Throat thickness = $\frac{5}{8}$

(Thickness of thinner plate)

(b) Double V-groove weld Throat thickness - thickness of tinner plate

Fig. 3.24

3.22.2 Effective Length of Fillet Weld

$$\text{Design stress for weld} = f_{wd} = \frac{f_y}{\gamma_{mw}}$$

$$\text{Design strength of weld for length} \quad L = P_{dw} = L, t_f\, f_{wd} = \frac{L t_t\, f_y}{\gamma_{mw}}$$

$$t_t = \text{Throat thickness}$$

Equating factored load $\qquad\qquad P_u = \text{Design strength } P_{dw}$

$$P_u = \frac{L \cdot t_t \cdot f_y}{\gamma_{mw}} = L \cdot t_t \cdot f_{wd}$$

$\therefore\quad$ Effective length $\qquad\qquad L = \dfrac{P_u}{t_t \cdot f_{wd}}$ $\qquad\qquad$ ... (3.15)

3.23 REDUCTION FACTOR FOR LONG JOINTS

- When the length of the welded joint L_j of a splice or end connection in a compression or tension element is greater than $150\, t_t$, the design capacity of the weld shall be reduced by the factor.

$$\beta_{tw} = 1.2 - \frac{0.2\, L_j}{150\, t_t} \leq 1.0 \qquad\qquad \text{... (3.16)}$$

where, $\qquad\qquad L_j = $ Length of the joint in the direction of the force transfer

$\qquad\qquad\qquad t_t = $ Throat size of the weld

3.24 ANALYSIS AND DESIGN OF FILLET WELDS

- A fillet weld may be subjected to direct, bending and shear stresses but the shear controls the design since the fillet always fails in shear at an angle of about 45° through the throat.

  1. Assume the size S of the weld based on thickness of members to be connected.
  2. Calculate the strength of fillet weld/mm using

$$p_q = f_{wd} \cdot L \cdot t_t \text{ where, } L = 1 \text{ mm.}$$

  3. Estimate the Pull (Tension) or Thrust (Compression) to be transmitted by the connections.
  4. Calculate effective length of weld using

$$L = \frac{P_{dw}}{f_{wd} \times t_t} \quad \text{or} \quad \frac{P_{dw}}{p_q}$$

  Adjust this length either by longitudinal fillet welds (parallel to direction of load) or as transverse fillet welds (perpendicular to direction of load) alongwith longitudinal fillet welds.

  5. If only a longitudinal fillet weld is provided check that length of each longitudinal fillet weld is greater than perpendicular distance between them.
  6. Provide end returns of length $= 2 s$ at each end of longitudinal fillet weld.

SOLVED EXAMPLES

Ex. 3.16: *A 80 mm × 8 mm plate is to be connected to a 120 × 8 mm plate in a lap joint to transmit a factored pull of 125 kN. Using 6 mm site welds, design the connection.*

Sol.:

  1. Size of weld, s = 6 mm (given).

   $\therefore$ Throat thickness $t_t = 0.7\, s = 0.7 \times 6 = 4.2$ mm.

  2. Design stress for site weld ($\gamma_{mw} = 1.5$)

$$f_{wd} = \frac{f_u}{\sqrt{3}\ \gamma_{mw}} = \frac{410}{\sqrt{3} \times 1.5} = 157.8 \text{ N/mm}^2$$

   $\therefore$ Design strength per mm length of weld,

$$p_q = f_{wd} \times 1 \times t_t = 157.8 \times 4.2 = 662.8 \text{ N/mm}$$

  3. Factored pull, $P_{dw} = 125$ kN $= 125 \times 10^3$ N.

4. Effective length of weld required,

$$L = \frac{P_{dw}}{p_q} = \frac{125 \times 10^3}{662.8} = 188.59 \simeq 189 \text{ mm}$$

Provide this length as longitudinal welds on both side. Therefore length of weld on each side

$$= \frac{189}{2} = 94.5 \text{ mm say } \textbf{95 mm}.$$

5. Minimum length of each longitudinal weld = Perpendicular distance between longitudinal weld = 80 mm.

6. Provide end return = 2 s = 2 × 6 = 12 mm on each side.

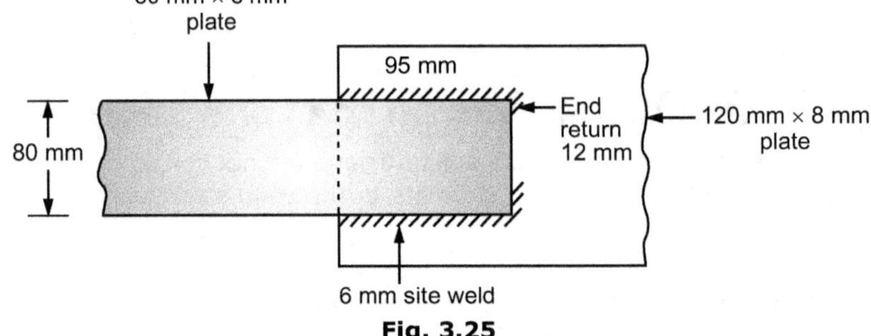

Fig. 3.25

Ex. 3.17: *Find safe load transmitted by a fillet welded joint between a flat 60 mm wide overlapping 100 mm over a gusset plate. Thickness of both plates is 10 mm. Weld is on all sides of overlap. Size of weld is 6 mm, which is provided at shop.*

Sol.:

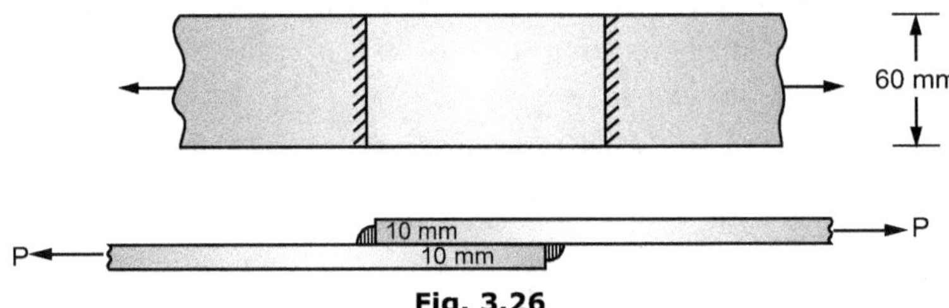

Fig. 3.26

As both the flats are of same width i.e. 60 mm only transverse welds as shown Fig. 3.26 can be provided.

1. Size of weld, s = 6 mm.

∴ Throat thickness, $t_t = 0.7 s = 0.7 \times 6 = 4.2$ mm.

2. Design stress for shop weld $(\gamma_{mw} = 1.25)$

$$f_{wd} = \frac{f_u}{\sqrt{3}\ \gamma_{mw}} = \frac{410}{\sqrt{3} \times 1.25} = 189.37 \text{ N/mm}^2$$

3. Effective length of weld, L = 60 + 60 = 120 mm

4. Factored load $P_{dw} = f_{wd} \times L \times t_t$

$$P_{dw} = 189.37 \times 120 \times 4.2 = 95442.9 \text{ N} = 95.44 \text{ kN}$$

5. Safe load $P = \dfrac{P_{dw}}{\gamma_f} = \dfrac{95.44}{1.5} = \textbf{63.62 kN}$

Ex. 3.18: *Design a suitable fillet weld to connect a tie bar 80 × 8 mm to a 10 mm thick gusset plate. Joint has to be designed for full strength of the tie bar and welding on all three sides.*

Sol.: 1. Design strength of 80 × 8 mm plate.

$$P_{dw} = \frac{f_y}{\gamma_{mo}} A_g = \frac{250}{1.10} \times 80 \times 8 = 145454 \text{ N} = 145.45 \text{ kN}$$

2. Size of weld

  Minimum size = 3 mm

  Maximum size = 8 − 1.5 = 6.5 mm

Provide 4 mm site weld

3. Design stress for site weld.

$$f_{wd} = \frac{f_u}{\sqrt{3}\ \gamma_{mw}} = \frac{410}{\sqrt{3} \times 1.50} = 157.80 \text{ N/mm}^2 \qquad (\gamma_{mw} = 1.5)$$

4. Design strength per mm length of weld

$$p_q = f_{wd} \times t_t = 157.80 \times (0.7 \times 4) = 441.84 \text{ N/mm}$$

5. Effective length of weld required

$$L = \frac{P_{dw}}{p_q} = \frac{145454}{441.84} = 329.20 \text{ mm say } \textbf{330 mm}$$

In such an arrangement the distance between longitudinal welds shall not exceed 16 t.

i.e.   16 × 8 = 128 mm

Let us provide two longitudinal and one transverse weld.

Length of transverse weld = 80 mm (< 128 mm).

Length of each longitudinal weld = $\dfrac{330 - 80}{2}$ = **125 mm**

                  (∡ transverse dist. = 80 mm)

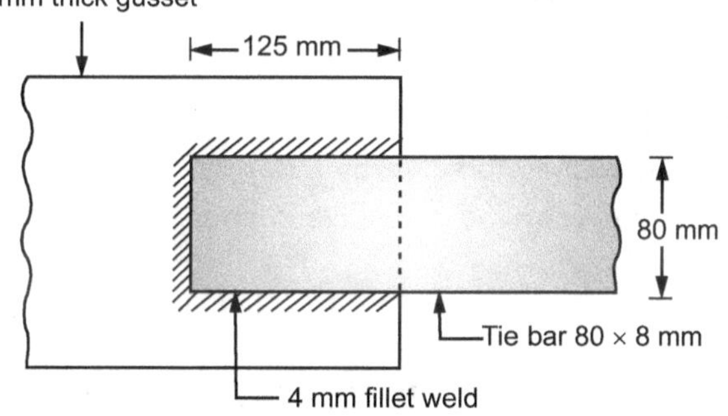

Fig. 3.27

Ex. 3.19: *A plate 150 mm × 10 mm is connected by 8 mm fillet weld. Find lap required to provide only longitudinal weld. Draw neat sketch and use I.S. specifications.*

Sol.: 1. Design strength of 150 mm × 10 mm plate.

$$P_{dw} = \frac{f_y}{\gamma_{mo}} A_g = \frac{250}{1.1} \times 150 \times 10 = 340909 \text{ N}$$

2. Size of weld = 8 mm (given)

As the type of weld is not given consider it as site weld.

3. Design stress for site weld

$$f_{wd} = \frac{f_u}{\sqrt{3}\ \gamma_{mw}} = \frac{410}{\sqrt{3} \times 1.5} = 157.80 \text{ N/mm}^2$$

4. Design strength/mm length of weld

$$p_q = f_{wd} \times t_t = 157.80 \times (0.7 \times 0.8) = 883.68 \text{ N/mm}$$

5. Effective length of weld required

$$L = \frac{P_{dw}}{p_q} = \frac{340909}{883.68} = 385.78 \text{ say } 390 \text{ mm}$$

As only longitudinal welds are to be provided length of each longitudinal weld

$$x = \frac{390}{2} = \textbf{195 mm} \quad (\nleq \text{transverse distance} = 150 \text{ mm})$$

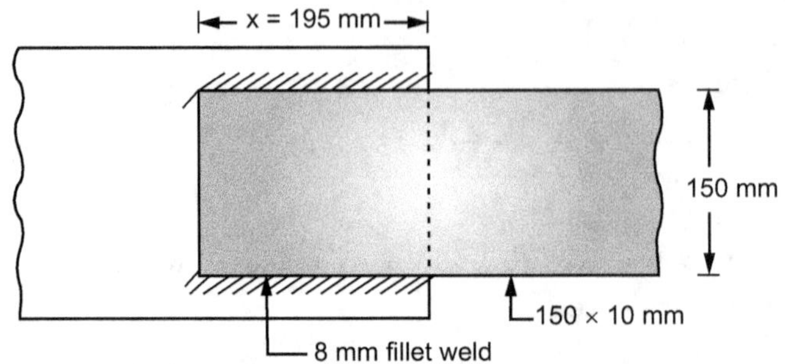

Fig. 3.28

Ex. 3.20: *A plate 150 mm × 10 mm is connected by 8 mm fillet weld. Find lap required to be provided for longitudinal weld only. Draw neat sketch showing lap length. Take, permissible stress in weld material as 108 N/mm².*

    Sol.: Flat size 150 mm × 10 mm

         Fillet weld 8 mm

         Permissible shear stress 108 N/mm²

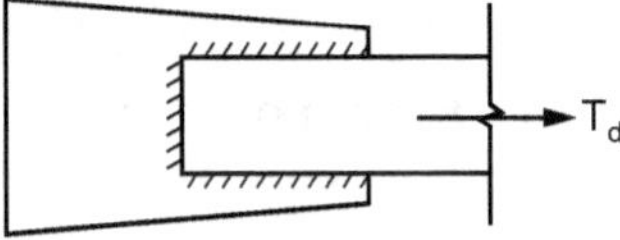

Fig. 2.29

(**Note:** Given data represents question is based on working stress method, not as per Limit State Method).

- External pull a tension force is also not mentioned

- Cannot be solved referring LSM.

Note: Further, if a student wishes to solve by LSM, he has to assume f_u has to calculate T_d (at least referring T_{dg} or T_{dn}), then only he can solve such as:

Assuming f_u = 410 MPa

Design strength of weld/mm

$$P_d = \frac{f_u}{\sqrt{3}\ \gamma_{mw}} \times t = \frac{410}{\sqrt{3} \times 1.5}(0.7 \times 8) = 883.73 \text{ N/mm}$$

For plate,

$$\Rightarrow \qquad T_{dg} = \frac{A_g \cdot f_y}{\gamma_{mo}} = \frac{(150 \times 10)\ 250}{1.1} = 340.90 \times 10^3 \text{ N}$$

$$\Rightarrow \qquad T_{dn} = \frac{0.9\ A_n\ f_u}{\gamma_{mo}}$$

$$= \frac{0.9\ (150 \times 10)\ 410}{1.25}$$

$$= 442.8 \times 10^3 \text{N}$$

$$\Rightarrow \qquad T_d = 340.90 \times 10^3 \text{ N (Minimum out of } T_{dg} \cdot T_{dn})$$

$$\Rightarrow \qquad \text{Weld length reqd.} = \frac{T_d}{P_d}$$

$$= \frac{340.90 \times 10^3}{883.73} = 385.75 \text{ mm say 390 mm}$$

$$\therefore \qquad \text{Required lap length} = \frac{390}{2} = 195 \text{ mm.}$$

Ex. 3.21: *A tie member 100 × 10 mm has to transmit an axial load of 100 kN. Design fillet weld and calculate necessary overlap by assuming welding on all four sides. Also draw a neat sketch of connection. Take permissible shear stress in weld material as 108 Mpa.*

Sol.: Given data:
(i)   Axial load = 200 kN
(ii)   Tie member 100 × 10 mm
(iii) Shear stress in weld material
      = 108 Mpa
Assume gusset plate of 12 mm thick.

Let s = size of weld

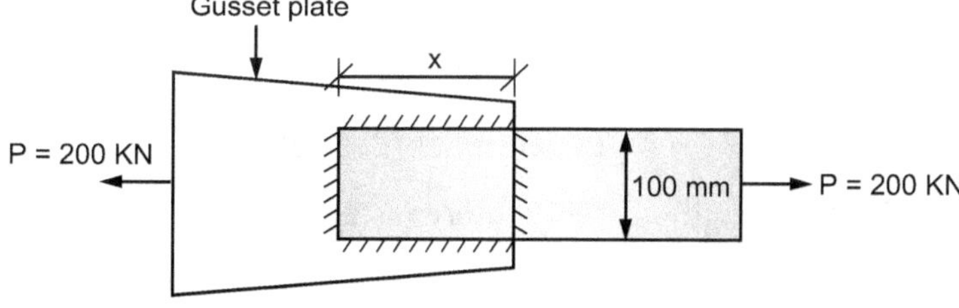

Fig. 3.30

∴   Minimum size of weld for 12 mm thicker plate = 5 mm
Also,
     Maximum size of weld for 10 mm thinner plate
           = Thickness of Thinner plate – 1.5 = 10 – 1.5 = 8.5 mm
∴   We provide size of weld (s) = 6 mm
∴   Throat thickness (t) = 0.7 × Size of weld = 0.7 × 6 = 4.2 mm
           Factored load (P_{wd}) = $200 \times 10^3 \times 1.5 = 300 \times 10^3$ N

For Fe 410 grade steel
          Ultimate stress (fu) = 410 N/mm^2
Let   x = Lap length
∴   Total length of weld (*l*) = x + x + 100 + 100 = 2x + 200
Now, we know that

$$\text{Design strength of fillet weld} = t \times l \times \frac{f_y}{\sqrt{3} \times \gamma_{mw}}$$

$$300 \times 10^3 = 4.2 \times (2x + 200) \times \frac{410}{\sqrt{3} \times 1.5}$$

$$\text{Assume } \gamma_{mw} = 1.5 \text{ (as shop or site weld note given)}$$

$$300 \times 10^3 = 662.79 \times (2x + 200)$$

$$452.62 = 2x + 200$$

$$x = 126.31 \text{ mm}$$

$$\text{Say} \boxed{x = 130 \text{ mm}}$$

Ex. 3.22: *Design a suitable filled weld of size 4 mm to connect a tie bar 80 × 8 mm to 10 mm thick gusset plate. Joint has to be designed for full strength of the tie bar and welding on all three sides. Draw a neat sketch showing lap length.*

Take - f_y = 250 N/mm^2, γ_{mo} = 1.10, f_u = 410 N/mm^2, γ_{mo} = 1.10.

Ans.   Given: f_y = 250 N/mm^2 γ_{mo} = 1.10

              f_u = 410 N/mm^2 γ_{mw} = 1.50

              Size of weld = 4 mm

Design strength of 80 × 8 mm plate

$$P_{dw} = \frac{f_y}{\gamma_{mo}} \, A_g = \frac{250}{1.10} \times 80 \times 8 = 145454.55 \text{ N} = 145.45 \text{ kN}$$

$$\boxed{P_{dw} = 145.45 \text{ kN}}$$

Design stress for site weld

$$f_{wd} = \frac{f_u}{\sqrt{3} \, \gamma_{mw}} = \frac{410}{\sqrt{3} \times 1.5} = 157.81 \text{ N/mm}^2$$

$$\boxed{f_{wd} = 157.81 \text{ N/mm}^2}$$

Design strength per mm length of weld

$$p_q = f_{wd} \times t_t = f_{wd} \times 0.7 \times 5 = 157.80 \times 0.7 \times 4 = 441.84 \text{ N/mm}$$

$$\boxed{\therefore \; p_q = 441.84 \text{ N/mm}}$$

Effective length of weld required

$$L = \frac{P_{dw}}{2} = \frac{145454.55}{441.84}$$

$$\boxed{\therefore \; L = 329.20 \text{ mm} \cong 330 \text{ mm}}$$

In such an arrangement the distance between longitudinal weld shall not exceed 16 t.

i.e. $16 \times 8 = 128$ mm

Let us provide two longitudinal and one transverse weld.

Length of transverse weld = 80 mm.

Length of Each longitudinal weld $= \dfrac{330 - 80}{2}$

= 125 mm

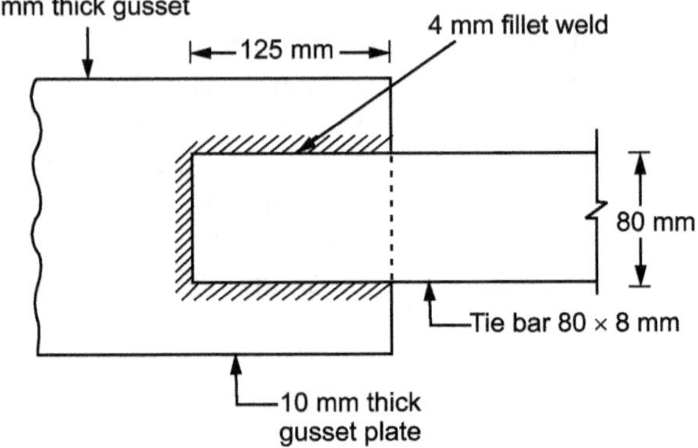

Fig. 3.31: Welded Connection of Lap Joint

Ex. 3.23: *Calculate the length of fillet weld required to connect an ISA 100 × 100 × 10 mm with gusset plate using 6 mm weld as shown in Fig. 3.33. The angle is subjected to factored axial force of 300 kN. $C_{xx} = C_{yy}$ for angle is 28.4 mm.*

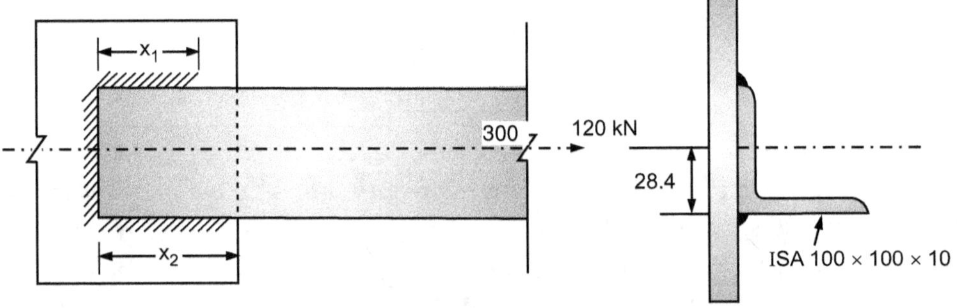

Fig. 3.32

Sol.:

1. $P_u = 300$ kN

2. Size of weld = 6 mm (consider site weld)

3. Design stress for site weld.

$$f_{wd} = \frac{f_u}{\sqrt{3} \, \gamma_{mw}} = \frac{410}{\sqrt{3} \times 1.5} = 157.80 \text{ N/mm}^2$$

4. Design strength per mm length of weld

$$p_q = f_{wd} \times t_t = 157.80 \times (0.7 \times 6) = 662.76 \text{ N/mm}$$

5. Effective length of weld required

$$L = \frac{P_u}{p_q} = \frac{300 \times 10^3}{662.76} = 452.65 \text{ say } 455 \text{ mm}$$

Let x_1 and x_2 be the lengths of weld at the upper and lower edges of the angle, such that

$$x_1 + x_2 + 100 = 455 \text{ mm.}$$

$\therefore \qquad x_1 + x_2 = 355$

Taking moment about the bottom weld

$$662.76\, x_1 \times 100 + 662.76 \times 100 \times 50 = 300 \times 10^3 \times 28.4$$

Dividing throughout by 66276

$$x_1 + 50 = 128.55$$

$\therefore \qquad x_1 = 78.55 \text{ say } \textbf{80 mm}$

$\therefore \qquad x_2 = 355 - 80 = \textbf{275 mm}$

Ex. 3.24: *A 75 mm × 50 mm × 8 mm angle is to be connected to a gusset plate by 6 mm fillet welds at extremities of the longer leg. Design the welded connection corresponding to the full tensile strength of angle. Assume shop welding, Gross Area of angle = 938 mm², C_{xx} = 25.2 mm.*

Sol.: 1. Full tensile strength of the angle

$$P = \frac{f_y\, A_g}{\gamma_{mo}} = \frac{250 \times 938}{1.10} = 213182 \text{ N}$$

2. Size of weld = 6 mm (shop welding is given)

3. Design stress of shop weld

$$f_{wd} = \frac{f_u}{\sqrt{3}\ \gamma_{mw}} = \frac{410}{\sqrt{3} \times 1.25} = 189.4 \text{ N/mm}^2$$

4. Design strength per mm length of weld

$$P_q = f_{wd} \times t_t = 189.4 \times (0.7 \times 6) = 795.48 \text{ N/mm}$$

5. Effective length of weld required

$$L = \frac{P}{P_q} = \frac{213182}{795.48} = 267.99 \text{ mm} \approx 270 \text{ mm.}$$

Let x_1 and x_2 be the lengths of longitudinal weld at upper and lower edges of angle.

$\therefore \qquad x_1 + x_2 = 270 \text{ mm}$

Taking moment about the bottom weld

$$795.48\, x_1 \times 75 = 213182 \times 25.2$$

$\therefore \qquad x_1 = \textbf{90 mm}$

$\therefore \qquad x_2 = 270 - 90 = \textbf{180 mm}$

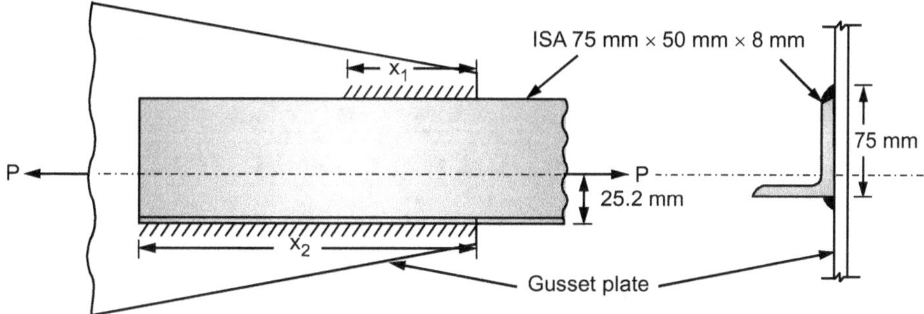

Fig. 3.33

Ex. 3.25: *The tension member of a truss consists of 2 angles 80 mm × 50 mm × 8 mm and is welded on either side of a gusset plate. The member is subjected to factored tensile force of 275 kN. Design the fillet weld connection. Assume shop welding for each angle C_{xx} = 27.3 mm.*

Sol.:

1. $P_u = 275$ kN.

2. Size of weld

    (i)   Minimum size = 3 mm

    (ii)   Maximum size $= \frac{3}{4}\,t = \frac{3}{4} \times 8 = 6$ mm

$\therefore$ Provide 5 mm size shop weld

3. Design stress of shop weld

$$f_{wd} = \frac{f_u}{\sqrt{3}\,\gamma_{mw}} = \frac{410}{\sqrt{3} \times 1.5} = 189.4 \text{ N/mm}^2$$

4. Design strength per mm length of weld

$$p_q = f_{wd} \times t_t$$
$$= 189.4 \times (0.7 \times 5) = 662.9 \text{ N/mm}$$

5. Effective length of weld required

$$L = \frac{P_u}{p_q} = \frac{275 \times 10^3}{662.9} = 414.84 \text{ mm say } 420 \text{ mm}$$

Length required on each angle $= \dfrac{420}{2} = 210$ mm

Let x_1 and x_2 be the length of weld at upper and lower edge of angle such that $x_1 + x_2 = 210$ mm. Taking moment about the bottom weld

$$662.9 \times x_1 \times 80 = \frac{275 \times 10^3}{2} \times 27.3$$

$\therefore$ $x_1 = 70.78$ mm $\approx$ **75 mm**

$\therefore$ $x_2 = 210 - 75$ mm $\approx$ **135 mm**

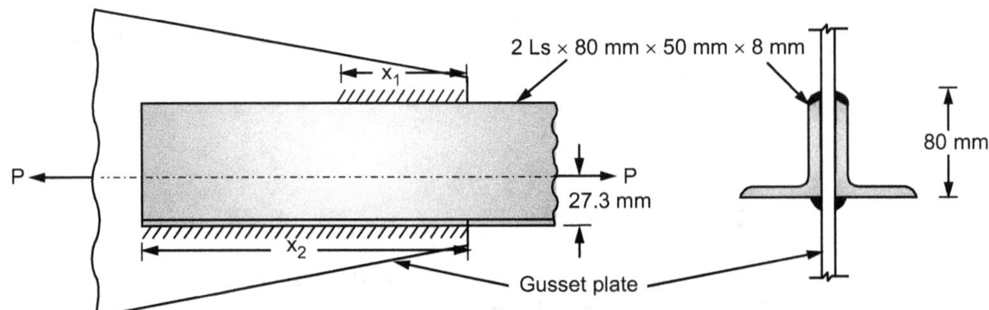

Fig. 3.34

3.25 ECCENTRIC BOLTED CONNECTIONS

- Uptil now, we have seen that, the resistance offered by a bolt was entirely to prevent a linear or translatory displacement of the connected plate or member. There are also circumstances in which the bolts provided for a connection may have to offer not only a resistance to prevent translatory displacement but also a resistance to prevent rotary displacement. A bracket connection is an example of this type of connection.

- There are two types of bracket connections:

    (i) Bolted connection subjected to moment in the plane of the connection.

    (ii) Bolted connection subjected to moment in a plane normal to the plane of the connection.

- Out of these two we will discuss case (i) as under:

Case (i): Bolted connection subject to moment in the plane of the connection

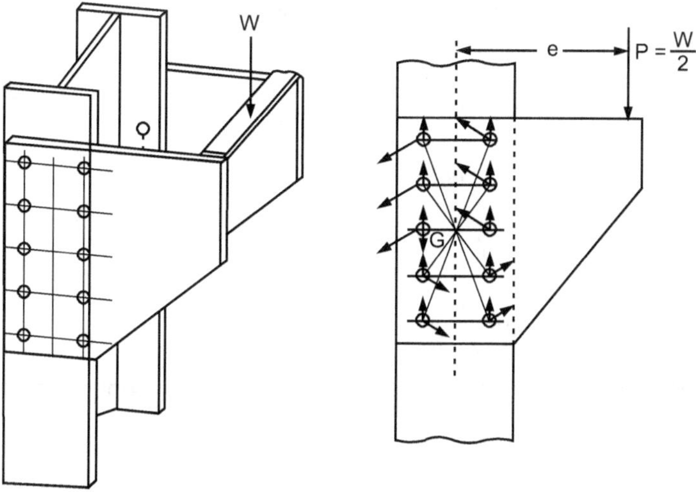

Fig. 3.35

* Fig. 3.35 shows an eccentric bolted connected for a bracket. It consists of two bracket plates bolted to the flanges of a rolled steel column. If a load W be applied to the bracket, a load $P = \dfrac{W}{2}$ is transmitted to each bracket plate.

* The line of action of the load P on the bracket plate does not pass through the centroid of the bolt group. The perpendicular distance between the centroid of the bolt group and the line of action of the load P is called the eccentricity of the load P.

* The bolts connecting the bracket plate and the flange of the column have to offer the following resistances.

    (i) Resistance against translation: The resistance is assumed to be uniform for all the bolts.

    If P be the load on one bracket plate resistance against translation per bolt $= \dfrac{P}{n}.$ ... (3.17)

    where, n = number of bolts on one bracket plate.

(ii) Resistance offered by the bolts against the rotation of the bracket plate: The load P being eccentric there is a tendency for the bracket plate to rotate about the centroid G of the bolt group. The bolts therefore have to offer a resistance to prevent such rotation. Such a resisting force offered by a bolt is called the *torsional shear* in the bolt.

It will be assumed that the torsional shear in a bolt directly proportional to the distance of the bolt from the centroid of the bolt group. The direction of this resisting force S is at right angles to the line joining G and the bolt.

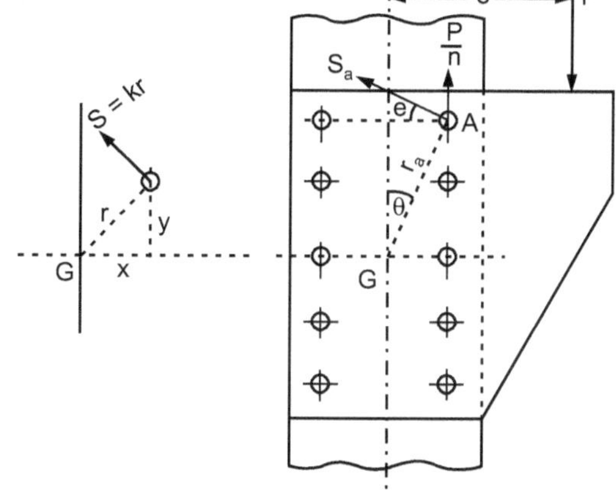

Fig. 3.36

- Let S be the torsional shear for a bolt distant r from G. The torsional shear S acts at right angles to the line joining G and the bolt. As per out assumption,

$$S = Kr$$

where   K = a constant of proportionality

Restoring moment provided by this bolt

$$= Sr = Kr^2$$

Total restoring moment provided by all the bolts

$$= \Sigma Kr^2 = K\Sigma r^2$$

If (x, y) are the co-ordinates of the rivet distant r and G

$$r^2 = x^2 + y^2$$
$$\Sigma r^2 = \Sigma x^2 + \Sigma y^2$$

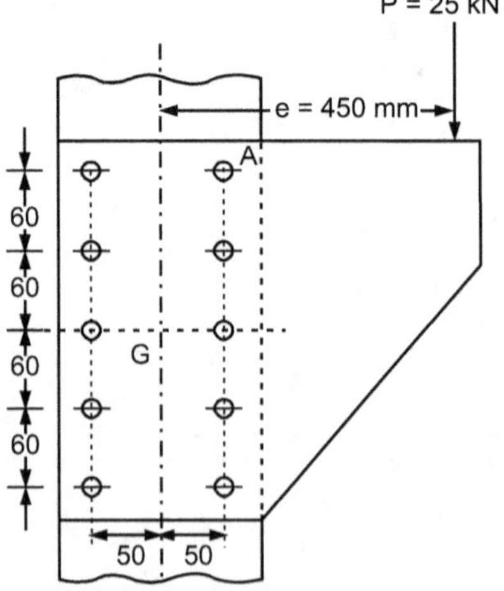

$\therefore$ Total restoring moment provided by all the bolts

$$= K\,[\Sigma x^2 + \Sigma y^2]$$

Equating the restoring moment to the turning moment on the connection,

$$= K\,[\Sigma x^2 + \Sigma y^2] = Pe$$

$\therefore$
$$K = \frac{Pe}{\Sigma x^2 + \Sigma y^2} \qquad \ldots (3.18)$$

Now consider the bolt A most distant from G. This bolt offers the maximum resistance.

$$\text{Resistance against translation} = \frac{P}{n}$$

(n = number of bolts on one bracket plate)

$$\text{Torsional shear} = S_a = Kr_a \qquad \text{(See Fig. 3.38)}$$

Total vertical component on the bolt $= V = \dfrac{P}{n} + S_a \sin\theta$

$$= \frac{P}{n} + Kr_a \sin\theta = \frac{P}{n} + Kx_a$$

Horizontal component on the bolt $= H = S_a \cos\theta = Kr_a \cos\theta = Ky_a$

$$= \sqrt{V^2 + H^2} \qquad \ldots (3.19)$$

Resultant resistance.

If P is the factored load on one bracket plate, then the resultant resistance offered by the bolt shall be less than the design strength of the bolt.

Ex. 3.26: *A line shaft transmits a load of 25 kN at an eccentricity of 450 mm across a bracket plate bolted to a stanchion. Two rows of bolts 100 mm part are provided with five bolts per row. The pitch of bolts in each row is 60 mm. Find the greatest forced induced in any bolt.*

Sol.: Resistance against translation per bolt $= \dfrac{P}{n} + \dfrac{25 \times 10^3}{10} = 2500$ V

Fig. 3.38 shows the arrangement of the bolts.

Fig. 3.38

With respect to the centroid G of the bolt group

$$\Sigma x^2 + \Sigma y^2 = 10(50)^2 + 4(120)^2 + 4(60)^2 \text{ mm}^2 = 97000 \text{ mm}^2$$

$$K = \frac{Pe}{\Sigma x^2 + \Sigma y^2} = \frac{25000 \times 450}{97000} = 116 \text{ N/mm}$$

Now the consider the bolt A

Resisting force against resolution $= S_a = Kr_a = 116\, r_a$

See Fig. 3.40.

Total vertical component on the bolt A

$$= V = 2500 + 116\, r_a \sin\theta$$

$$= 2500 + 116 \times 50 = 8300$$

Horizontal component on the bolt A

$$= H = 116\, r_a \cos\theta$$

$$= 116 \times 120 = 13920 \text{ N}$$

Resultant resistance offered by the bolt A

$$= \sqrt{V^2 + H^2} = \sqrt{8300^2 + 13920^2}$$

$$= 1620 \text{ N}$$

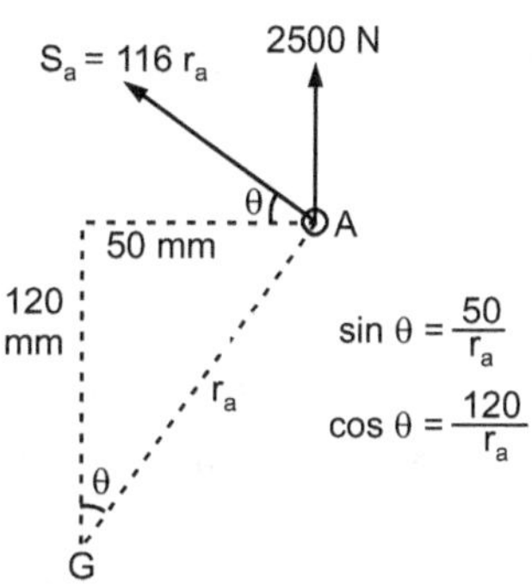

Fig. 3.39

This is the greatest force transmitted to a bolt.

Ex. 3.27: *A working load of 160kN is applied to a bracket plate at an eccentricity of 300 mm. Sixteen bolts of 20 mm diameter are arranged in two rows with 8 bolts per row. The rows are 200 mm part. The pitch of bolts in each vertical row is 80 mm. The thickness of the bracket plate is 12.5 mm. Investigate the safety of the design.*

Sol.: Fig. 3.40 shows the bracket connection. Let G be the centroid of the group of welds.

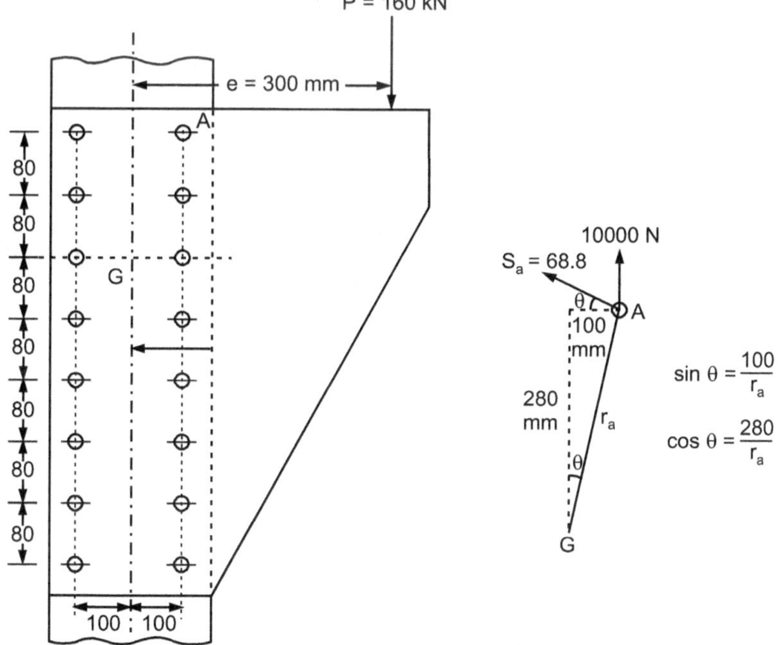

Fig. 3.40

With respect to G,

$$\Sigma x^2 + \Sigma y^2 = 16(100)^2 + 4(280)^2 + 4(200)^2 + 4(120)^2 + 4(40)^2 \text{ mm}^2$$

$$= 697600 \text{ mm}^2$$

Consider the bolt marked A

$$\text{Resistance against translation} = \frac{P}{n} = \frac{160 \times 10^3}{16} = 10000 \text{ N}$$

$$\text{Torsional shear} = S_a = Kr_a$$

$$K = \frac{Pe}{\Sigma x^2 + \Sigma y^2} = \frac{160 \times 10^3 \times 300}{697600} = 68.8 \text{ N/mm}$$

$$\therefore \qquad S_a = 68.8 \, r_a$$

Total vertical component
$$V = 10000 + 68.8 \, r_a \sin\theta$$
$$= 10000 + 68.8 \times 100 = 16880 \text{ N}$$

Horizontal component
$$H = 68.8 \, r_a \cos\theta$$
$$= 68.8 \times 280 = 19264 \text{ N}$$

Resultant resistance
$$= \sqrt{(16880^2 + 19200)^2} = 25613 \text{ N}$$

$\therefore$ Factored shear load on the bolt $A = 1.5 \times 25613 = 38420$ N

$$\text{Bolt diameter} = d = 20 \text{ mm}$$
$$\text{Bolt hole diameter} = d_o = 20 + 2 = 22 \text{ mm}$$

Bolts are in single shear.

$$\text{Design shearing strength of bolt} = \frac{1}{\gamma_{mb}} \cdot \frac{f_u}{\sqrt{3}} (0.8) \frac{\pi d^2}{4}$$

$$\frac{1}{1.25} \times \frac{400}{\sqrt{3}} (0.8) \frac{\pi \times 20^2}{4} = 45274 \text{ N}$$

$$\text{Design bearing strength of bolt} = \frac{1}{\gamma_{mb}} [2.5 \, K_b \, dt \, f_u]$$

K_b is the least of the following:

(i) $\dfrac{e}{3d_o} = \dfrac{40}{3 \times 22} = 0.606$

(ii) $\dfrac{p}{3d_o} - 0.25 = \dfrac{80}{3 \times 22} - 0.25 = 0.962$

(iii) $\dfrac{f_{ub}}{f_u} = \dfrac{400}{410} = 0.975$

(iv) 1

$\therefore$ $K = 0.606$

$\therefore$ Design bearing strength of bolt

$$\frac{1}{1.25} [2.5 \times 0.606 \times 20 \times 12.5 \times 400]$$

$$= 121200 \text{ N}$$

$$\therefore \qquad \text{Bolt value} = 45274 \text{ N}$$

But factored load on the bolt $= 38420$ N

$\therefore$ The design is safe.

Ex. 3.28: *Fig. 3.42 shows an eccentric bolted bracket connection, supporting a load of 75 kN at an eccentricity of 200 m. Find the size of the bolt required. Thickness of the bracket plate = 10 mm.*

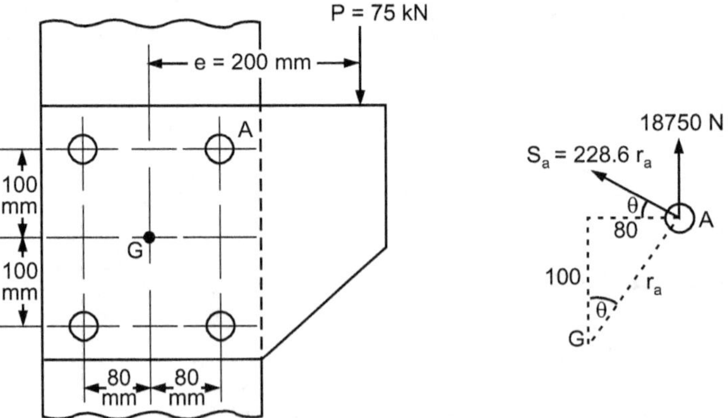

Fig. 3.41

Sol.: With respect to the centroid G of the bolt group.

$$\Sigma x^2 + \Sigma y^2 = 4(80)^2 + 4(100)^2 = 65600 \text{ mm}^2$$

$$K = \frac{Pe}{\Sigma x^2 + \Sigma y^2} = \frac{75 \times 10^3 \times 200}{65600} = 228.6 \text{ N/mm}$$

Consider the bolt marked A.

$$\text{Resistance against translation} = \frac{P}{n} = \frac{75 \times 10^3}{4} = 18750 \text{ N}$$

Torsional shear on bolt $\qquad A = S_a = K\, r_a = 228.6\, r_a$

Total vertical component $\qquad V = 18750 + 228.6\, r_a \sin\theta$

$$= 18750 + 228.6 \times 80 = 37038$$

Horizontal component $\qquad H = 228.6\, r_a \cos\theta = 228.6 \times 100 = 22860 \text{ N}$

Resultant force on the bolt $\qquad A = \sqrt{37038^2 + 22860^2} = 43525$

Factored shear force on the bolt $\quad A = 1.5 \times 43425 = 65287.5 \text{ N}$

Let the diameter of the bolt be d mm.

Design shear strength of the bolt

$$\frac{1}{\gamma_{mb}} \cdot \frac{f_u}{\sqrt{3}}\,(0.8)\,\frac{\pi d^2}{4} = 65287.5$$

$$\frac{1}{1.25} \cdot \frac{400}{\sqrt{3}}\,(0.8)\,\frac{\pi d^2}{4} = 65287.5$$

$$116.09\, d^2 = 65287.5$$

$$d = 23.71 \text{ mm}$$

Provide 24 mm diameter bolt

Check for bearing strength

$$\text{Design bearing strength} = \frac{1}{\gamma_{mb}}\,[2.5\, K_b\, dt\, f_u]$$

K_b is the least of the following:

(i) $\dfrac{e}{3d_o} = \dfrac{40}{3 \times 22} = 0.606$

(ii) $\dfrac{p}{3d_o} - 0.25 = \dfrac{100}{3 \times 26} - 0.25 = 1.03$

(iii) $\dfrac{f_{ub}}{f_u} = \dfrac{400}{410} = 0.976$

(iv) 1

$\therefore\ K = 0.606$

$\therefore\ $ Design bearing strength of bolt

$$\frac{1}{1.25}\,[2.5 \times 0.606 \times 24 \times 10 \times 400]$$

$$= 116352 \text{ N}$$

Factored load on the bolt = 65287.5 N

$\therefore\quad$ The design is safe.

Ex. 3.29: *A factored load of 150 kN is applied to a bracket at an eccentricity of 350 mm from the axis of a column. This load is transmitted to the flanges of the column with 2 rows of 20 mm diameter bolts for each bracket plate. The rows are 120 mm apart and the pitch of bolts is 75 mm. Investigate the safety of the design.*

Sol.: Load on one bracket plate.

$$P = \frac{150}{2} = 75 \text{ kN}$$

Consider the bolts on one bracket plate.

With the respect to the centroid G of the bolt group

$$\Sigma x^2 + \Sigma y^2 = 8(60)^2 + 4(112.5)^2 + 4(37.5)^2 = 85050 \text{ mm}^2$$

Consider the bolt marked A

$$\text{Resistance against translation} = \frac{P}{n} = \frac{75000}{80} = 9375 \text{ N}$$

$$K = \frac{P_e}{\Sigma x^2 + \Sigma y^2} = \frac{75000 \times 350}{80050}$$

$$= 308.642 \text{ N/mm}$$

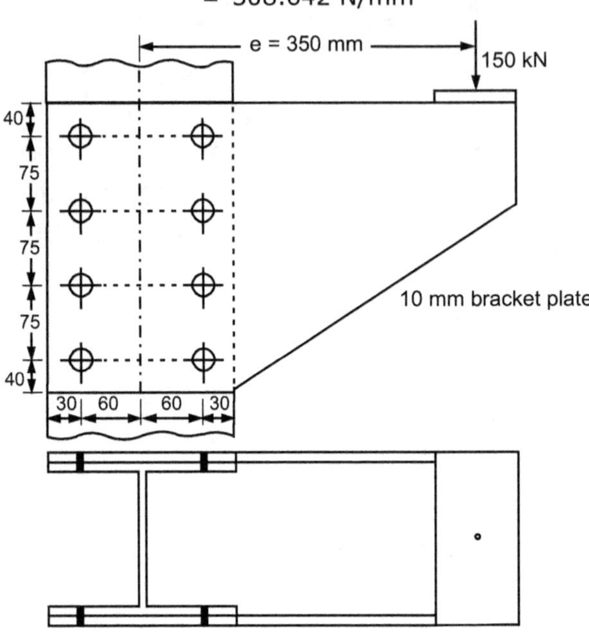

Fig. 3.42

$$\text{Torisonal shear for bolt A} = S_a = Kr_a = 308.642 \, r_a$$

Total vertical component on the bolt A

$$= V = 9375 + 308.642 \, r_a \sin \theta$$

$$= 9375 + 308.642 \times 60 = 27893.5 \text{ N}$$

Horizontal component of the bolt A

$$= H = 308.642 \, r_a \cos \theta$$

$$= 308.642 \times 112.5 = 34722.2 \text{ N}$$

Resultant resistance of the bolt

$$= \sqrt{27893.5^2 + 34722.2^2}$$

$$= 44538.5 \text{ N}$$

$$\text{Bolt diameter} = d = 20 \text{ mm}$$

$$\text{Bolt hole diameter} = d_o = 20 + 2 = 22 \text{ mm}$$

The bolts are in single shear.

Design strength of the bolt in single shear

$$\frac{1}{\gamma_{mb}}\left[\frac{f_u}{\sqrt{3}}(0.78)\frac{\pi d^2}{4}\right] = \frac{1}{1.25}\left[\frac{400}{\sqrt{3}}(0.78)\frac{\pi 20^2}{4}\right] \text{ N}$$

$$= 45274 \text{ N}$$

Design strength of the bolt in bearing $= \dfrac{1}{\gamma_{mb}}[2.5 \, K_b \, dt \, f_u]$

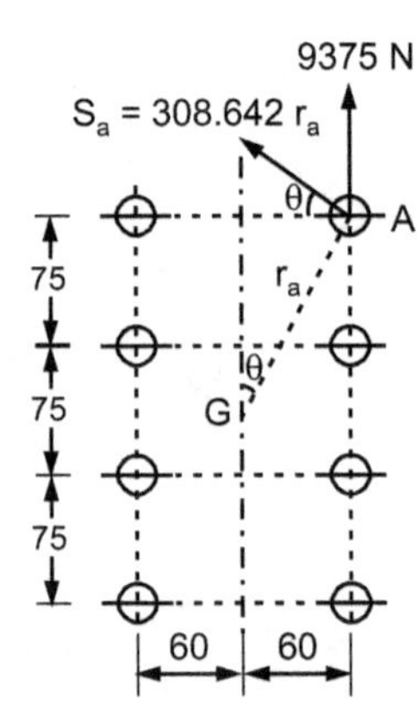

Fig. 3.43

K_b is the least of the following:

(i) $\dfrac{e}{3d_o} = \dfrac{40}{3 \times 22} = 0.606$

(ii) $\dfrac{p}{3d_o} - 0.25 = \dfrac{75}{3 \times 22} - 0.25 = 0.886$

(iii) $\dfrac{f_{ub}}{f_u} = \dfrac{400}{410} = 0.975$

(iv) 1

$\therefore$ The design is safe.

$\therefore$ $K_b = 0.606$

$\therefore$ Design bearing strength of bolt

$$= \frac{2.5 \times 0.606 \times 20 \times 10 \times 400}{1.25} = 96960 \text{ N}$$

Maximum load on the bolt = 44538.5 N

3.26 ECCENTRIC WELDED CONNECTIONS

- A bracket connection is an example of eccentric connection.

- There are two types of bracket connections viz.

   (i) Welded connection subjected to moment in the plane of the weld.

   (ii) Welded connection subjected to moment in a plane normal to the plane of the weld.

Case (i) Welded bracket connection subjected to moment in the plane of the weld

- Consider the bracket connection shows in Fig. 3.44. It consists of two bracket plates welded to the flanges of a steel column. If a load W be applied to the bracket a load $P = \dfrac{W}{2}$ is transmitted to each bracket plate. The line of action of the load P does not pass through the centroid of the weld group.

- Let G be the centroid of the weld lengths

   Let $\qquad$ e = Eccentricity of the load

   $\qquad$ = Distance between G and the line of action of the load P

   $\qquad$ = $\bar{x} + a$ (See Fig. 3.44) $\hfill$ ... (3.20)

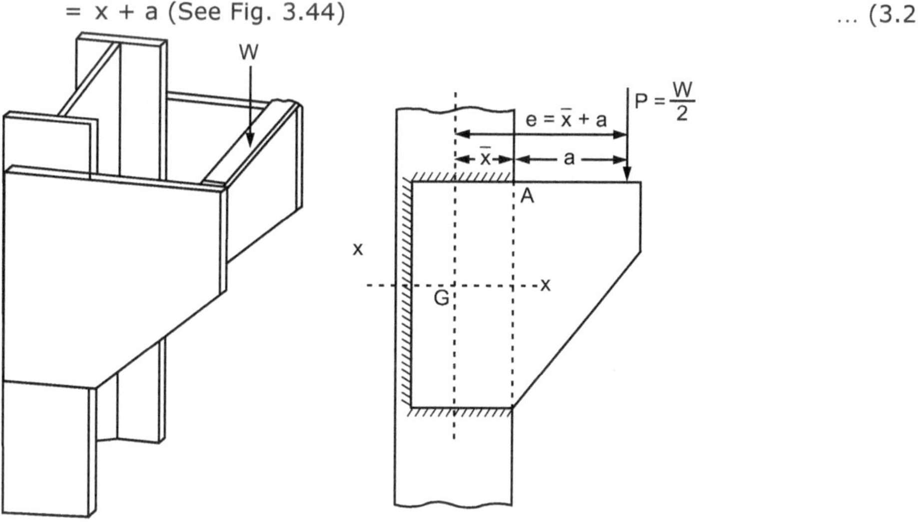

Fig. 3.44

- The weld has to offer the following resistances:

   (a) Resistance against translation: This resistance is assumed to be uniform over the whole length of the weld.

   $\therefore$ Resistance against translation per unit length of the weld = $\dfrac{P}{L}$.

   where, L = Total length of the weld on the bracket plate.

(b) Resistance against the rotation of the bracket plate: The force of resistance per unit length of the weld at any point of the weld length against rotation of the bracket plate is assumed to be proportional to the distance of the point from the centroid of the weld group.

Consider an elemental length dl of weld at any point Z of the weld line distant r from the centroid G.

Resistance offered against rotation by the elemental weld length

$$= Krdl \text{ acting normal to GZ}$$

Restoring moment offered by the elemental weld length

$$= (Krdl) r = Kdlr^2$$

Total restoring moment offered by the whole weld

$$\Sigma Kdlr^2 = \Sigma Kdlr^2 = KI_p$$

$$I_p = \Sigma dlr^2 = \text{Polar moment of inertia of the weld length}$$

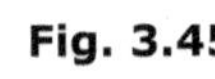

Fig. 3.45

But $I_p = I_{xx} + I_{yy}$

where, I_{xx} = Moment of inertia of the weld length about the axis XX in the plane of the weld through G

I_{yy} = Moment of inertia of the weld length about the axis YY in the plane of the weld through G

Total restoring moment offered by the weld length

$$= K(I_{xx} + I_{yy})$$

External moment on the connection = P_e

Equating the restoring moment to the external moment,

$$K(I_{xx} + I_{yy}) = Pe$$

$\therefore$ $$K = \frac{Pe}{I_{xx} + I_{yy}} \qquad \text{... (3.21)}$$

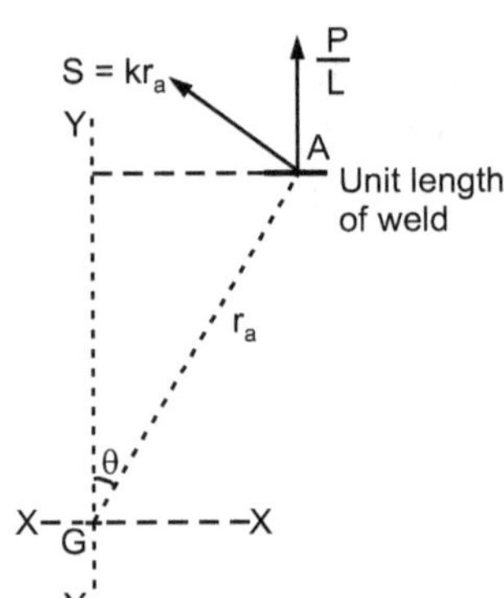

Fig. 3.46

From the above relation, the constant K can be determined for any given arrangement of weld length. The maximum force of resistance against rotation per unit length of the weld is offered at the point A most distant from G.

Considering unit length (say 1 mm length of weld) at A,

Resistance against translation $= \dfrac{P}{L}$

Resistance against rotation $= S = Kr_a$ (acting at right angles of GA)

See Fig. 3.46.

Let θ be the inclination of GA with the YY axis.

Total vertical force of resistance per unit length of weld at A

$$= V = \frac{P}{L} + S \sin \theta \qquad \text{... (3.22)}$$

Horizontal force of resistance per unit length of weld at A

$$= H = S \cos \theta \qquad \text{... (3.23)}$$

Resultant resistance offered by unit length of weld at A

$$= R = \sqrt{V^2 + H^2} \qquad \text{... (3.24)}$$

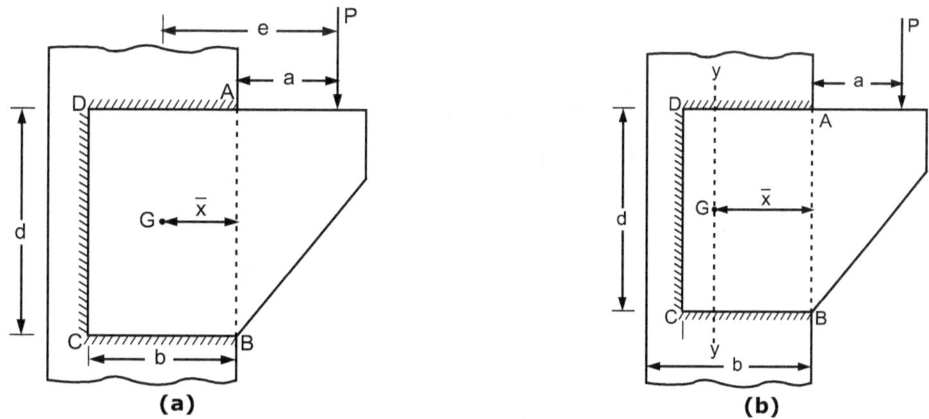

Fig. 3.47

Position of centroid G of the weld group for two usual arrangement of weld lengths are given below:

(a) When the bracket plate is welded to the column flange as shown in Fig. 3.47 (a). In this case the welding is done on all the four sides of the rectangle ABCD

$$\bar{x} = \frac{b}{2} \quad \text{and} \quad e = \bar{x} + a$$

(b) When the bracket plate is welded to the column flange as shown in Fig. 3.47 (b). In this case the welding is done only on three sides AD, DC and CD of the rectangle ABCD.

$$\bar{x} = \frac{b(b + d)}{b + (b + d)} \quad \text{and} \quad e = \bar{x} + a$$

Moment of inertia of a weld strength (See Fig. 3.48 and 3.49)

Moment of inertia of a weld line of length d about the axis XX = $I_{xx} = \dfrac{d^3}{12}$.

Moment of inertia of a weld line about the axis EF = $I_{ef} = I_{xx} + d(\bar{y})^2$ parallel to the XX.

Moment of inertia of a weld line about its longitudinal axis = 0.

Moment of inertia of a weld line of length d about an axis parallel to the weld length at a distance x_1, from the weld line = dx_1^2.

Fig. 3.48   **Fig. 3.49**

Ex. 3.30: *The bracket plate shown in Fig. 3.50 welded to the flange of an ISMB 200 column has to support a factored load of 120 kN. Determine the size of the weld required. Assume shop welding.*

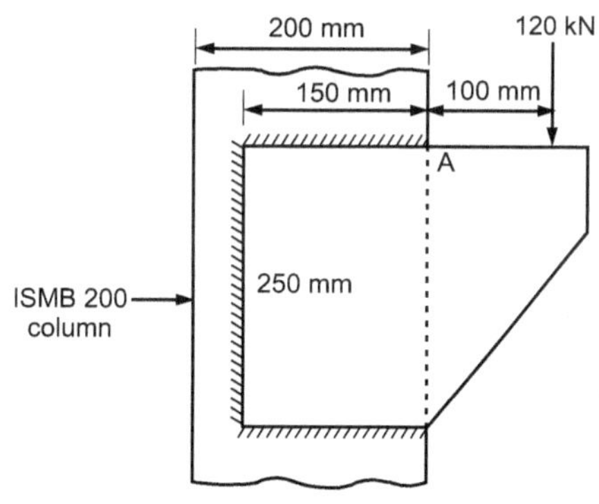

Fig. 3.50

Sol.: Total length of weld = L = 2 (150) + 250 = 550 mm

Resistance against translation per mm length of weld

$$\frac{P}{L} = \frac{120 \times 10^3}{550} = 218.18 \text{ N/mm}$$

Horizontal distance of the centroid G of the weld group from A.

$$\bar{x} = \frac{b(b + d)}{b + (b + d)} = \frac{150 \times 400}{150 + 400} = 109.1 \text{ mm}$$

Eccentricity $e = a + \bar{x} = 100 + 109.1 = 209.1$ mm
See Fig. 3.51

$$I_{xx} = 2 \times 150 \times 125^2 + \frac{250^3}{12} = 5.9896 \times 10^6 \text{ mm}^3$$

$$I_{yy} = 2\left[\frac{150^3}{12} + 150(109.1 - 75)^2\right] + 250 \times 40.9^2 = 1.3295 \times 10^6 \text{ mm}^3$$

$$I_{xx} + I_{yy} = (5.9896 + 1.3295)\,10^6 = 7.3191 \times 10^6 \text{ mm}^3$$

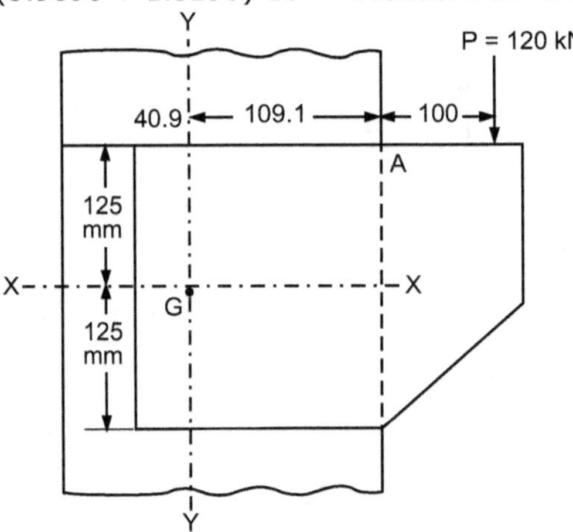

Fig. 3.51

At any point of the weld line distant r from G resistance per mm length of weld against rotation

$$= S = Kr$$

where, $K = \dfrac{Pe}{I_{xx} + I_{yy}} = \dfrac{120 \times 10^3 \times 209.1}{7.3191 \times 10^6} = 3.428$

Consider a 1 mm length of weld at A

Resistance against rotation per mm length of weld a A = S_a
= Kr_a = 3.428 r_a N/mm

See Fig. 3.52.

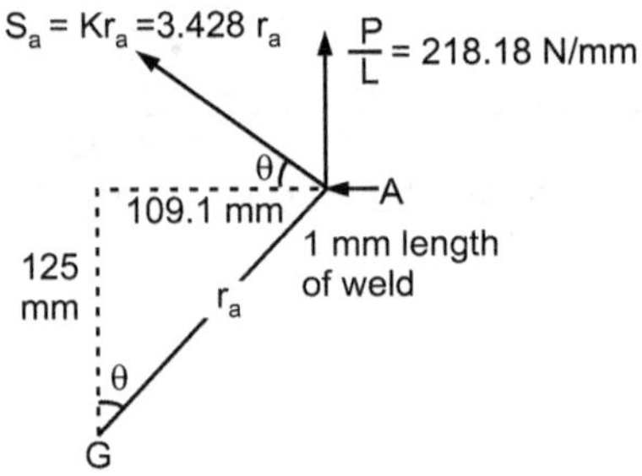

Total vertical component per mm length of weld a A

$$= V = \frac{P}{L} + S_a \sin \theta$$

$$= 218.18 + 3.428\, r_a \sin \theta$$

$$= 218.18 + 3.428 \times 109.1 = 592.17 \text{ N/mm}$$

Fig. 3.52

Horizontal component per mm length of weld at A

$$= H = S_a \cos \theta$$

$$= 3.428\, r_a \cos \theta$$

$$= 3.428 \times 125$$

$$= 428.5 \text{ N/mm}$$

Resultant resistance per mm length of weld at A

$$= R = \sqrt{592.17^2 + 428.50^2} = 730.94 \text{ N/mm}^2$$

Let the size of the weld be s mm

$$\text{Design stress for the weld} = f_{wd} = \frac{f_u}{\sqrt{3}\gamma_{mw}} = \frac{410}{\sqrt{3} \times 1.25} = 189.58\ S$$

$$\therefore \qquad 132.58\ s = 730.94$$

$$\therefore \qquad s = 5.51\ mm \qquad\qquad\qquad \text{Provide 6 mm weld.}$$

Ex. 3.31: *Fig. 3.53 shows a bracket consisting of two bracket plates welded to the flanges of an ISHB 225 column. The factored load on the bracket is 400 kN. Each bracket plate is welded to the corresponding flange of the column as shown in the Fig. Determine the size of the weld. Assume shop welding.*

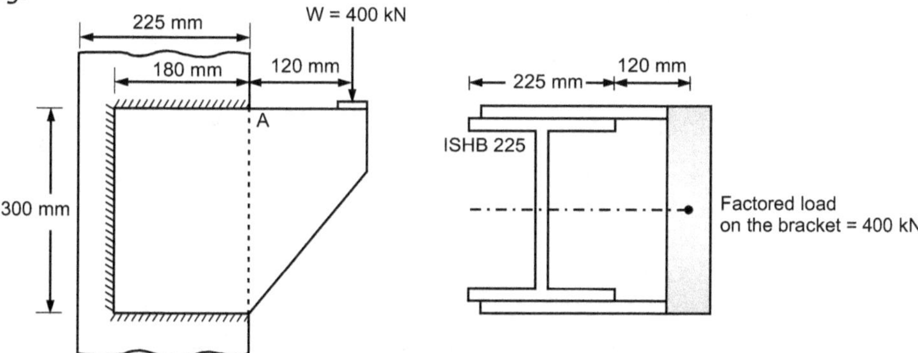

Fig. 3.53

Sol.: Consider one bracket plate

$$\text{Factored load on one bracket plate} = P = \frac{400}{2} = 200\ kN$$

$$\text{Total length of weld} = L = 2\ (180) + 300 = 660\ mm$$

Resistance against translation per mm length of the weld

$$= \frac{P}{L} = \frac{200 \times 10^3}{660} = 303.3\ N/mm$$

Horizontal distance of the centroid G of the weld group from A

$$= \bar{x} = \frac{b(b + d)}{b + (b + d)} = \frac{180 \times 480}{180 + 480} = 130.91\ mm$$

$$= e = \bar{x} + a = 130.91 + 120 = 250.91\ mm$$

Eccentricity

See Fig. 3.54

$$I_{xx} = 2(180)\ 150^2 + \frac{300^3}{12} = 10.35 \times 10^6\ mm^3$$

$$I_{yy} = 2\left[\frac{180^3}{12} + 180\ (130.81 - 80)^2\right] + 300 \times 49.09^2$$

$$= 2.297 \times 10^6\ mm^3$$

$$I_{xx} + I_{yy} = (10.350 + 2.297)\ 10^6 = 12.647 \times 10^6\ mm^3$$

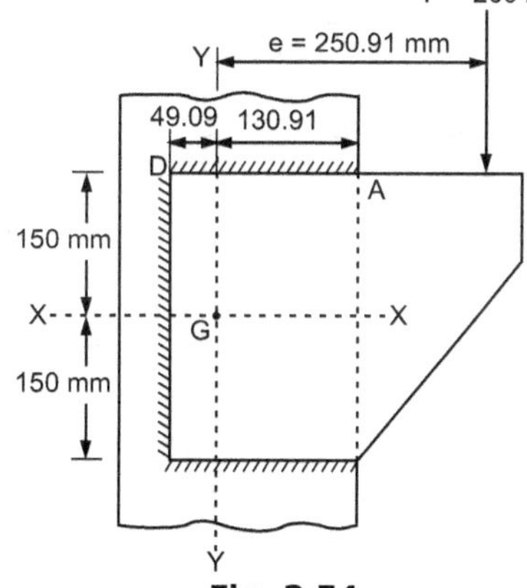

Fig. 3.54

At any point of the weld line distant r from G resistance per mm length of weld against rotation

$$= S = Kr$$

where,

$$K = \frac{Pe}{I_{xx} + I_{yy}} = \frac{200 \times 10^3 \times 250.91}{12.647 \times 10^6} = 3.968$$

Consider a 1 mm length of weld at A.

Resistance against rotation per mm length of weld at A.

$$= S_a = Kr_a = 3.968 \, r_a \text{ see Fig. 3.55.}$$

Total vertical component per mm length of weld at A

$$= V = \frac{P}{L} + S_a \sin \theta$$

$$= 303.03 + 3.968 \, r_a \sin \theta$$

$$= 303.03 + 3.968 \times 130.91$$

$$= 822.48 \text{ N/mm}$$

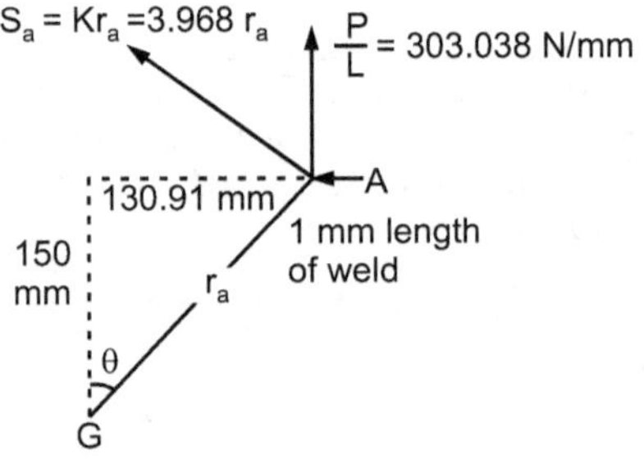

Fig. 3.55

Horizontal component per mm length of weld

$$= H = S_a \cos \theta = 3.968 \, r_a \cos \theta$$

$$= 3.968 \times 150 = 595.2 \text{ N/mm}$$

Resultant resistance per mm length of weld at A

$$= R = \sqrt{822.48^2 + 595.2^2} = 1015.25 \text{ N/mm}$$

Design stress for the weld $= f_{wd} = \dfrac{f_u}{\sqrt{3}\gamma} = \dfrac{410}{\sqrt{3} \times 1.25} = 189.4 \text{ N/mm}^2$

Let the size of the weld S mm.

Design strength of the weld per mm length

$$= 189.4 \times 1 \times 0.7 \, S = 1015.25 \qquad \therefore S = 7.66 \text{ mm}$$

Provide 8 mm weld.

Ex. 3.32: *Determine the maximum resistance offered by the weld per mm length in the eccentric bracket connection shown in Fig. 3.57. Load on the bracket plate is 250 kN. Find also the size of weld required. Assume shop welding.*

Sol.: Geometrical properties of weld length

Total length of the weld $= L = 200 + 200 + 300 = 700$ mm

$$\bar{x} = \frac{2(200 \times 100)}{700} = 57.143 \text{ mm}$$

Moment of inertia of the weld length

$$I_x = \frac{300^3}{12} + 2[200 \times 150^2] = 11.25 \times 10^6 \text{ mm}^3$$

$$I_y = 2\left[\frac{200^3}{12} + 200\,(100 - 57.143)^2\right] + 300 \times 57.143 \text{ mm}^3$$

$$= 3.048 \times 10^6 \text{ mm}^3$$

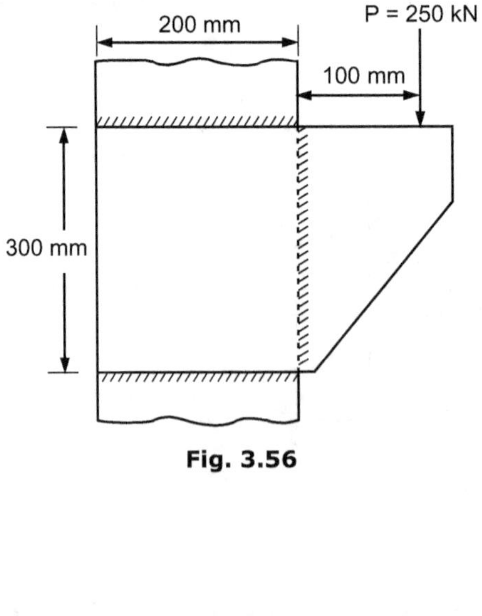

Fig. 3.56

$$I_p = I_x + I_y = (11.250 + 3.048)\,10^6\ mm^3 = 14.298 \times 10^6\ mm^3$$

$$K = \frac{Pe}{I_p}$$

$$= \frac{250 \times 10^3 \times (100 + 57.143)}{14.298 \times 10^6} = 2.748$$

Consider a 1 mm length of weld at A. See Fig. 3.57.

$$\text{Resistance against translation} = \frac{P}{L} = \frac{250 \times 10^3}{700} = 357.143\ N$$

$$\text{Resisting force against rotation} = S_a = Kr_a = 2.748\ r_a$$

$$\text{Total vertical component} = V = 357.143 + 2.748$$

$$= 357.43 + 2.748 \times 57.143\ N = 514.172\ N/mm$$

$$\text{Horizontal component} = H = 2.748\ r_a \cos\theta = 2.748 \times 150 = 412.20\ N/mm$$

Resultant resistance per mm length of weld

$$= \sqrt{514.172^2 + 412.20^2} = 659\ N/mm$$

This is the maximum resistance offered per mm length of the weld.

Let the size of the weld be s mm.

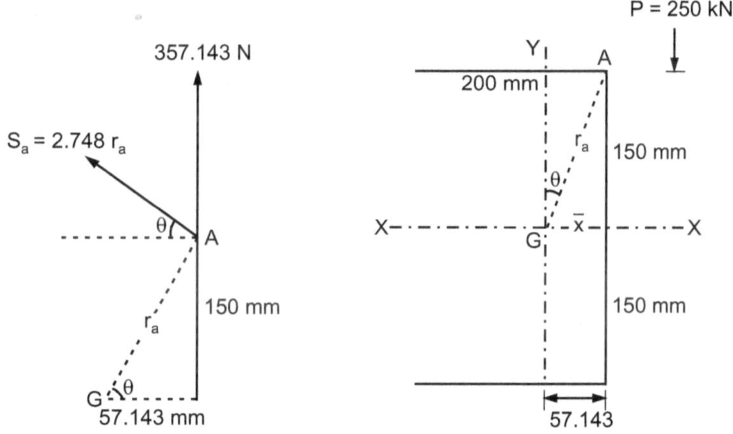

Fig. 3.57

Design strength of the weld per mm length of the weld

$$\frac{f_u}{\sqrt{3}\gamma_{mw}}\,[0.7\ s] = 659$$

$$\frac{410}{\sqrt{3} \times 1.25}\,[0.7\ s] = 659$$

$$\therefore \qquad\qquad s = 4.97\ mm$$

Provide 5 mm weld.

Ex. 3.33: *Determine the greatest load P per bracket plate that can be resisted by the bracket connection shown in Fig. 3.58 based on the strength of the weld if 8 mm fillet welds are used. Assume shop welding.*

Sol.: Total length of weld = L = 100 + 100 + 200 = 400 mm

Corresponding to factored load P

Resistance against translation per mm length of weld

$$= \frac{P}{L} = \frac{P}{400}\ N/mm$$

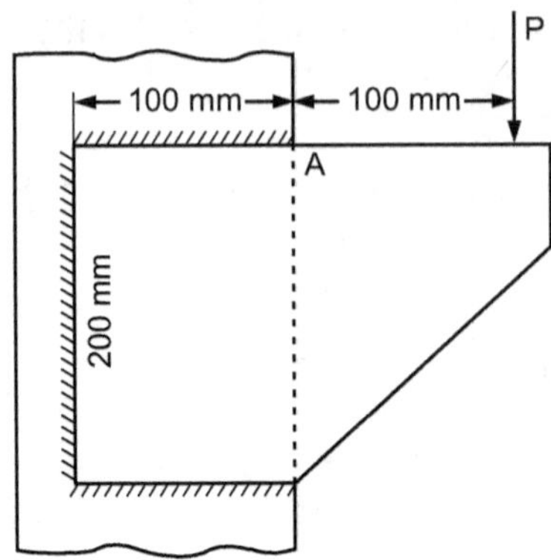

Fig. 3.58

For the given geometry of the weld group.

Horizontal distance of the centroid G of the weld group from A

$$= \bar{x} = \frac{b(b + d)}{b + (b + d)} = \frac{100 \times 300}{100 + 300} = 75 \text{ mm}$$

Eccentricity $\quad e = \bar{x} + a = 75 + 100 = 175 \text{ mm}$

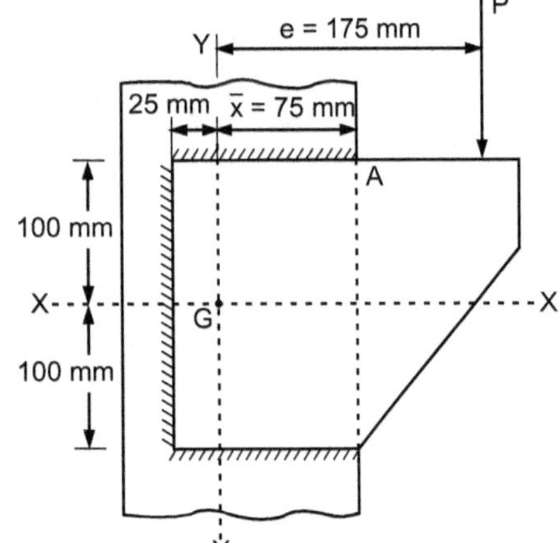

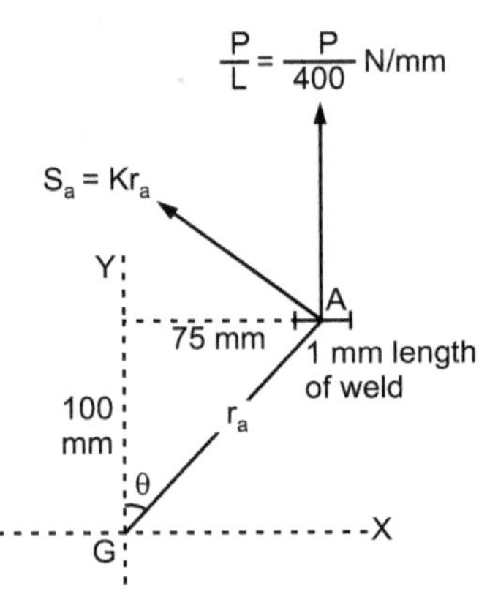

Fig. 3.59

Moment of inertia of weld length:

$$I_{xx} = \frac{200^3}{12} + 2\,(100 \times 100^2) = 2.667 \times 10^6 \text{ mm}^3$$

$$I_{yy} = 2\left[\frac{100^3}{12} + 100\,(75 - 50)^2\right] + 200 \times 25^2 = 0.417 \times 10^6 \text{ mm}^3$$

$$I_{xx} + I_{yy} = (2.667 + 0.417)\,10^6 \text{ mm}^3 = 3.084 \times 10^6 \text{ mm}^3$$

Resistance against translation per mm length of weld $= \dfrac{P}{400}$ N/mm

Resistance against rotation per mm length of weld at A

$$= S_a = Kr_a, \text{ where, } K = \frac{Pe}{I_{xx} + I_{yy}} = \frac{P \times 175}{3.084 \times 10^6}$$

$$\therefore \qquad S_a = \left[\frac{175\,P}{3.084 \times 10^6}\right] r_a$$

Total vertical component per mm length of weld at A $= \dfrac{P}{400} + S_a \sin\theta$

$$= V = \frac{P}{400} + \left[\frac{175\,P}{3.084 \times 10^6}\right] r_a \sin\theta$$

$$a = \frac{P}{400} + \frac{175\,P}{3.084 \times 10^6} \times 75 = 0.00675\,P \text{ N/mm}$$

Horizontal component per mm length of weld at $A = H = S_a \cos \theta$

$$= H = \left[\frac{175\ P}{3.084 \times 10^6}\right] r_a \cos \theta$$

$$= \frac{175\ P}{3.084 \times 10^6} \times 100 = 0.00567\ P\ \text{N/mm}$$

Resultant resistance per mm length of weld at A

$$= \sqrt{(0.00675\ P)^2 + 0.00567\ P^2)} = 0.008815\ P\ \text{N/mm}$$

$$\text{Design of stress of weld} = \frac{f_u}{\sqrt{3}\gamma_{mw}} = \frac{410}{\sqrt{3} \times 1.25} = 189.4\ \text{N/mm}^2$$

Design strength of 8 mm weld per mm length

$$= 189.4 \times 1 \times 0.7 \times 8 = 1060.6\ \text{N/mm}$$

$\therefore \qquad\qquad 0.008815\ P = 1060.6$

$\therefore \qquad\qquad\qquad P = 120317\ \text{N} = 120.317\ \text{kN}$

Important Points

(A) BOLTED CONNECTIONS

- A bolt may be defined as a metal pin with a head at one end and a shank threaded at the other end to receive a nut.
- Bolts are classified as:
  (i) Black bolt
  (ii) High strength bolt
- Types of bolted joints are:
  (i) Lap joints
  (ii) Butt joints
- Bolted joint may fail under:
  (i)   Shear failure
  (ii) Tensile failure
  (iii) Bearing failure
- Net tensile stress area of bolt at threads = $0.78 \times$ plain shank area of bolts.
- Pitch of bolts
  (i) Minimum pitch = 2.5 d; where d = nominal diameter of bolt
  (ii) Maximum pitch = 32 t or 200 mm which is less
- Edge distance $\approx$ 2d
- Shearing strength of bolt

$$V_{dsb} = \frac{f_u}{\sqrt{3}} \frac{(n_n \cdot A_{nb} + n_s \cdot A_{sb})}{\gamma_{mb}}$$

- Bearing strength of bolt

$$V_{dpb} = \frac{2.5\ k_b \cdot d \cdot t_p \cdot f_u}{\gamma_{mb}}$$

  where, k_b is least of $\left[\dfrac{e}{3d_o};\ \dfrac{p}{3d_o} - 0.25;\ \dfrac{f_{ub}}{f_u};\ 1.0\right]$

- Tensile strength of plate $T_{dn} = \dfrac{0.9\ f_u \cdot A_n}{\gamma_{m1}}$

- Bolt value is minimum of shearing strength and bearing strength of bolt.

- Efficiency of joint = $\dfrac{\text{Minimum Actual Strength of Joint}}{\text{Gross Strength of Solid Plate}} \times 100$

(B) WELDED CONNECTIONS

- When the various members of a structure are connected together by means of welds, the connection is called as welded connection.
- Types of welds:
  - (i) Butt or Groove Welds
  - (ii) Fillet Welds
- Welded joints are more advantageous due to:
  - (i) Avoidance of holes
  - (ii) More rigidity and stiffness
  - (iii) Light in weight and economical
  - (iv) Better finish and appearance
  - (v) Reduction in noise
  - (vi) Less maintenance
  - (vii) Air tightness and water tightness
- Design stress of fillet weld

$$f_{wd} = \frac{f_u}{\sqrt{3}\,\gamma_{mw}}$$ where γ_{mw} = 1.25 for shop welds = 1.50 for site welds

- Design strength of fillet weld

$$P_{dw} = f_{wd} \cdot L \cdot t_t$$

- Effective length of weld

$$L = \frac{P_{dw}}{p_q} \text{ or } \frac{P_{dw}}{f_{wd} \times t_t}$$

Practice Questions

(A) BOLTED CONNECTIONS

1. Define 'bolt value'. How and why is it calculated?
2. Explain how pitch is decided for bolted joints.
3. State two modes of failure of bolted joints.
4. Define pitch distance of gauge distance of bolts.
5. Draw typical plan and section elevations of double bolted lap joint and single cover double bolted butt joint.
6. Find the bolt value of 16 mm diameter 4.6 grade bolt in a lap joint. The thickness of plate is 10 mm. Use I.S. specifications.
7. 12 mm thick plates are connected using double bolted lap joint using 16 mm diameter bolts of 4.6 grade at a pitch of 80 mm. Calculate strength and efficiency of joint.
8. Fig. 3.60 shows a joint a lower chord of a roof truss. The members are connected with 16 mm diameter bolts of 4.6 grade to the gusset plate 12 mm thick. The forces shown are factored forces.

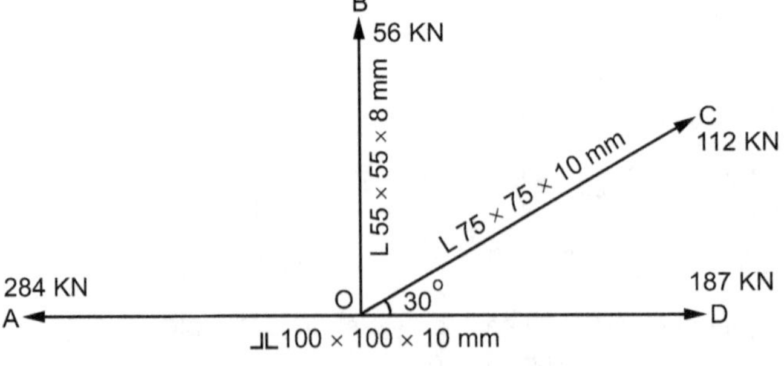

Fig. 3.60

9. Two plates 12 mm × 60 mm are connected in a lap joint with 4 bolts of 16 mm diameter as shown in Fig. 3.61. Determine the strength of the joint. Ultimate strength of bolt material = 400 MPa and ultimate strength of plate material = 410 MPa.

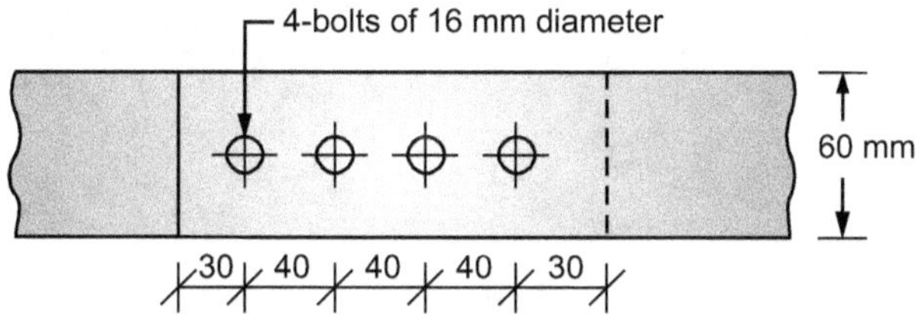

Fig. 3.61

[Ans.: 115900 N]

10. Design a lap joint connecting two plates 120 mm × 8 mm to transmit a factored load of 120 kN. Use 12 mm diameter bolts of grade 4.6 and plates of grade 410.

[Ans.: Provide 8 bolts at pitch = 30 mm]

(B) WELDED CONNECTIONS

11. State the advantages and disadvantages of welded connections over bolted connections. Write any four points of differences.

12. Define effective length of weld. State how to calculate length of weld required.

13. State the relation between throat thickness and size of fillet weld.

14. Draw a neat sketch of welded single 'V' butt joint.

15. Draw a neat sketch of double 'V' butt joint in weld.

16. State two types of welded joints.

17. Drawing a neat sketch show the size and throat of a welded lap joint. What is the relation between them?

18. Find out the design strength of the welded joint as shown in Fig. 3.62.

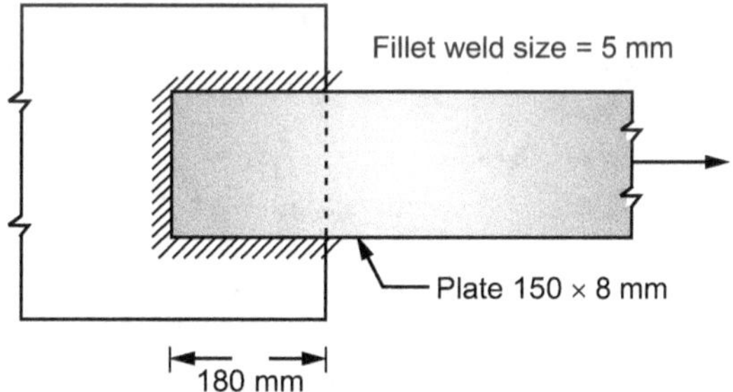

Fig. 3.62

19. A flat 120 × 10 mm is to be connected to a gusset plate using 6 mm fillet weld as shown in Fig. 3.63. Calculate the minimum lap required to full strength of flat.

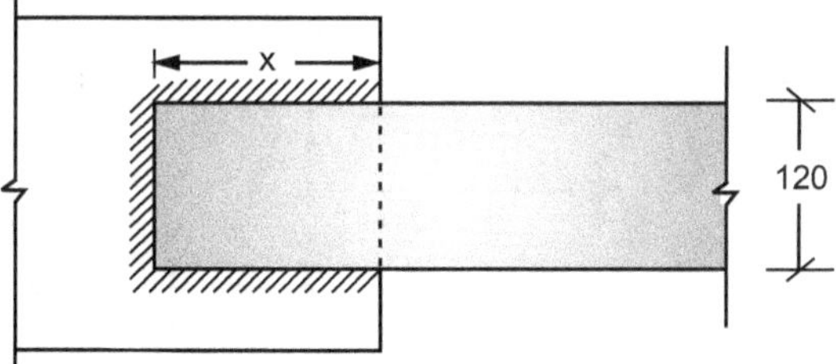

Fig. 3.63

20. Calculate minimum lap required to connect a flat 100 × 16 mm with gusset plate using 10 mm fillet weld as shown in Fig. 3.64 if it is subjected to force equal to its full strength in tension.

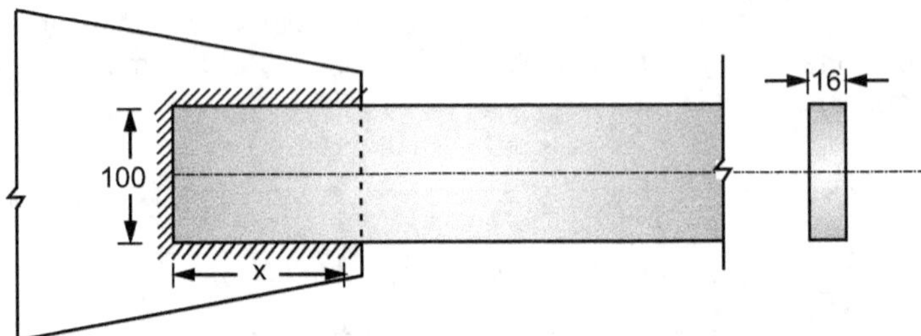

Fig. 3.64

21. Calculate the size of weld required for the joint shown in Fig. 3.65. Use I.S. specifications.

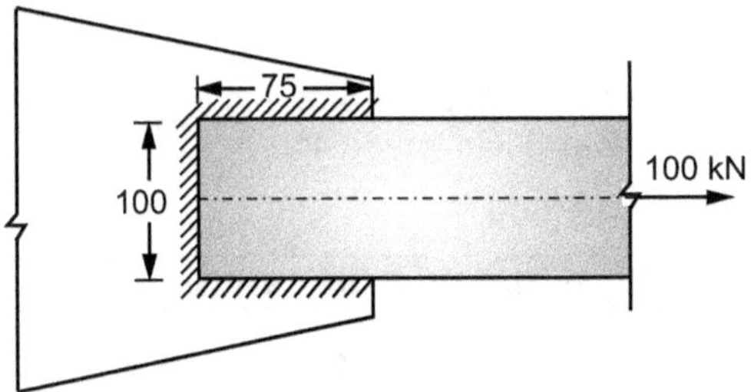

Fig. 3.65

22. Enlist any four advantages of welded connection. Draw a neat labelled sketch of a fillet weld. Find safe load transmitted by a fillet welded joint between a flat 60 mm wide overlapping 100 mm over a gusset plate. Thickness of both plates is 10 mm. Weld is on all sides of overlap. Size of weld is 6 mm.

23. A tie member 75 mm × 8 mm has to transmit an axial load of 90 kN. Design the fillet weld and calculate the necessary overlap. Consider the welding on all four sides.

24. A 100 mm wide and 12 mm thick tie plate has been connected to a gusset plate by a 10 mm fillet weld. The lap of the plate over gusset plate is 120 mm. Fillet welding has been done on two longitudinal edges and not at ends. Find out the strength of the fillet welded joint.

25. A tie in a truss consists of a pair of angles ISA 50 × 50 × 6 mm welded on either side of a 8 mm thick gusset plate. Design the welded joint.

26. A plate of 60 mm × 8 mm in cross-section is connected to a gusset plate. What is the length of weld required for full strength of plate, if size is weld is 6 mm? Assume design stresses, as per I.S. specifications.

27. Find length of 8 mm thick fillet weld to connect a tie 100 mm × 12 mm to a 16 mm thick gusset plate. Consider full strength of plate.

28. Design a suitable fillet weld to connect a tie bar 80 × 10 mm to a 12 mm thick gusset plate. Use I.S. specifications for stresses.

29. Calculate length of fillet weld required to connect an ISA 100 × 100 × 10 mm with gusset plate using 6 mm weld as shown in Fig. 3.66. The angle is subjected to an axial force of 120 kN. Use I.S. specifications for design stresses. $C_{xx} = C_{yy}$ for angle is 28.4 mm.

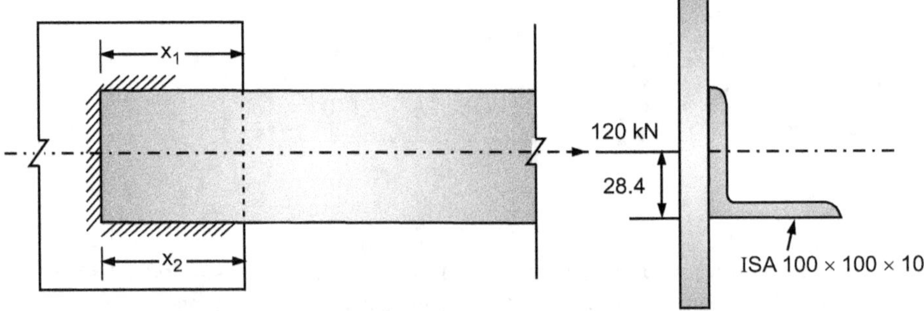

Fig. 3.66

30. Design a suitable fillet weld to connect a tie bar 100×12 mm to a 16 mm thick gusset plate. The design stresses in the tie bar and fillet weld shall be assumed as per IS specifications.

31. Calculate the size of fillet weld required for joint shown in Fig. 3.67 to carry an axial load of 200 kN.

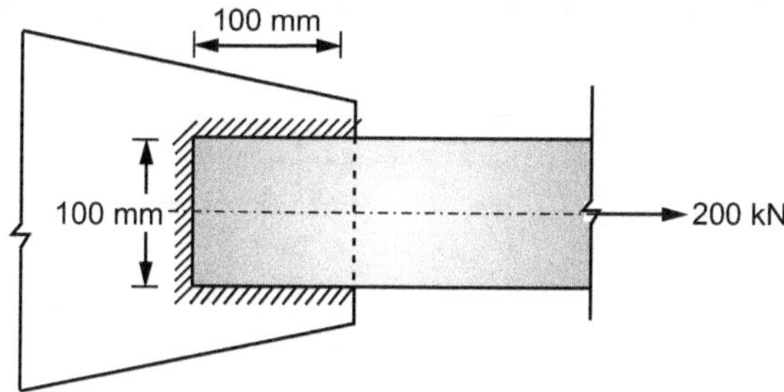

Fig. 3.67

32. Calculate minimum pitch to be provided to connect 12 mm thick plates using 10 mm diameter rivets. Double riveted lap joint is to be provided.

33. A flat is connected with a gusset plate using 6 mm fillet weld as shown in Fig. 3.68. Calculate minimum lap length 'x' to carry a force of 100 kN. Take τ_{vf} for weld as 100 MPa.

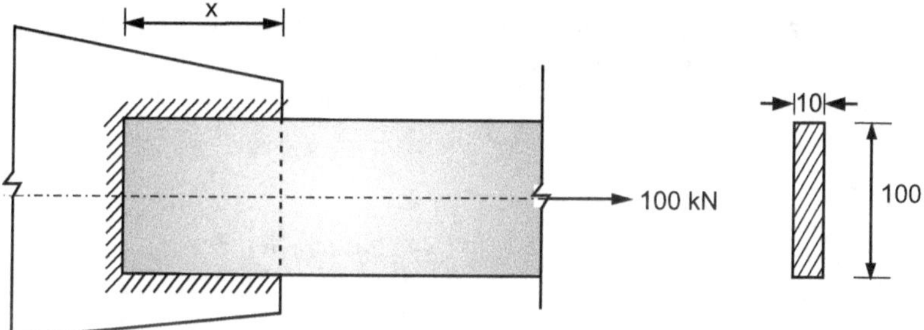

Fig. 3.68

(C) PROBLEMS ON ECCENTRIC CONNECTIONS

34. A factored load of 40 kN is applied to a 10 mm thick bracket plate bolted to the flange of a column with 20 mm diameter bolts, Eight bolts are provided with 4 bolts in each of the two vertical rows as shown in Fig. 3.69. The eccentricity of the load from the centroid of the bolt group is 300 mm. Determine the greatest resistance offered by a bolt.

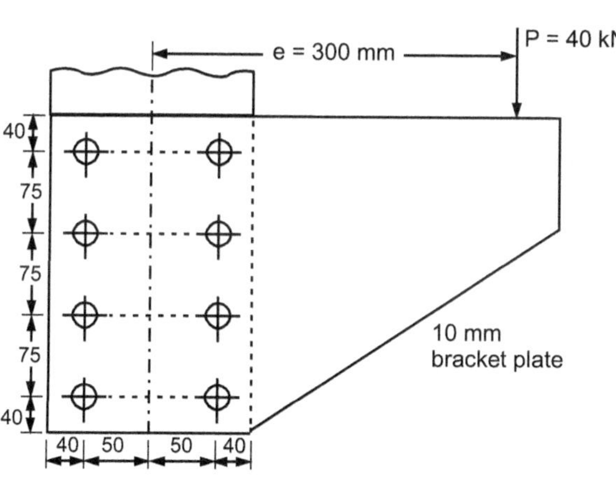

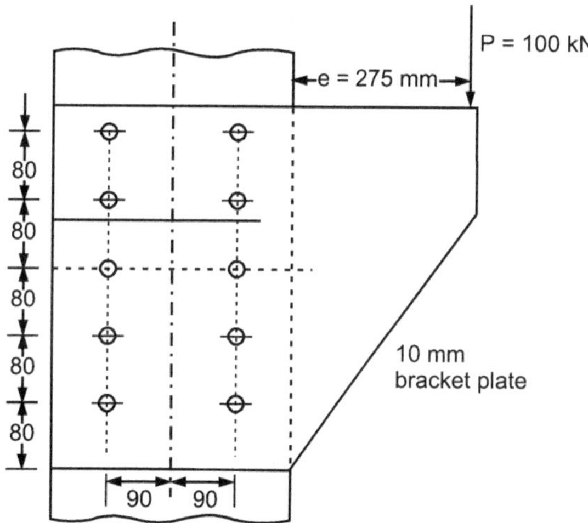

Fig. 3.69   **Fig. 3.70**

35. Investigate the safety of the design of the bracket connection shown in Fig. 3.70. The factored load on the bracket plate is 100 kN at an eccentricity of 275 mm. The bolts are 20 mm in diameter.

36. The bracket plate shown in Fig. 3.71 welded to the flange of a ISHB 250 column has to support a factored load of 150 kN. Determine the size of the weld required. Assume shop welding.

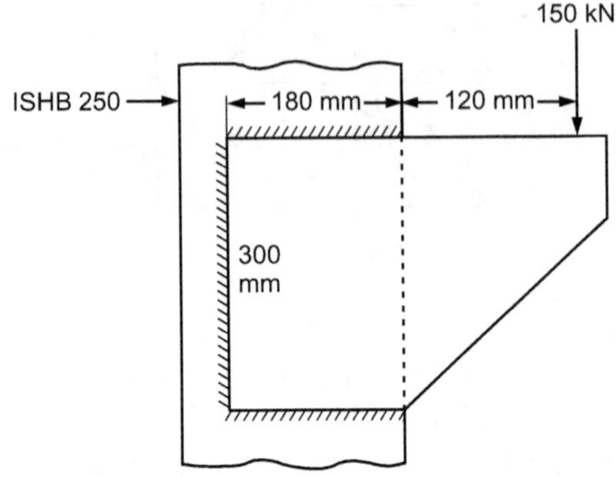

Fig. 3.71

37. Fig. 3.72 shows as bracket consisting of two bracket plates welded to the flanges of a ISHB 350 column. The factored load on the bracket is 350 kN. Each bracket plate is welded to the corresponding flange of the column as shown in Fig. 3.73. Determine the size of the weld require. Assume shop welding.

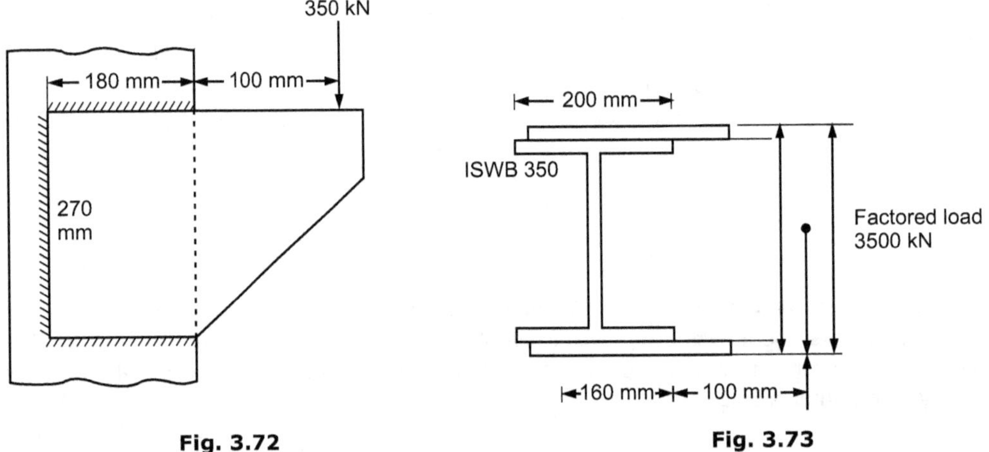

Fig. 3.72   Fig. 3.73

∎∎∎

Chapter **4**

DESIGN OF TENSION MEMBER BY L.S.M.

- Tension Members - Effective Length and Effective Sectional Area of Tension Members - Design Strength of Tension Members against Yielding of Gross Section Requirements: Deflection Limits, Vibration, Durability and Fire Resistance, against Rupture of Critical Section and due to Block Shear. Problems on Determination of Design Strength of given Members and Designing Tension Members using Rolled Steel Sections for given Loads - Design of Bolted/Riveted and Welded Connections for Tension Members - Problems.

About this Chapter

After reading this chapter students can understand:
- Design of Tension Member
- Types of Section Used as Tension Member
- Net Sectional Area in Chain or Staggered Bolting
- Strength of Tension Members

4.1 INTRODUCTION

- A tension member is a structural member which transmits a direct axial pull applied at its ends.

- Tension members can sustain load upto the ultimate load without fracture, but the elongation of the member at this load will be more (upto 18%) and hence the member may become unserviceable. Therefore in design of tension members, **the yield load is generally taken as failure load.**

4.2 TYPES OF SECTIONS USED

- The various forms of tension members are as follows:

  (i) Wires and Cables

  (ii) Bars and Rods

  (iii) Plates and Flat Bars

  (iv) Single and Built-up Sections

- Wire ropes are exclusively used for hoisting purposes and as guy wires in steel stacks and towers. Cables are used as floor suspenders in suspension bridges and as stay cables in cable stayed bridges.

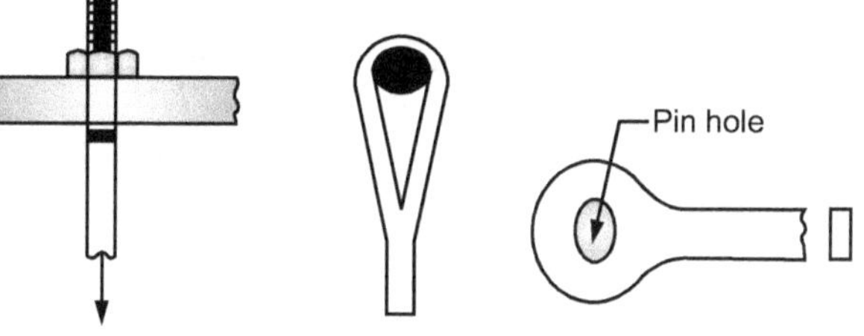

Fig. 4.1: Connections of bars and rods

- Bars and rods are often used in bracing systems, sag rods and to support girts in industrial buildings (See Fig. 4.1).

- Plates and flat bars are used in transmission towers, foot bridges, columns, lacing flats and batten plates.

- Single sections such as angles, channels and I-sections are used as open sections. Built-up sections such as double angles, double channels with or without additional plates are used as compound sections. (See Fig. 4.2).

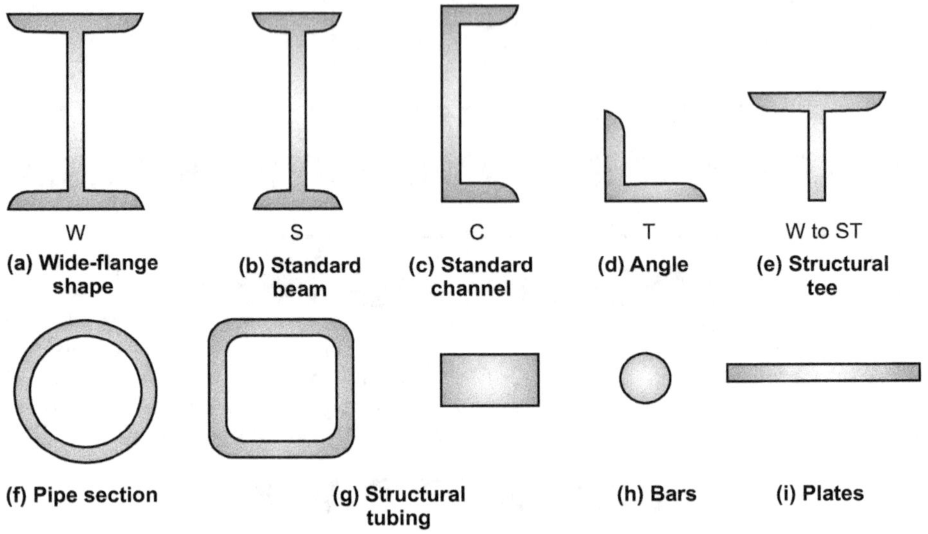

W	S	C	T	W to ST
(a) Wide-flange shape	**(b) Standard beam**	**(c) Standard channel**	**(d) Angle**	**(e) Structural tee**

(f) Pipe section	**(g) Structural tubing**	**(h) Bars**	**(i) Plates**

(a) Single section Tension Members

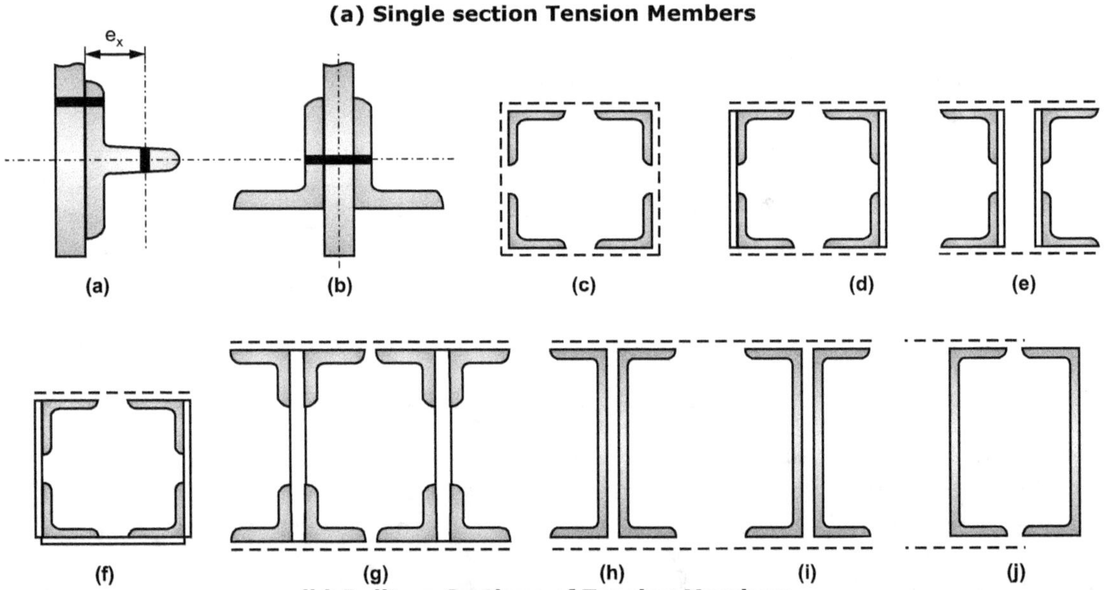

(a)	(b)	(c)	(d)	(e)

(f)	(g)	(h)	(i)	(j)

(b) Built-up Sections of Tension Members

Fig. 4.2

4.3 NET SECTIONAL AREA

- The net sectional area of a tension member is the gross-sectional area of the member minus the sectional area of the maximum number of holes.
- Generally, a tension member without bolt holes can resist loads upto the ultimate load without failure. The presence of holes reduces strength of the tension member even if the hole is occupied by the bolt.

(a) Chain bolting:

- Refer to the plate shown in Fig. 4.3 subjected to pull T and provided with chain bolting (i.e. holes in one line across the section). The possibility of failure can be along the section ABCD. Net area at the section is equal to gross area minus the area of bolt holes B and C.

$$A_n = A_g - \text{Sectional area of holes}$$
$$A_n = (b - n \cdot d_h) \times t \qquad ... (4.1)$$

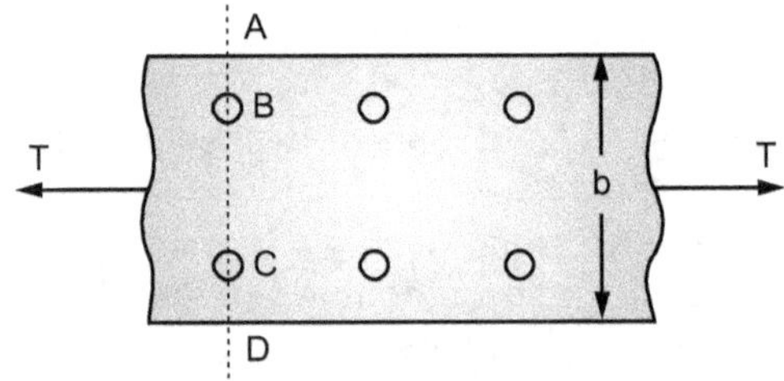

Fig. 4.3: Chain Bolting

where, A_n = Net sectional area of plate
              A_g = Gross-sectional area of plate
              b = Width of plate
              n = Number of bolts
              d_h = Diameter of bolt hole (d_h = d + 2 mm in case of directly punched holes)
              t = Thickness of plate

(b) Zig-Zag Bolting:
- Refer to the plate shown in Fig. 4.4 subjected to a pull T and provided with zig-zag bolting or staggered bolting. Staggering improves the load carrying capacity of the member for a given row of bolts.
- For computation of critical net area, calculate area of section along ABC less area of one bolt hole and then the area of a section along ABDE less two bolt holes. The least of the two value would be the critical value. The net area for staggered pattern shown in Fig. 4.4 (a) is taken as.

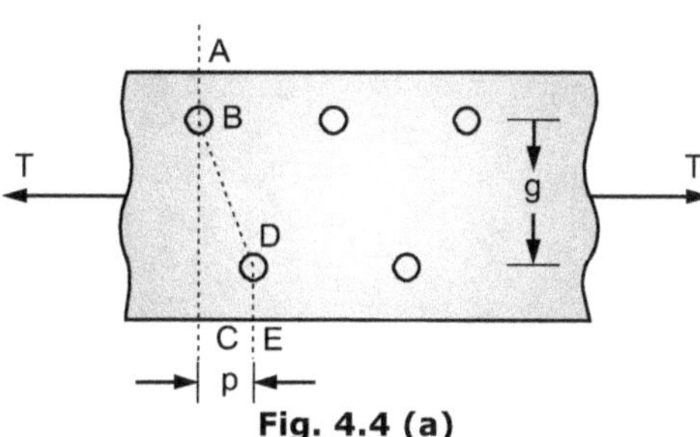

Fig. 4.4 (a)

$$A_n = [b - n\, d_h + (\Sigma p^2/4g)] \times t \qquad \ldots (4.2)$$

where,   g = Gauge distance i.e. distance measured along right angles to the direction of pull.
              p = Pitch length parallel to pull.

Other symbols have the same meaning as applied for chain bolting.

A typical case of a plate subjected to pull T and connected to gusset plate using bolts is shown in Fig. 4.4 (b) in staggered pattern.

The net area along the section ABCDE is

$$A_n = \left[b - 3d + \frac{p_1^2}{4g_1} + \frac{p_2^2}{4g_2}\right] \cdot t \ldots (4.3)$$

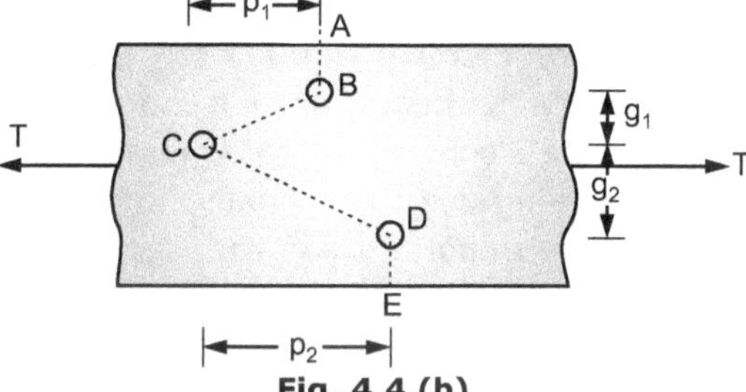

Fig. 4.4 (b)

4.4 DESIGN STRENGTH OF TENSION MEMBERS

- The design strength of members under axial tension shall be minimum of following three failures:
  (i)   due to yielding of the gross-section.
  (ii) rupture of net section.
  (iii) failure due to block shear.

4.4.1 Gross-Section Yielding

- A tension member can fail by reaching one of following two limit states i.e. excessive deformation or fracture. Generally, a tension member can sustain loads upto ultimate load without failure. But it deforms considerably in longitudinal direction before fracture and becomes unserviceable. Hence to prevent excessive deformation due to yielding, the stress on gross-section shall be less than yield stress.

$$\frac{T}{A_g} < f_y$$

$$T < A_g \cdot f_y$$

or $$T_{dg} = \frac{A_g \cdot f_y}{\gamma_{mo}}$$

where, T = Factored tensile force

              T_{dg} = Design strength
              A_g = Gross-sectional area
              γ_{mo} = Partial safety factor (= 1.1.)

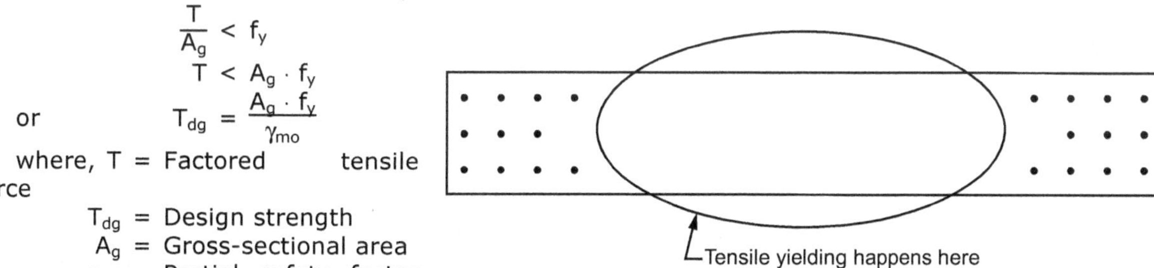

Fig. 4.5: Gross-section Yielding

4.4.2 Net Section Rupture

(a) For Plates and Threaded Rods:

- A tension member is usually connected to other members by bolts or welds. The fibres adjacent to the bolt hole yield due to stress concentration. However, the ductility of steel permits the initially yielded zone to deform without fracture. At this stage the entire net section reaches the ultimate stress f_u. To prevent failure of tension member by net section rupture;

$$T < A_n \cdot f_u$$

$$T_{dn} = \frac{T}{\gamma_{m1}} = \frac{A_n f_u}{\gamma_{m1}}$$

where, T_{dn} = Design strength in rupture

    f_u = Ultimate stress

    γ_{m1} = Partial safety factor in rupture of cross-section (= 1.25)

- As there is no reserve of strength of any kind beyond ultimate strength, a factor = 0.9 shall be multiplied to equation of T_{dn} which is based on statistical equation determined from experimental results.

$$\therefore \quad T_{dn} = 0.9 \frac{A_n f_u}{\gamma_{m1}} \qquad \ldots (4.5)$$

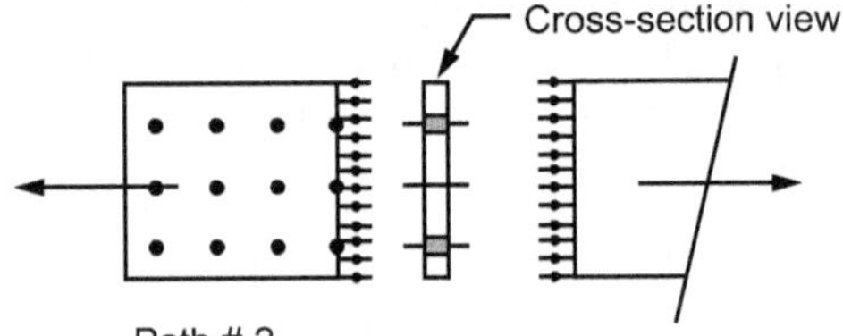

(b) For Angles:

- While transferring the tensile force from gusset plate to tension member through one leg by bolts or welds, the connected leg of section (such as angle, channel) may be subjected to more stress than the outstanding leg and finally the stress distribution becomes uniform over the section away from the connection. Thus one part lags behind the other, this is called as *shear lag*.

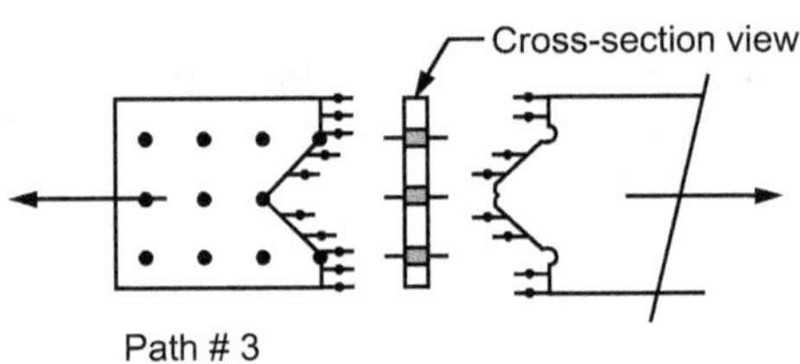

Fig. 4.6

- The tearing strength of an angle section connected through one leg is affected by shear lag also. Thus, the design strength, T_{dn} governed by tearing at net section is given by;

$$T_{dn} = 0.9 \frac{A_{nc} f_u}{\gamma_{m1}} + \beta \frac{A_{go} f_y}{\gamma_{mo}} \qquad \ldots (4.6)$$

where, $\beta = 1.4 - 0.076 \dfrac{w}{t} \times \dfrac{f_y}{f_u} \times \dfrac{b_s}{L_c} \le 0.9 \dfrac{f_u}{f_y} \times \dfrac{\gamma_{mo}}{\gamma_{m1}} \ge 0.7$

where, b_s = Shear lag width

    (See Fig. 4.7)

    L_c = Length of end connection (the distance between the outer-most bolts in the joint along the length or length of the weld along the load direction).

    t = Thickness of angle leg

    A_{nc} = Net area of connecting leg

    A_{go} = Gross-area of outstanding leg

  $b_s = w + g - t$ $b_s = w$

Fig. 4.7: Shear Lag Width

Approximate rupture strength

$$T_{dn} = \alpha \cdot \frac{A_n f_u}{\gamma_{m1}} \qquad \ldots (4.7)$$

where, α = Reduction factor = 0.6 for one or two bolts

    = 0.7 for three bolts = 0.8 for four or more bolts or equivalent weld length

4.4.3 Block Shear

- This type of failure occurs due to tearing of a segment or block of the material at the end of a member. It occurs along a path involving tension on one plane and shear on a perpendicular plane as shown in Fig. 4.8.

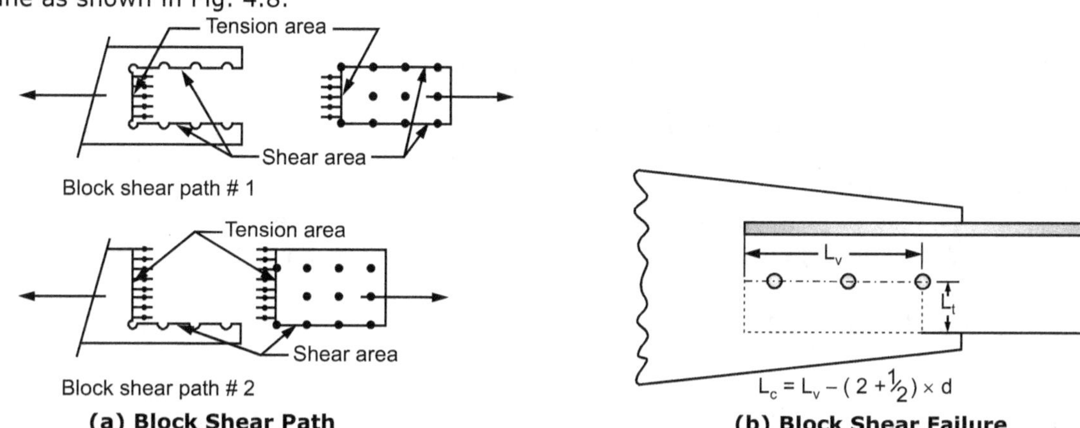

Fig. 4.8

- The block shear strength at an end connection shall be taken least of following two

  (i)   For shear yield and tension fracture

$$T_{db1} = \frac{A_{vg}\,f_y}{\sqrt{3}\cdot\gamma_{mo}} + 0.9\,\frac{A_{tn}\cdot f_u}{\gamma_{m1}} \qquad \ldots (4.8)$$

  (ii)   For shear fracture and tension yield

$$T_{db2} = \frac{A_{tg}\cdot f_y}{\gamma_{mo}} + 0.9\,\frac{A_{vn}\cdot f_u}{\sqrt{3}\cdot\gamma_{m1}} \qquad \ldots (4.9)$$

where, A_{tg} and A_{tn} = Minimum gross and net area in tension from the hole to the toe of angle or next last row of bolts in plates perpendicular to the line of force respectively.

$$A_{tg} = L_t \cdot t \quad \text{and} \quad A_{tn} = (L_t - 0.5\,d_h) \times t$$

A_{vg} and A_{vn} = Minimum gross and net area in shear along the line of action for force respectively

$$A_{vg} = L_v \cdot t$$

and

$$A_{vn} = \left[L_v - \left(2 + \frac{1}{2}\right) d_h \right] \times t$$

4.5 SLENDERNESS RATIO (SR)

Effective length of tension member:

- It is the unsupported length of the member between centres of the end connection.

- It is defined as the ratio of its unsupported length 'L' to it least radius of gyration 'k_{min}'. IS : 800 – 2007 limits the values of slenderness ratio for tension member as given in table 4.1.

Table 4.1: Maximum Slenderness Ratios for Tension Member (I.S. 800)

Sr. No.	Type of Tension Member	Maximum Slenderness Ratio
(i)	A tension member in which a reversal of direct stress due to loads other than wind or seismic forces occur.	180
(ii)	A member normally acting as a tie in a roof truss or bracing system but subject to possible reversal of stress resulting from the action of wind or seismic forces.	350
(iii)	Tension members other than pretensioned members.	400

4.6 STEPWISE PROCEDURE FOR DESIGN OF TENSION MEMBERS SUBJECTED TO AXIAL LOAD

1. Determine gross area required from its yield strength

$$A_g = \frac{T}{f_y/\gamma_{mo}} = \frac{T}{f_y} \cdot \gamma_{m0}$$

where, f_y = yield strength of material

       γ_{mo} = 1.1

       T = Factored tensile load

2. From steel table select suitable rolled steel section providing area matching with calculated gross area in step 1.

3. Calculate number of bolts required to connect the member of gusset plate/another member.

4. Calculate design strength T_d of trial section. It should be minimum of T_{dg}, T_{dn} and T_{db}

   where,

   (i)   Gross section yielding

$$T_{dg} = \frac{A_g\, f_y}{\gamma_{m_0}}$$

   (ii)   Net section Rupture

      (a)   For plates and threaded rods $T_{dn} = \dfrac{0.9\, A_n\, f_u}{\gamma_{m_1}}$

      (b)   For angles $T_{dn} = \dfrac{\alpha\, A_n\, f_u}{\gamma_{m_1}}$ and

$$T_{dn} = 0.9\, \frac{A_{nc}\, f_u}{\gamma_{m_1}} + \beta\, \frac{A_{go} \cdot f_y}{\gamma_{mo}}$$

   (iii) Block shear failure

      (a)   For shear yield and tension fracture

$$T_{db_1} = \frac{A_{vg}\, f_y}{\sqrt{3}\, \gamma_{mo}} + 0.9\, \frac{A_{tn} \cdot f_u}{\gamma_{m_1}}$$

      (b)   For shear fracture and tension yield

$$T_{db_2} = \frac{A_{tg} \cdot f_y}{\gamma_{mo}} + 0.9\, \frac{A_{vn} \cdot f_u}{\sqrt{3}\, \gamma_{m_1}}$$

Minimum of T_{db_1} and T_{db_2} is considered as block shear strength.

Solved Examples

Ex. 4.1: *A plate 180 mm × 10 mm is connected to a gusset plate with 16 mm diameter bolts as shown in Fig. 4.7. Find the minimum net area of the plate section to act as a tension member.*

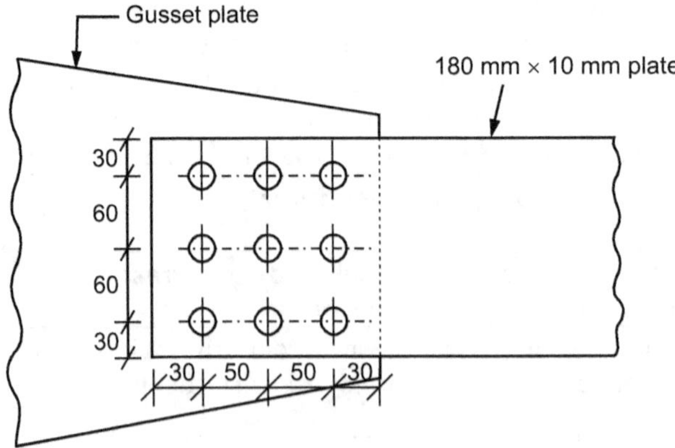

Fig. 4.9

Sol.: Diameter of bolt hole, $d_h = 16 + 2 = 18$ mm.

Minimum net area of plate section

$A_n = (b - nd_h) \cdot t = (180 - 3 \times 18) \times 10 =$ **1260 mm²**.

Ex. 4.2: *Two flat plates 180 mm × 20 mm are connected in a lap joint using 6 bolts of 20 mm diameter as shown in Fig. 4.8. Determine the strength of a plate in tension. Take $f_y = 250$ N/mm² and $f_u = 410$ N/mm².*

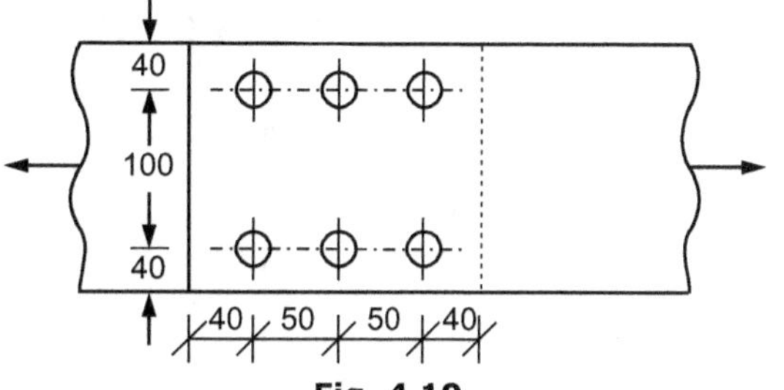

Fig. 4.10

Sol.: Diameter of bolt hole,

$$d_h = 20 + 2 = 22 \text{ mm.}$$
$$A_n = (b - n \, d_h) \cdot t$$
$$= (180 - 2 \times 22) \times 20 = 2720 \text{ mm}^2$$

(i) Design strength governed by gross-section yielding

$$T_{dg} = \frac{A_g \cdot f_y}{\gamma_{m_0}} = \frac{(180 \times 20) \times 250}{1.10} = 818182 \text{ N}$$

(ii) Design strength to governed by net section rupture

$$T_{dn} = \frac{0.9 \, A_n \, f_u}{\gamma_{m_1}} = \frac{0.9 \times 2720 \times 410}{1.25} = 802944 \text{ N}$$

∴ Design strength of plate in tension (minimum of T_{dg} and T_{dn})

$$T_{dn} = 802944 \text{ N} = \textbf{802.94 kN}$$

Ex. 4.3: *Find the minimum net area of the 240 mm × 10 mm plate in the bolted connection shown in Fig. 4.9. Bolts are 16 mm in diameter.*

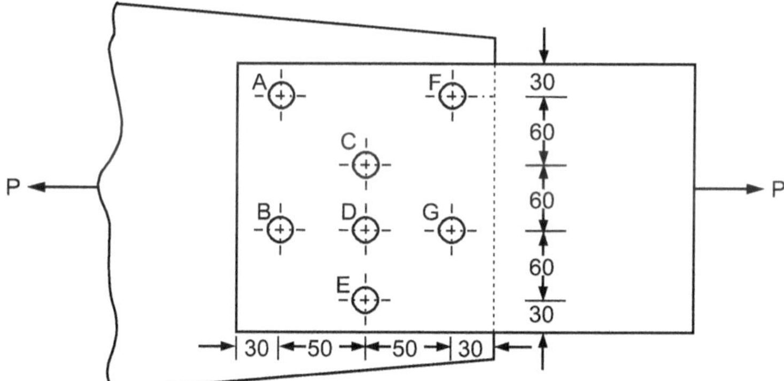

Fig. 4.11

Sol.: Diameter of bolt hole, $d_h = 16 + 2 = 18$ mm.

For any path, net area is given by equation (4.2)

$$A_n = \left[b - n \cdot d_h + \Sigma \frac{p^2}{4g} \right] \cdot t$$

(i) For path AB and FG (2 holes).

$$A_n = [240 - 2 \times 18] \times 10 = 2040 \text{ mm}^2$$

(ii) For path CDE (3 holes)

$$A_n = [240 - 3 \times 18] \times 10 = 1860 \text{ mm}^2$$

(iii) For path ACDE (4 holes) and FCDE

$$A_n = \left[240 - 4 \times 18 + \frac{50^2}{4 \times 60} \right] \times 10 = 1784.2 \text{ mm}^2$$

(iv) For path ACG (3 holes)

$$A_n = \left[240 - 3 \times 18 + 2 \times \frac{50^2}{4 \times 60} \right] \times 10 = 2068.3 \text{ mm}^2$$

∴   Minimum net area = **1784.2 mm²**

Ex. 4.4: *Two flat plates 180 mm × 20 mm are connected in a lap joint using 6 bolts of 20 mm diameter as shown in Fig. 4.10. Determine the design strength in tension of the plates. Take $f_y = 250$ N/mm² and $f_u = 410$ N/mm².*

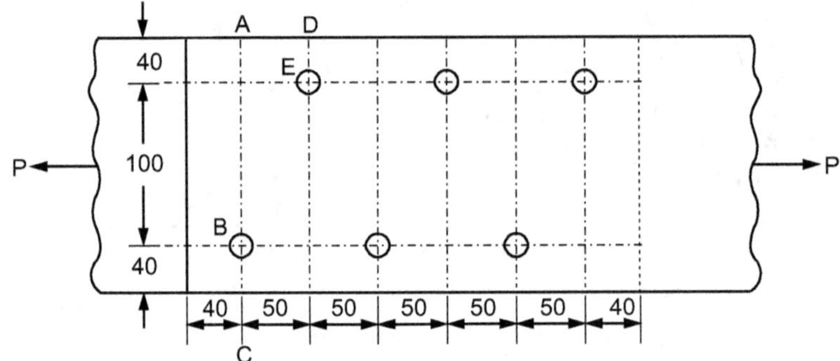

Fig. 4.12

Sol.: Diameter of bolt hole

$$d_h = 20 + 2 = 22 \text{ mm.}$$

Net area along any path,

$$A_n = \left[b - n\, d_h + \Sigma \frac{p^2}{4g} \right] t$$

(i) For path ABC (1 hole)

$$A_n = [180 - 1 \times 22] \times 20$$
$$= 3160 \text{ mm}^2$$

(ii) For path DEBC (2 holes)

$$A_n = \left[180 - 2 \times 22 + \frac{50^2}{4 \times 100} \right] \times 20$$

$$= 2845 \text{ mm}^2$$

∴   Minimum net area $= 2845 \text{ mm}^2$

Design Strength of Plate

(i) Design strength governed by gross-section yielding

$$T_{dg} = \frac{A_g \cdot f_y}{\gamma_{m_0}}$$

$$= \frac{(180 \times 20) \times 250}{1.10}$$

$$= 818182 \text{ N}$$

(ii) Design strength governed by net section rupture

$$T_{dn} = \frac{0.9\, A_n\, f_u}{\gamma_{m_1}}$$

$$= \frac{0.9 \times 2845 \times 410}{1.25}$$

$$= 839844 \text{ N}$$

∴ Minimum of T_{dg} and T_{dn}, $T_d = 818182$ N

Thus, design tensile strength of plate is **818.18 kN.**

Ex. 4.5: *In a bolted connection of an angle ISA 150 × 75 × 8 mm at a joint, 6 bolts of 20 mm diameter are provided as shown in Fig. 4.11. Determine the minimum net area of the angle.*

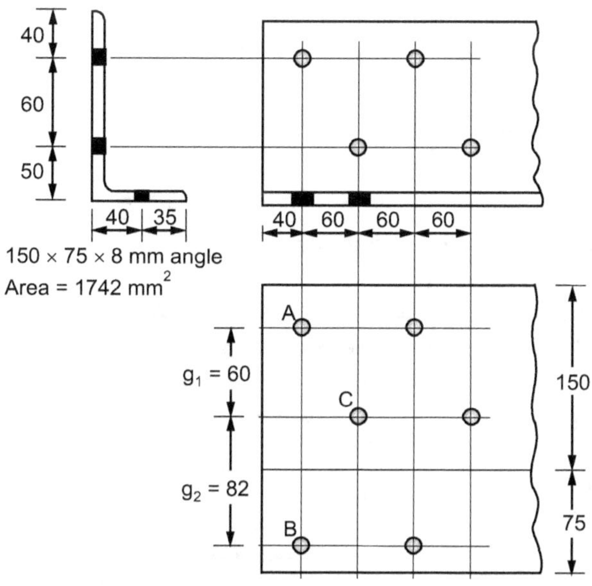

Fig. 4.13

Sol.: Diameter of bolt hole d_h = 20 + 2 = 22 mm. To analyse the angle it may be considered as flattened forming a plate as shown in bottom figure in Fig. 4.13.

Here $\quad g_1$ = 60 mm

$\quad g_2$ = 50 + 40 − t = 50 + 40 − 8 = 82 mm

(i) Net area along path AB (2 holes)

$\quad A_n$ = Area of angle − 2 × Bolt hole area

$\quad A_n$ = 1742 − 2 × 22 × 8 = 1390 mm²

(ii) Net area along path ACB (3 holes)

$$= \text{Area of angle} - 3 \times \text{bolt hole area} + \left[\frac{p_1^2}{4g} + \frac{p_2^2}{4g}\right] t$$

$$A_n = 1742 - 3 \times 22 \times 8 + \left[\frac{60^2}{4 \times 60} + \frac{60^2}{4 \times 82}\right] \times 8$$

$$= 1742 - 528 + (15 + 10.97561) \times 8$$

$$= 1742 - 528 + 207.8049 = 1421.8 \text{ mm}^2$$

∴ Minimum net area = **1390 mm²**

Ex. 4.6: *A single angle 100 mm × 75 mm × 10 mm is used as a tension member of a truss. The longer leg of the angle is connected to the gusset plate with 4 bolts of 20 mm diameter. Determine the net effective area of the angle.*

If the ends of the longer leg of angle are welded to gusset plate, determine the net effective area of angle.

Sol.: (i) When the angle is bolted to the gusset plate.

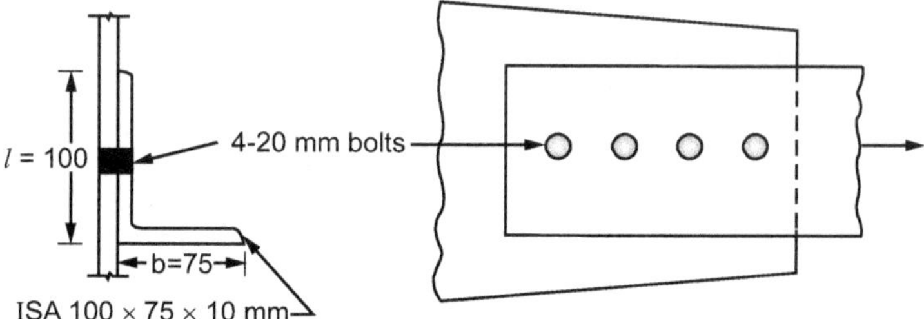

Fig. 4.14

$$\text{Net area of connected leg} = \left(l - d_h - \frac{t}{2}\right)t = \left(100 - 22 - \frac{10}{2}\right) \times 10$$

$$A_1 = 730 \text{ mm}^2$$

$$\text{Area of outstanding leg} = \left(b - \frac{t}{2}\right) \times t$$

$$A_2 = \left(75 - \frac{10}{2}\right)10 = 700 \text{ mm}^2$$

Net area of angle, $A = A_1 + A_2 = 730 + 700 = 1430 \text{ mm}^2$

Reduction factor $\alpha = 0.8$ (for 4 or more bolts)

∴ Effective net area, $A_n = 0.8 \times 1430 = \mathbf{1144 \text{ mm}^2}$

(ii)   When the angle is welded to the gusset plate

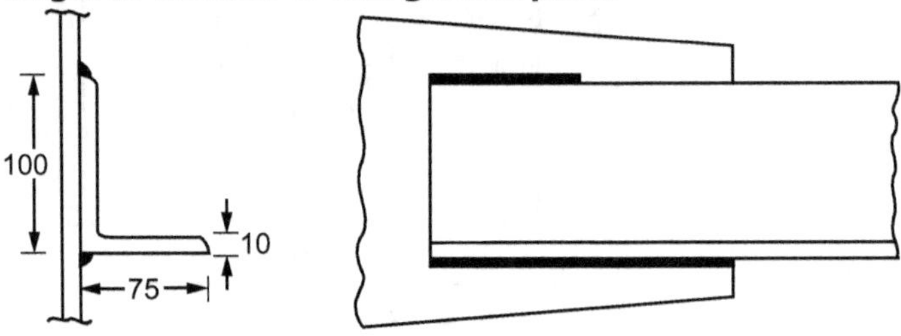

Fig. 4.15

$$\text{Net area of connected leg, } A_1 = \left(l - \frac{t}{2}\right)t = \left(100 - \frac{10}{2}\right) \times 10 = 950 \text{ mm}^2$$

$$\text{Area of outstanding leg, } A_2 = \left(b - \frac{t}{2}\right)t = \left(75 - \frac{10}{2}\right)10 = 700 \text{ mm}^2$$

Net area of angle, $A = A_1 + A_2 = 950 + 700 = 1650 \text{ mm}^2$

Reduction factor $\alpha = 0.8$ (for angle leg welded)

∴ Effective net area of angle $= 0.8 \times 1650 = \mathbf{1320 \text{ mm}^2}$

Ex. 4.7: *A single angle 150 × 75 × 10 mm is used as a tension member. Connected to 12 mm thick gusset plate at ends with 5 of 18 mm diameter bolts. Bolts are pitched at 50 mm. Find Net area, if (i) longer leg connected to gusset plate (ii) shorter leg connected to gusset plate.*

Ans. (i) Case I:

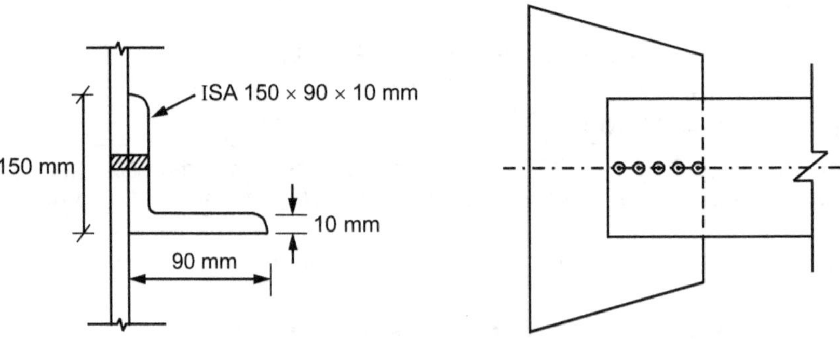

Fig. 4.16

Diameter of bolt $= 18$ mm

Diameter of hole $(d_h) = 18 + 2 = 20$ mm

∴ Net sectional area of connected leg (A_{nc})

$$= \left(150 - \frac{t}{2}\right)t - d_h \times t = \left(150 - \frac{10}{2}\right) \times 10 - 20 \times 10 = 1250 \text{ mm}^2$$

Gross area of outstanding leg (A_{go})

$$= \left(75 - \frac{t}{2}\right) \times t = \left(75 - \frac{10}{2}\right) \times 10 = 700 \text{ mm}^2$$

∴ Net sectional area $(A_n) = A_{nc} + A_{go} = 1250 + 700 = 1950 \text{ mm}^2$

Now, find net effective area A_{ne}

∴ Net effective area $(A_{ne}) = \alpha \times A_n = 0.8 \times 1950 = 1560$ mm^2

∴ Shear lug factor $\alpha = 0.8$ for No. of bolts ≥ 4

(ii) Case - II:

When shorter leg connected to Gusset plate.

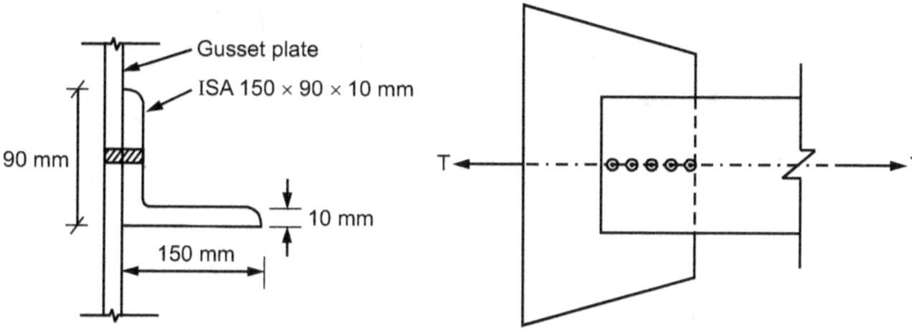

Fig. 4.17

Diameter of bolt $= 18$ mm

Diameter of hole (dh) $= 18 + 2 = 20$ mm

Net sectional area of connected leg

$$= \left(75 - \frac{t}{2}\right) \times t - dh \times t = \left(75 - \frac{10}{2}\right) 10 - 20 \times 10 = 500 \text{ mm}^2$$

Gross Area of outstanding leg (A_{go})

$$= \left(150 - \frac{t}{2}\right) \times t = \left(150 - \frac{10}{2}\right) \times 10 = 1450 \text{ mm}^2$$

∴ Net sectional area $(A_n) = 500 + 1450 = 1950$ mm^2

∴ $A_{ne} = \alpha \times A_n = 0.8 \times 1950 = 1560$ mm^2

Shear factor $\alpha = 0.8$ for number of bolt ≥ 4

Ex. 4.8: *The longer leg of a single angle 90 mm × 60 mm × 10 mm is connected to the gusset plate with 3 bolts of 20 mm diameter at a pitch of 60 mm. For this tension member determine the block shear strength.*

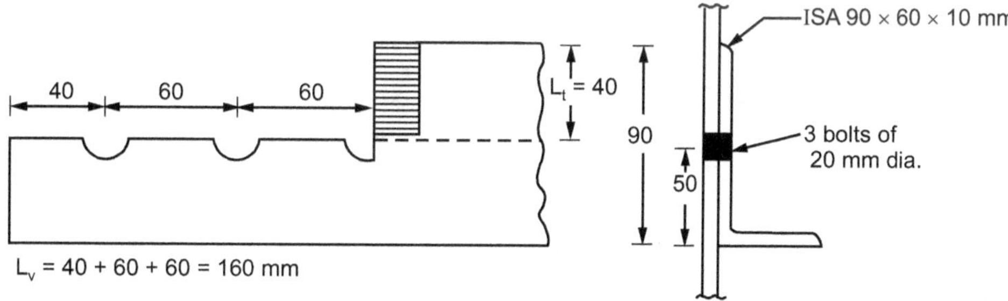

Fig. 4.18

Sol.: Diameter of bolt hole, $d_h = 20 + 2 = 22$ mm

Referring Fig. 4.17.

A_{vg} = Minimum gross area in shear along bolt line = L_v, T

$\quad = (160) \times 10 = 1600$ mm^2

A_{vn} = Minimum net area in shear along bolt line

$$= \left[L_v - \left(2 + \frac{1}{2}\right) d_h\right] t = (160 - 2.5 \times 22) \times 10 = 1050 \text{ mm}^2$$

A_{tg} = Minimum gross area in tension from bolt hole to the toe of angle perpendicular to line of force

$\quad = L_t \cdot t = 40 \times 10 = 400$ mm^2

A_{tn} = Minimum net area in tension from bolt hole to the toe of angle perpendicular to line of force

$\quad = (L_t - 0.5\, d_n)\, t = (40 - 0.5 \times 22) \times 10 = 290$ mm^2

Design tensile strength governed by block shear

$$T_{db_1} = \frac{A_{vg}\, f_y}{\sqrt{3}\, \gamma_{m_0}} + \frac{0.9\, A_{tn} \cdot f_u}{\gamma_{m_1}} = \frac{1600 \times 250}{\sqrt{3} \times 1.10} + \frac{0.9 \times 290 \times 410}{1.25} = 295553 \text{ N}$$

or $$T_{db_2} = \frac{A_{tg}\, f_y}{\gamma_{m_0}} + \frac{0.9\, A_{vn} \cdot f_u}{\sqrt{3}\, \gamma_{m_1}} = \frac{400 \times 250}{1.10} + \frac{0.9 \times 1050 \times 410}{\sqrt{3} \times 1.25} = 269864 \text{ N}$$

$\therefore$ Design tensile strength governed by block shear will be minimum by T_{db_1} and T_{db_2}

i.e. T_{db} = 269864 N = **269.86 kN**

Ex. 4.9: *The longer leg of a single angle 100 × 75 × 8 mm is connected to the gusset plate with 3 bolts in a line of 20 mm diameter at a pitch of 60 mm, for this tension member. Determine block shear strength.*

Sol.: Given: ISA = 100 × 75 × 8 mm, d = 20 mm, d_o = 22 mm, P = 60 mm, 3 Bolts in a line of 20 mm in diameter, End distance e = 2d = 2 × 20 = 40 mm

$$L_v = (60 + 60 + 40) = 160 \text{ mm}$$

Assume $\quad\quad\quad\quad L_t = 40 \text{ mm}$

1. $\quad\quad\quad\quad A_{tg} = (L_t \times t) = 40 \times 8 = 320 \text{ mm}^2$

2. $\quad\quad\quad\quad A_{tn} = (L_t - 0.5\, d_0) \times t = 232 \text{ mm}^2$

3. $\quad\quad\quad\quad A_{vg} = (L_v \times t) = (160 \times 10) = 1280 \text{ mm}^2$

4. $\quad\quad\quad\quad A_{vn} = (L_v - 2.5\, d_0) \times t = (160 - 2.5 \times 22) \times 8 = 840 \text{ mm}^2$

To calculate Block shear strength:

$$T_{db_1} = \frac{A_{vg} \times f_y}{\sqrt{3} \times \gamma_{m_0}} + \frac{0.9 \times A_{tn} \times f_u}{\gamma_{m_1}}$$

$$\gamma_{m_0} = 1.10$$

$$\gamma_{m_1} = 1.25$$

$$f_{ub} = 400 \text{ N/mm}^2$$

$$f_u = 410 \text{ N/mm}^2$$

$$T_{db_1} = \frac{1280 \times 250}{\sqrt{3} \times 1.10} + \frac{0.9 \times 232 \times 410}{1.25} = 236.43 \text{ kN}$$

$$T_{db_2} = \frac{0.9\, A_{vn} \times f_u}{\sqrt{3} \times \gamma_{m_1}} + \frac{A_{tg} \times f_y}{\gamma_{m_0}} = \frac{0.9 \times 840 \times 410}{\sqrt{3} \times 1.25} + \frac{232 \times 250}{1.10} = 195.88 \text{ kN}$$

Block shear strength of member is 195.88 kN

Ex. 4.10: *For the tension member shown in Fig. 4.19. Determine the block shear strength.*

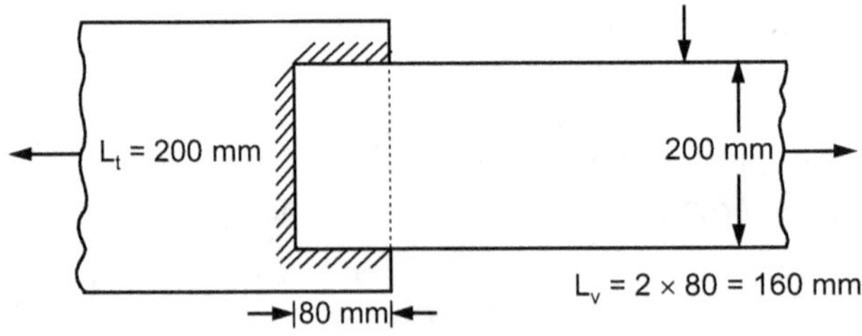

Fig. 4.19

Sol.: $\quad\quad\quad A_{vg} = L_v \cdot t = 160 \times 10 = 1600 \text{ mm}^2$

$\quad\quad\quad\quad\quad\quad A_{vn} = 1600 \text{ mm}^2$

$\quad\quad\quad\quad\quad\quad A_{tg} = L_t \cdot t = 200 \times 10 = 2000 \text{ mm}^2$

$\quad\quad\quad\quad\quad\quad A_{tn} = 2000 \text{ mm}^2$

Block shear strength

$$T_{db_1} = \frac{A_{vg}\, f_y}{\sqrt{3}\; \gamma_{m_0}} + \frac{0.9\, A_{tn} \cdot f_u}{\gamma_{m_1}} = \frac{1600 \times 250}{\sqrt{3} \times 1.10} + \frac{0.9 \times 2000 \times 410}{1.25} = 800345 \text{ N}$$

or

$$T_{db_2} = \frac{A_{tg}\, f_y}{\gamma_{m_0}} + \frac{0.9\, A_{vn} \cdot f_u}{\sqrt{3}\; \gamma_{m_1}} = \frac{2000 \times 250}{1.10} + \frac{0.9 \times 1600 \times 410}{\sqrt{3} \times 1.25} = 727239 \text{ N}$$

$\therefore$ Block shear strength

$$T_{db} = \text{Least of } T_{db_1} \text{ and } T_{db_2} = 727239 \text{ N} = \textbf{727.24 kN}$$

Ex. 4.11: *The longer leg of a single angle ISA 125 × 75 × 8 mm is connected to a 10 mm thick gusset plate by 4 bolts of 18 mm diameter arranged as shown in Fig. 4.16. Determine the design tensile strength of the angle. Take f_y = 250 N/mm² and f_u = 410 N/mm².*

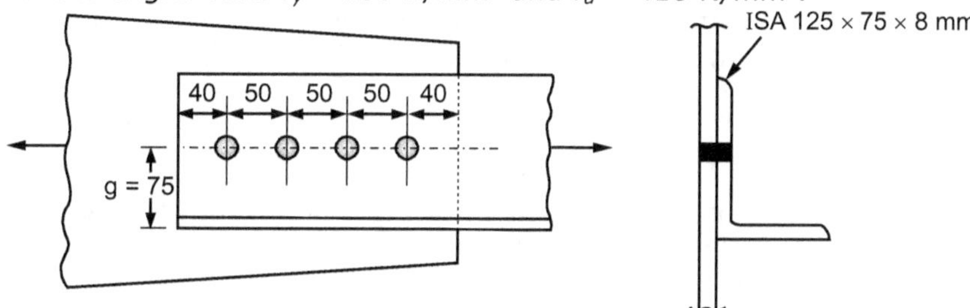

Fig. 4.20

Sol.: Bolt hole diameter $d_h = 18 + 2 = 20$ mm.
From steel table area of ISA 125 × 75 × 8, $A_g = 1538$ mm².
Net area of connected leg,

$$A_{nc} = \left[l - d_h - \frac{t}{2} \right] t = \left[125 - 20 - \frac{8}{2} \right] \times 8 = 808 \text{ mm}^2$$

Gross area of outstanding leg,

$$A_{go} = \left[b - \frac{t}{2} \right] t = \left[75 - \frac{8}{2} \right] 8 = 568 \text{ mm}^2$$

(i) Design tensile strength governed by yielding of gross section

$$T_{dg} = \frac{A_g\, f_y}{\gamma_{m_0}} = \frac{1538 \times 250}{1.10} = 349545 \text{ N} = 349.54 \text{ kN}$$

(ii) Design tensile strength governed by net section rupture

$$T_{dn} = \frac{0.9\, A_{nc}\, f_u}{\gamma_{m_1}} + \frac{\beta\, A_{go} \cdot f_y}{\gamma_{m_0}}$$

where,

$$\beta = 1.4 - 0.076\, \frac{w}{t} \times \frac{f_y}{f_u} \times \frac{b_s}{L_c} \leq 0.9\, \frac{f_u}{f_y} \cdot \frac{\gamma_{m_0}}{\gamma_{m_1}} \geq 0.7$$

$$w = \text{Outstand leg width} = 75 - \frac{8}{2} = 71 \text{ mm}$$

$$b_s = w + g - t = 71 + 75 - 8 = 138 \text{ mm}$$

L_c = Length of end connection
   = Distance between outermost bolt holes
   = 50 + 50 + 50 = 150 mm

$\therefore$

$$\beta = 1.4 - 0.076 \times \frac{71}{8} \times \frac{250}{410} \times \frac{138}{150} = 1.02$$

$$0.9\, \frac{f_u}{f_y} \cdot \frac{\gamma_{m_0}}{\gamma_{m_1}} = 0.9 \times \frac{410}{250} \times \frac{1.10}{1.25} = 1.29$$

Thus, $\beta \leq 1.29 \geq 0.7$

$\therefore$ $\beta = 1.02$ is acceptable

$\therefore$

$$T_{dn} = \frac{0.9 \times 808 \times 410}{1.25} + \frac{1.02 \times 568 \times 250}{1.10} = 370194 \text{ N} = 370.19 \text{ kN}$$

(iii) Design tensile strength by block shear

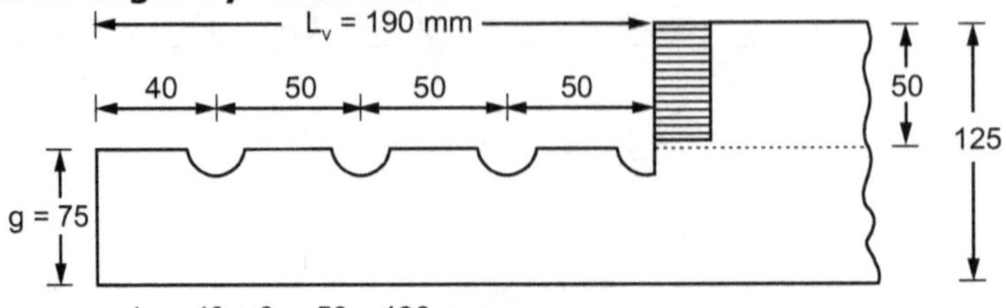

Fig. 4.21

$L_v = 40 + 3 \times 50 = 190$

A_{vg} = Minimum gross area in shear along bolt line = $(L_v) \cdot t$

$\quad = [190] \times 8 = 1520 \text{ mm}^2$

A_{vn} = Minimum net area in shear along bolt line = $\left[L_v - \left(3 + \frac{1}{2}\right) d_h \right] t$

$\quad = [190 - 3.5 \times 20] \times 8 = 960 \text{ mm}^2$

A_{tg} = Minimum gross area in tension from bolt hole to toe of angle perpendicular to the line of force $L_t \cdot t$

$\quad = 8 \times 50 = 400 \text{ mm}^2$

A_{tn} = Minimum net area in tension from bolt hole to toe of angle perpendicular to the line of force = $(L_t - 0.5 \, d_h) \cdot t$

$\quad = (50 - 0.5 \times 20) \times 8 = 320 \text{ mm}^2$

Design tensile strength governed by block shear

$$T_{db_1} = \frac{A_{vg} \, f_y}{\sqrt{3} \, \gamma_{m_0}} + \frac{0.9 \, A_{tn} \cdot f_u}{\gamma_{m_1}} = \frac{1520 \times 250}{\sqrt{3} \times 1.1} + \frac{0.9 \times 320 \times 410}{1.25} = 293912 \text{ N}$$

or $\quad T_{db_2} = \dfrac{A_{tg} \, f_y}{\gamma_{m_0}} + \dfrac{0.9 \, A_{vn} \cdot f_u}{\sqrt{3} \, \gamma_{m_1}} = \dfrac{400 \times 250}{1.1} + \dfrac{0.9 \times 960 \times 410}{\sqrt{3} \times 1.25} = 254525 \text{ N}$

$\therefore \quad T_{db}$ = Minimum of T_{db_1} and T_{db_2} = 254525 N = 254.525 kN

$\therefore \quad T_{db}$ = Minimum of T_{db_1} and T_{db_2} = 254525 N = 254.525 kN

$\therefore$ Design tensile strength of angle

$\quad$ = Least of the strengths (i), (ii) and (iii) = **254.525 kN**

Ex. 4.12: *Design a suitable unequal angle section as a tie member to carry. A tensile factored load of 215 kN to be connected to a 20 mm diameter bolt to 12 mm gusset plate. The design strength of 20 mm diameter bolt = 45.3 kN.*

Sol.: $\quad$ Factored load, T = 215 kN

$\quad$ Diameter of bolt hole $d_h = 20 + 2 = 22$ mm

1. Approximate gross sectional area required

$$A_g \text{ required} = \frac{T}{f_y} \cdot \gamma_{m_0} = \frac{215 \times 10^3}{250} \times 1.1 = 946 \text{ mm}^2$$

2. From steel stable select ISA 100 × 75 × 6 mm having

$$A_g = 1014 \text{ mm}^2$$

3. Number of bolts required $= \dfrac{\textbf{Factored load}}{\textbf{Design strength of bolt}} = \dfrac{215}{45.3} = 4.74$

$\quad \therefore$ Provide 5 bolts of 20 mm diameter.

4. Check for design strength T_d of trial section

(i) Design tensile strength governed by gross-section yielding

$$T_{dg} = \frac{A_g \, f_y}{\gamma_{m_0}} = \frac{1014 \times 250}{1.1} = 230455 \text{ N} = 230.45 \text{ kN}$$

(ii) Design tensile strength governed by net section rupture

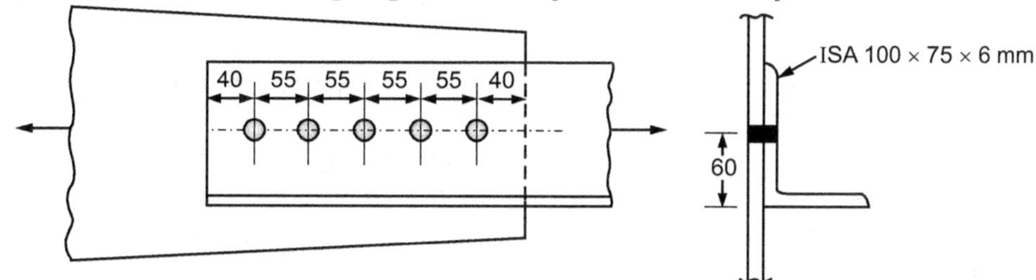

Fig. 4.22

$$T_{dn} = \frac{0.9\, A_{nc}\, f_u}{\gamma_{m_1}} + \frac{\beta\, A_{go} \cdot f_y}{\gamma_{m_0}}$$

where $\beta = 1.4 - 0.076 \dfrac{w}{t} \times \dfrac{f_y}{f_u} \times \dfrac{b_s}{L_c} \leq 0.9 \dfrac{f_u}{f_y} \cdot \dfrac{\gamma_{m_0}}{\gamma_{m_1}} \geq 0.7$

Net area of connected leg,

$$A_{nc} = \left[l - d_h - \frac{t}{2} \right] t = \left[100 - 22 - \frac{6}{2} \right] \times 6 = 450 \text{ mm}^2$$

Gross area of outstanding leg,

$$A_{go} = \left[b - \frac{t}{2} \right] t = \left[75 - \frac{6}{2} \right] \times 6 = 432 \text{ mm}^2$$

Outstanding leg width,

$$w = b - \frac{t}{2} = 75 - \frac{6}{2} = 72 \text{ mm}$$

$$b_s = w + g - t = 72 + 60 - 6 = 126 \text{ mm}$$

L_c = Distance between outermost bolt holes = $4 \times 55 = 220$ mm

$\therefore \qquad \beta = 1.4 - 0.076 \times \dfrac{72}{6} \times \dfrac{250}{410} \times \dfrac{126}{220} = 1.08$

$$0.9 \frac{f_u}{f_y} \times \frac{\gamma_{m_0}}{\gamma_{m_1}} = 0.9 \times \frac{410}{250} \times \frac{1.10}{1.25} = 1.29$$

Thus, $\beta = 1.08 \leq 1.29$ and ≥ 0.7

$\therefore \qquad \beta = 1.08$ is acceptable

$\therefore \qquad T_{dn} = \dfrac{0.9\, A_{nc}\, f_u}{\gamma_{m1}} + \dfrac{\beta\, A_{go}\, f_y}{\gamma_{m_0}} = \dfrac{0.9 \times 450 \times 410}{1.25} + \dfrac{1.08 \times 432 \times 250}{1.10}$

$\qquad\qquad = 238876$ N $= 238.87$ kN

(iii) Design tensile strength governed by block shear

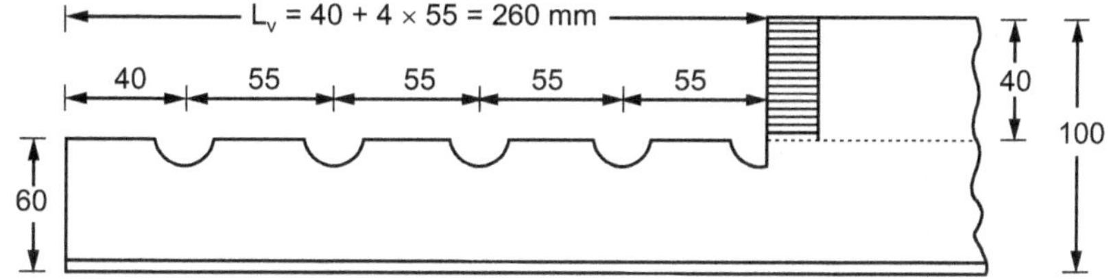

Fig. 4.23

A_{vg} = Minimum gross area in shear along bolt line = $L_v \cdot t$

$\qquad = [260] \times 6 = 1560 \text{ mm}^2$

A_{vn} = Minimum net area in shear along bolt line = $\left[L_v - \left(4 + \frac{1}{2} \right) d_h \right] \cdot t$

$\qquad = [260 - 4.5 \times 22] \times 6 = 966 \text{ mm}^2$

A_{tg} = Minimum gross-area in tension from bolt hole to toe of angle perpendicular to line of force = $L_t \cdot t$

$\qquad = 40 \times 6 = 240 \text{ mm}^2$

$$A_{tn} = \text{Minimum net area in tension from bolt hole to toe of angle perpendicular to the line of force} = (L_t - 0.5\, d_h) \cdot t$$

$$= [40 - 0.5 \times 22] \times 6 = 174 \text{ mm}^2$$

$$T_{db_1} = \frac{A_{vg}\, f_y}{\sqrt{3}\,\gamma_{m_0}} + \frac{0.9\, A_{tn} \cdot f_u}{\gamma_{m_1}} = \frac{1560 \times 250}{\sqrt{3} \times 1.10} + \frac{0.9 \times 174 \times 410}{1.25} = 256062 \text{ N}$$

or $$T_{db_2} = \frac{A_{tg}\, f_y}{\gamma_{m_0}} + \frac{0.9\, A_{vn} \cdot f_u}{\sqrt{3}\,\gamma_{m_1}} = \frac{240 \times 250}{1.10} + \frac{0.9 \times 966 \times 410}{\sqrt{3} \times 1.25} = 219184 \text{ N}$$

$\therefore \quad T_{db} = \text{Minimum of } T_{db_1} \text{ and } T_{db_2} = 219184 \text{ N} = 219.18 \text{ kN}$

$\therefore$ The design tensile strength of the angle

$$= \text{Least of } T_{dg},\, T_{dn},\, T_{db} = \mathbf{219.18} \text{ kN} < T\ (= 215 \text{ kN})$$

$\therefore \quad$ Design is safe.

Ex. 4.13: *Design a tension member consisting of single unequal angle section to carry a tension load of 340 kN. Assume single row 20 mm bolted connection. The length of member is 2.4 m. Take $f_u = 410$ MPa, $\alpha = 0.80$.*

Section available (mm)	Area (mm^2)
ISA 100 × 75 × 8	1336
ISA 125 × 75 × 8	1538
ISA 150 × 75 × 8	1748

Sol.: Approximate gross-area required A_g

$$\text{Required } A_g = \frac{1.1 \times T_d}{f_y} = \frac{1.1 \times 340 \times 10^3}{250} = 1496 \text{ mm}^2$$

Try ISA 125 × 75 × 8 mm giving $A_g = 1538$ mm^2, $r_{min} = 16.1$ mm. Assuming longer leg connected, check the strength of the section.

(i) Design strength due to yielding of gross-section

$$T_{dg} = \frac{A_g \times f_y}{\gamma_{mo}} = \frac{1538 \times 250}{1.10} = 349545.4 \text{ N}$$

$$\boxed{T_{dg} = 349.54 \text{ kN}}$$

(ii) Design strength due rupture of critical-section

$$T_{dn} = \alpha\, A_n \frac{f_u}{\gamma_{m1}} \quad \text{(Approximate rupture strength)}$$

$$A_n = A_{nc} + A_{go}$$

$$A_{nc} = \left(B_1 - d_n - \frac{t}{2}\right) \times t = \left(125 - 22 - \frac{8}{2}\right) \times 8$$

$$A_{nc} = 792 \text{ mm}^2$$

$$A_{go} = \left(B_2 - \frac{t}{2}\right) t = \left(75 - \frac{8}{2}\right) 8 = 568 \text{ mm}^2$$

$$A_n = A_{nc} + A_{go}$$

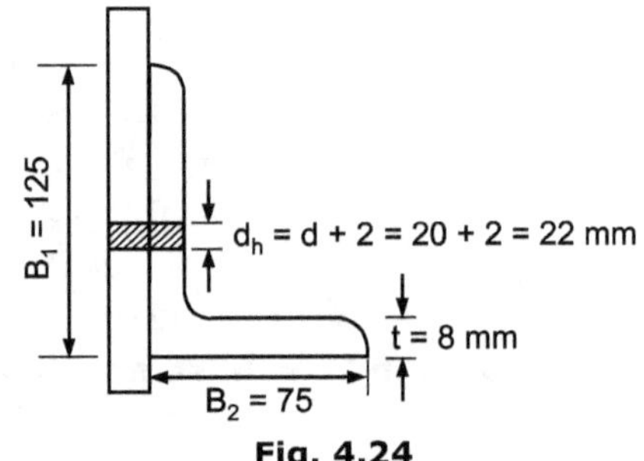

Fig. 4.24

$$A_n = 792 + 568$$

$$A_n = 1360 \text{ mm}^2$$

Considering more than four bolts in a row $\alpha = 0.8$.

$$T_{dn} = \frac{0.8 \times 1360 \times 410}{1.25}$$

$$\boxed{T_{dn} = 356.864 \text{ kN}}$$

Design of bolts:

Capacity of bolts in single shear = 45.3 kN

$$\text{Capacity of bolt in bearing} = \frac{20 \times 8 \times 410}{1000} = 65.6 \text{ kN}$$

Least bolt value = 45.3 kN (min. of two above)

$$\text{Number of bolts required} = \frac{340}{45.3} = 7.5 \text{ say } 8$$

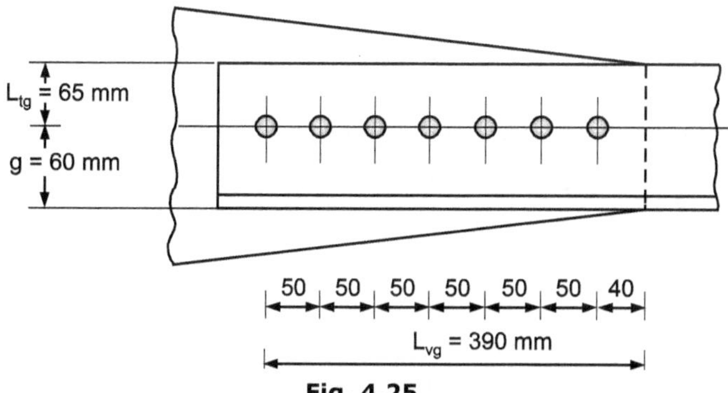

Fig. 4.25

(iii) Design strength due to block shear

$$\text{Assuming edge distance} = 40 \text{ mm}$$
$$g = 60 \text{ mm}$$
$$\text{Spacing of bolts} = 50 \text{ mm}$$
$$\text{Average} = L_{vg} \times t = 390 \times 8 = 3120 \text{ mm}^2$$
$$A_{vg} = 3120 \text{ mm}^2$$
$$A_{vn} = \{L_{vg} - [\text{No. of bolts} - 0.5\, d_h]\} \times t$$
$$A_{vn} = \{390 - [(8 - 0.5)\, 22]\} \times 8 = 1800 \text{ mm}^2$$
$$A_{vn} = 1800 \text{ mm}^2$$
$$A_{tg} = L_{tg} \times t = 65 \times 8 = 520 \text{ mm}^2$$
$$A_{tg} = 520 \text{ mm}^2$$
$$A_{tn} = (L_{tg} - 0.5\, d_n) \times t$$
$$A_{tn} = [65 - (0.5 \times 22)] \times 8$$
$$A_{tn} = 432 \text{ mm}^2$$
$$T_{db1} = \frac{A_{vg}\, f_y}{(\sqrt{3} \times \gamma_{mo})} + 0.9\, A_{tn} \frac{f_u}{\gamma_{m1}} = \frac{3120 \times 250}{(\sqrt{3} \times 1.10)} + 0.9 \times 432 \times \frac{410}{1.25}$$
$$= 409393.8 + 127526.5 = 536920.2 \text{ N}$$
$$T_{db1} = 536.92 \text{ kN}$$
$$T_{db2} = \left(\frac{0.9\, A_{vn}\, f_u}{\sqrt{3} \times \gamma_{m1}}\right) + \frac{A_{tg}\, f_y}{\gamma_{mo}} = \frac{0.9 \times 1800 \times 410}{(\sqrt{3} \times 1.25)} + \frac{520 \times 250}{1.10}$$
$$= 306780.8 + 118181.8 = 424962.8 \text{ N}$$
$$T_{db2} = 424.962 \text{ kN}$$
$$T_{db} = \text{Lesser than } T_{db1} \text{ and } T_{db2}$$
$$T_{db} = 424.96 \text{ kN}$$

∴   The tensile strength of angle = Lesser of T_{dg}, T_{dn} and T_{db}

$$(349.54, \ 356.86 \text{ and } 424.96)$$
$$= 349.54 \text{ kN}$$

This is greater than required 340 kN

Check for slenderness ratio $\lambda = \dfrac{l}{\gamma_{min}} = \dfrac{2400}{16.1} = 149.06 < 250$

Ex. 4.14: *Design a tie member using suitable equal angle section to carry a tensile factored load of 200 kN. The connection are with 20 mm dia. bolts and 12 mm thick gusset plate. Design strength of 20 mm dia. bolts = 45.3 kN, f_y = 250 MPa, f_u = 410 MPa, α = 0.8, Sections available.*

ISA - mm	Area mm²
90 × 90 × 8	1137
100 × 75 × 6	1014
125 × 75 × 6	116

Sol.: Given: $P_u = 200$ kN, $d = 20$ mm, $\therefore d_o = 20 + 2 = 22$ mm

Design strength of bolt = 45.3 kN, $f_y = 250$ N/mm^2

$f_u = 410$ N/mm^2, $\alpha = 0.8$

(i) Approximate Gross cross-sectional area required

$$A_{g\,reqd.} = \frac{T}{f_y} \times \gamma_{m0} = \frac{200 \times 10^3}{250} \times 1.1$$

$$A_g = 880 \text{ mm}^2$$

(ii) Select ISA 100 × 75 × 6 mm having

$$A_g = 1014 \text{ mm}^2$$

(iii)   No. of bolts reqd. $= \dfrac{\text{Factored load}}{\text{Design strength of Bolt}} = \dfrac{200}{45.3} = 4.41 \simeq 05$

Provide 5 bolts of 20 mm diameter.

(iv)   Check for design strength

(a) Design tensile strength governed by gross section yielding

$$T_{dg} = \frac{A_g f_y}{\gamma_{m0}} = \frac{1014 \times 250}{1.1}$$

$$T_{dg} = 230.45 \text{ kN}$$

(b) Design tensile strength governed by Net section rupture

$$T_{dn} = \frac{0.9 f_u A_n}{\gamma_{m_1}}$$

Net Area of connected leg

$$A_{nc} = \left[100 - 22 - \frac{6}{2}\right] \times 6 = 450 \text{ mm}^2$$

Net Area of outstanding leg

$$A_{go} = \left[75 - \frac{6}{2}\right] \times 6 = 432 \text{ mm}^2$$

Net Area $A_n = A_{nc} + A_{go} = 450 + 432 = 882 \text{ mm}^2$

$$T_{dn} = \frac{0.9 f_u A_n}{\gamma_{m_1}} = \frac{0.9 \times 410 \times 882}{1.25}$$

$$T_{dn} = 260.36 \text{ kN}$$

(c) Design tensile strength governed by block shear

$$P = 2.5\, d = 2.5 \times 20 = 50 \text{ mm}$$

$$e = 1.7\, d_o = 1.7 \times 22 = 37.4 \simeq 40 \text{ mm}$$

If P and e considered different marks given accordingly.

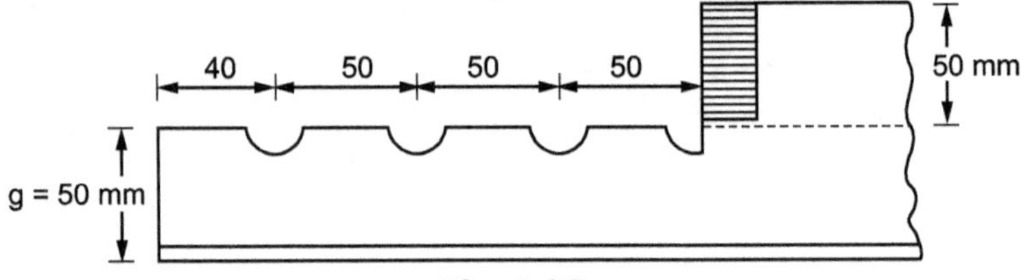

Fig. 4.26

$A_{vg} = [4 \times 50 + 40] \times 6 = 1440 \text{ mm}^2$

$A_{vn} = [4 \times 50 + 40 - 4.5 \times 22] \times 6 = 846 \text{ mm}^2$

$A_{tg} = 50 \times 6 = 300 \text{ mm}^2$

$A_{tn} = [50 - 0.5 \times 22] \times 6 = 234 \text{ mm}^2$

$$T_{db_1} = \frac{A_{vg} \cdot f_y}{\sqrt{3}\, \gamma_{m0}} + \frac{0.9\, A_{tn} \cdot f_u}{\gamma_{m_1}} = \frac{1440 \times 250}{\sqrt{3} \times 1.1} + \frac{0.9 \times 234 \times 410}{1.25}$$

$$= 188.97 + 69.07 = 258.04 \text{ kN}$$

$$T_{db_2} = \frac{A_{tg} \cdot f_y}{\gamma_{m_0}} + \frac{0.9\,A_{vn} \cdot f_u}{\sqrt{3}\,\gamma_{m_1}} = \frac{300 \times 250}{1.1} + \frac{0.9 \times 846 \times 410}{\sqrt{3} \times 1.25}$$

$$= 68.18 + 144.19 = 212.37 \text{ kN}$$

$$T_{db} = \text{Minimum of } T_{db_1} \text{ and } T_{db_2}$$

The design tensile strength of the angle

$$= \text{Least of } T_{dg},\, T_{dn},\, T_{db} = 212.37 \text{ kN} > 200 \text{ kN}$$

Hence Design is safe.

Ex. 4.15: *Determine the design strength of a single angle tension member consisting of single angle 80 mm × 50 mm × 8 mm. The shorter leg of the angle is connected to the gusset plate by 5 mm weld as shown in Fig. 4.27.*

Sol.: From Steel table area of ISA 80 × 50 × 8 mm, $A_g = 978 \text{ mm}^2$.

1. Design strength governed by yielding of gross-section

$$T_{dg} = \frac{A_g\, f_y}{\gamma_{m_0}} = \frac{978 \times 250}{1.10} = 222273 \text{ N}$$

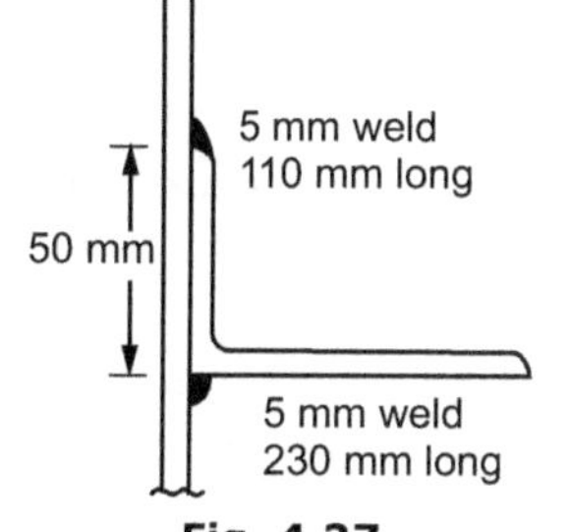

Fig. 4.27

2. Design strength governed by net section rupture

$$T_{dn} = \frac{0.9\,A_{nc}\,f_u}{\gamma_{m_1}} + \frac{\beta\,A_{go} \cdot f_y}{\gamma_{m_0}}$$

Area of connected leg, $A_{gc} = A_{nc} = \left(50 - \dfrac{8}{2}\right) \times 8 = 368 \text{ mm}^2$

Area of outstanding leg; $A_{go} = \left(80 - \dfrac{8}{2}\right) \times 8 = 608 \text{ mm}^2$

$$\beta = 1.4 - 0.076\,\frac{w}{t} \times \frac{f_y}{f_u} \times \frac{b_s}{L_c}$$

w = Outstand leg width = 80 mm

$b_s = w = 80$ mm

L_c = Average length of bottom and top welds = $\dfrac{230 + 110}{2} = 170$ mm

$$\therefore \quad \beta = 1.4 - 0.076 \times \frac{80}{8} \times \frac{250}{410} \times \frac{80}{170} = 1.18$$

$$0.9\,\frac{f_u}{f_y} \times \frac{\gamma_{m_0}}{\gamma_{m_1}} = 0.9 \times \frac{410}{250} \times \frac{1.10}{1.25} = 1.30$$

$$\therefore \quad \beta = 1.18 < 1.30 > 0.70$$

$$\therefore \quad \beta = 1.18 \text{ is acceptable}$$

$$\therefore \quad T_{dn} = \frac{0.9 \times 368 \times 410}{1.25} + \frac{1.18 \times 608 \times 250}{1.10} = 271688 \text{ N}$$

3. Design strength by block shear

A_{vg} = Minimum gross area in shear along weld line

$\quad$ = Total length of weld × t = 340 × 8 = 2720 mm²

A_{vn} = Minimum net area in shear along weld line = 2720 mm²

A_{tg} = Minimum gross-area in tension of connected leg = 50 × 8 = 400 mm²

A_{tn} = Minimum net area in tension of connected leg = 400 mm²

$$T_{db_1} = \frac{A_{vg}\, f_y}{\sqrt{3}\,\gamma_{m_0}} + \frac{0.9\,A_{tn} \cdot f_u}{\gamma_{m_1}} = \frac{2720 \times 250}{\sqrt{3} \times 1.10} + \frac{0.9 \times 400 \times 410}{1.25} = 474987 \text{ N}$$

or $$T_{db_2} = \frac{A_{tg}\, f_y}{\gamma_{m_0}} + \frac{0.9\,A_{vn} \cdot f_u}{\sqrt{3}\,\gamma_{m_1}} = \frac{400 \times 250}{1.10} + \frac{0.9 \times 2720 \times 410}{\sqrt{3} \times 1.25} = 554489 \text{ N}$$

$$\therefore \quad T_{db} = \text{Minimum of } T_{db_1} \text{ and } T_{db_2} = 474987 \text{ N}$$

$\therefore$ Design tensile strength of angle

$$= \text{Minimum of } T_{dg},\, T_{dn},\, T_{db} = 222273 \text{ N} = \mathbf{222.27\ kN}$$

Ex. 4.16: *A tension member of a truss consists of two angles 75 × 50 × 6 which are provided on either sides of a 10 mm thick gusset plate. 20 mm diameter bolts are used in one row for connecting the member to the gusset plate. Determine the design tensile strength of member and also number of bolts required to develop the design tensile strength.*

Sol.: Diameter of bolt hole,

$$d_h = 20 + 2 = 22 \text{ mm}$$
$$A_{nb} = 245 \text{ mm}^2$$

Gross area of 2 angles 75 × 50 × 6, $A_g = 2 \times 716 = 1432 \text{ mm}^2$.

1. Design strength governed by gross section yielding

$$T_{dg} = \frac{A_g \, f_y}{\gamma_{m_0}} = \frac{1432 \times 250}{1.1} = 325454 \text{ N}$$

2. Design strength governed by net section rupture

Net area of section,

$$A_n = 1432 - 2\,(22 \times 6) = 1168 \text{ mm}^2$$

Assuming bolts ≥ 4 in connection, $\alpha = 0.8$

Approximate rupture strength.

$$T_{dn} = \alpha \cdot \frac{A_n \, f_u}{\gamma_{m_1}} = 0.8 \times \frac{1168 \times 410}{1.25} = 306483 \text{ N}$$

∴   Design tensile strength = Minimum of T_{dg} and T_{dn} = 306.48 kN

$$\text{Design shear stress for a bolt} = \frac{f_{yb}}{\sqrt{3}\ \gamma_{mb}} = \frac{400}{\sqrt{3} \times 1.25} = 185 \text{ N/mm}^2$$

Design shear strength of bolt in double shear

$$= 2 \cdot A_{nb} \times \text{Design shear stress} = 2 \times 245 \times 185$$
$$= 90650 \text{ N} = 90.65 \text{ kN}$$

∴   Number of bolts required $= \dfrac{306.48}{90.65} = 3.38 \approx 4$

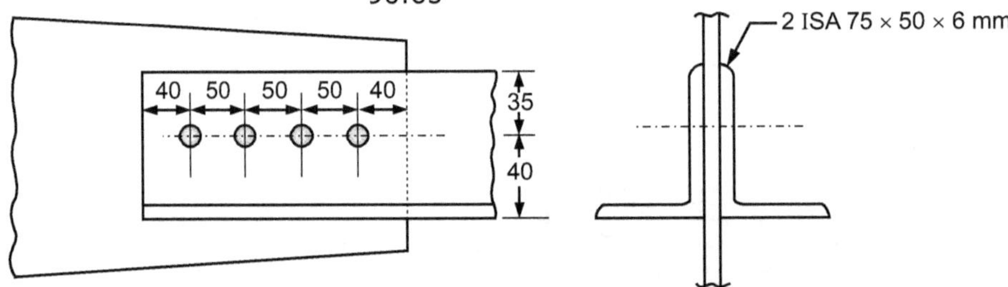

Fig. 4.28

Minimum pitch, $p = 2.5\,d = 2.5 \times 20 = 50$

Edge distance, $e \approx 2d = 2 \times 20 = 40$

3. Design strength governed by block shear

Consider one angle

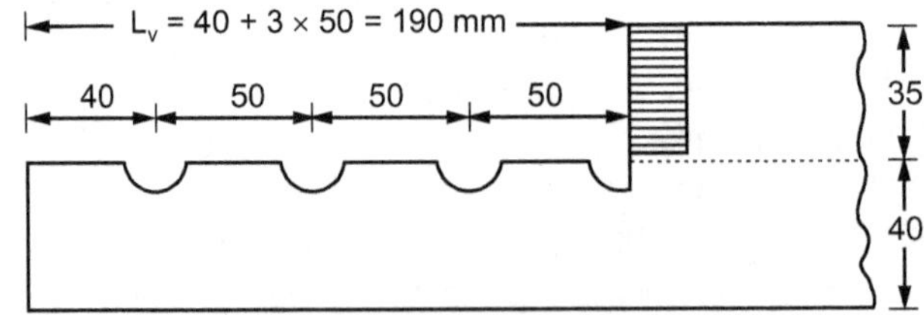

Fig. 4.29

A_{vg} = Minimum gross area in shear along bolt line = $L_v \cdot t$

$$= [190] \times 6 = 1140 \text{ mm}^2$$

A_{vn} = Minimum net area in shear along bolt line

$$= \left[L_v - \left(3 + \frac{1}{2}\right) d_h \right] t = [190 - 3.5 \times 22] \times 6 = 678 \text{ mm}^2$$

$$A_{tg} = \text{Minimum gross-area in tension from bolt hole to toe of angle perpendicular to line of force} = L_t \cdot t$$
$$= 35 \times 6 = 210 \text{ mm}^2$$

$$A_{tn} = \text{Minimum net area in tension from bolt hole to the toe of angle perpendicular to the line of force} = [L_t - 0.5\, d_h]\, t$$
$$= [35 - 0.5 \times 22] \times 6 = 144 \text{ mm}^2$$

$$T_{db_1} = \frac{A_{vg}\, f_y}{\sqrt{3}\ \gamma_{m_0}} + \frac{0.9\, A_{tn} \cdot f_u}{\gamma_{m_1}} = \frac{1140 \times 250}{\sqrt{3} \times 1.10} + \frac{0.9 \times 144 \times 410}{1.25} = 192095 \text{ N}$$

or
$$T_{db_2} = \frac{A_{tg}\, f_y}{\gamma_{m_0}} + \frac{0.9\, A_{vn} \cdot f_u}{\sqrt{3}\ \gamma_{m_1}} = \frac{210 \times 250}{1.10} + \frac{0.9 \times 678 \times 410}{\sqrt{3} \times 1.25} = 163281 \text{ N}$$

$\therefore \qquad T_{db} = \text{Minimum of } T_{db_1} \text{ and } T_{db_2} = 163281 \text{ N}$

For 2 angles $\quad T_{db} = 2 \times 163281 = 326562 \text{ N} = 326.56 \text{ kN}$

$\therefore$ The design tensile strength for double angle section

$$= \text{Minimum of } T_{dg},\ T_{dn},\ T_{db} = \textbf{306.48 kN}$$

Ex. 4.17: *The tension member of a truss consists of 2 ISA 70 × 70 × 6 mm connected on the same side of a 10 mm thick gusset plate. Determine the design tensile strength of member and also number of 20 mm dia. bolts required to develop design tensile strength.*

Sol.: Diameter of bolt hole $d_h = 20 + 2 = 22$ mm.

From steel table, gross area of 2 angles $A_g = 2 \times 806 = 1612 \text{ mm}^2$.

1. Design tensile strength governed by gross-section yielding

$$T_{dg} = \frac{A_g\, f_y}{\gamma_{m_0}} = \frac{1612 \times 250}{1.10} = 366363 \text{ N} = 366.36 \text{ kN}$$

2. Design tensile strength governed by net section rupture

$$T_{dn} = \alpha \frac{A_n\, f_u}{\gamma_{m_1}}$$

Here,
$$A_n = 2\left[\left(70 - 22 - \frac{6}{2}\right) + \left(70 - \frac{6}{2}\right)\right] \times 6 = 1344 \text{ mm}^2$$

$$\alpha = 0.8 \text{ (Consider minimum 4 bolts)}$$

$\therefore \qquad T_{dn} = 0.8 \times \dfrac{1344 \times 410}{1.25} = 352665 \text{ N} = 352.66 \text{ kN}$

$\therefore \qquad$ Design tensile strength $= \text{Minimum of } T_{dg} \text{ and } T_{dn} = 352.66 \text{ kN}$

Design shear stress for bolt $= \dfrac{f_{ub}}{\sqrt{3}\ \gamma_{mb}} = \dfrac{400}{\sqrt{3} \times 1.25} = 185 \text{ N/mm}^2$

Design shear strength of bolt in single shear

$$= 185 \times A_{nb} = 185 \times \left(0.78 \times \frac{\pi}{4} \times 20^2\right) = 45333 \text{ N} = 45.33 \text{ kN}$$

Number of bolts required $= \dfrac{352.66}{45.33} = 7.8 \approx 8$

Provide 8 bolts as shown in Fig. 4.30 with min. pitch.

$p = 2.5\, d = 2.5 \times 20 = 50$ mm and edge distance, $e = 40$ mm.

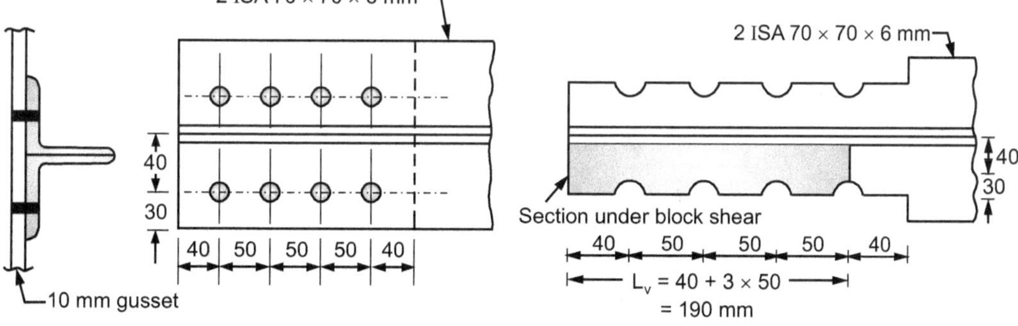

Fig. 4.30

3. Design tensile strength governed by block shear

For single angle ISA 70 × 70 × 6.

$$A_{vg} = \text{Minimum gross area in shear along bolt line} = L_v \cdot t$$
$$= [190] \times 6 = 1140 \text{ mm}^2$$

$$A_{vn} = \text{Minimum net area in shear along bolt line}$$
$$= \left[L_v - \left(3 + \frac{1}{2} \right) d_h \right] \cdot t = [190 - 3.5 \times 22 \times 6] = 678 \text{ mm}^2$$

A_{tg} = Minimum gross-area in tension from bolt hole to toe of angle perpendicular to line of force = $L_t \cdot t$
$$= 30 \times 6 = 180 \text{ mm}^2$$

A_{tn} = Minimum net area in tension from bolt hole to toe of angle perpendicular to the line of force = $[L_t - 0.5\, d_h] \cdot t$
$$= \left[30 - \frac{22}{2} \right] \times 6 = 114 \text{ mm}^2$$

$$T_{db_1} = \frac{A_{vg}\, f_y}{\sqrt{3}\; \gamma_{m_0}} + \frac{0.9\, A_{tn} \cdot f_u}{\gamma_{m_1}} = \frac{1140 \times 250}{\sqrt{3} \times 1.10} + \frac{0.9 \times 114 \times 410}{1.25} = 183239 \text{ N}$$

or
$$T_{db_2} = \frac{A_{tg}\, f_y}{\gamma_{m_0}} + \frac{0.9\, A_{vn} \cdot f_u}{\sqrt{3}\; \gamma_{m_1}} = \frac{180 \times 250}{1.10} + \frac{0.9 \times 678 \times 410}{\sqrt{3} \times 1.25} = 156463 \text{ N}$$

Minimum of T_{db_1} and T_{db_2} = 156463 N

∴   For double angle
$$T_{db} = 2 \times 156463 = 312926 \text{ N} = 312.92 \text{ kN}$$

∴   Design tensile strength
$$= \text{Minimum of } T_{dg},\, T_{dn} \text{ and } T_{db} = \mathbf{312.92 \text{ kN}}$$

Ex. 4.18: *A tension member consists of two angles ISA 75 × 75 × 8 mm bolted to 10 mm thick gusset plate one on each side using single row of bolt and tack bolted. Determine the maximum load that the member can carry.*

Take : (i)   Area of angle = 1140 mm²

        (ii)   Gauge distance as per IS clause.

Sol.: 2ISA 75 × 75 × 8 mm with 10 mm thick Gusset plate.

Assume 20 mm diameter 4.6 Grade bolts. If students consider other diameter bolts credit should be given to students accordingly.

$$A_g = 1140 \text{ mm}^2$$

Dia. of bolt hole
$$d_h = 20 + 2 = 22 \text{ mm}$$

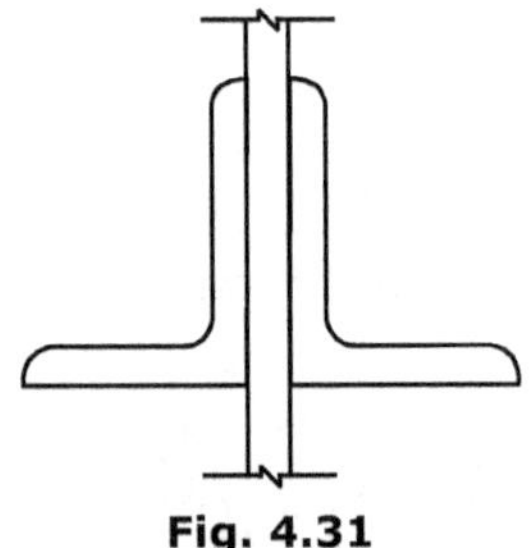

Fig. 4.31

Gross area of two angles
$$A_g = 2 \times 1140 \text{ mm}^2 = 2280 \text{ mm}^2$$

1. Design strength governed by gross-section yield
$$T_{dg} = \frac{A_g \cdot f_y}{\gamma_{m0}} = \frac{2280 \times 250}{1.1}$$
$$T_{dg} = 518.18 \text{ kN}$$

2. Design strength governed by Net Section Rupture

Net Area $A_n = 2280 - 2\,(22 \times 8) = 1928 \text{ mm}^2$

Assuming bolts ≥ 4 in connection $\alpha = 0.8$

Approximate Rupture strength
$$T_{dn} = \alpha \frac{A_n \cdot f_u}{\gamma_{m1}} = \frac{0.8 \times 1928 \times 410}{1.25}$$
$$T_{dn} = 505.90 \text{ kN}$$

∴ Design tensile strength = Minimum of T_{dg} and T_{dn} = 505.90 kN

$$\text{Design shear stress for a bolt} = \frac{f_{ub}}{\sqrt{3}\ \gamma_{mb}} = \frac{400}{\sqrt{3} \times 1.25} = 184.75 \text{ N/mm}^2$$

$$\text{Net Area of bolt } A_{nb} = 0.78 \times \frac{\pi}{4} \times 20^2 = 245.04 \text{ mm}^2$$

Design shear strength of bolt in double shear

$$= 2 \cdot A_n b \times 184.75 = 2 \times 245 \times 184.75 = 90.52 \text{ kN}$$

$$\therefore \quad \text{Number of bolts required} = \frac{505.90}{90.52} = 5.58 \approx 6 \text{ Nos.}$$

$$\text{Minimum pitch} = p = 2.5\ d = 2.5 \times 20 = 50 \text{ mm}$$

$$\text{Edge distance} = e = 1.7\ d_o = 1.7 \times 22 = 37.4 \approx 40 \text{ mm}$$

3. Design strength governed by block shear

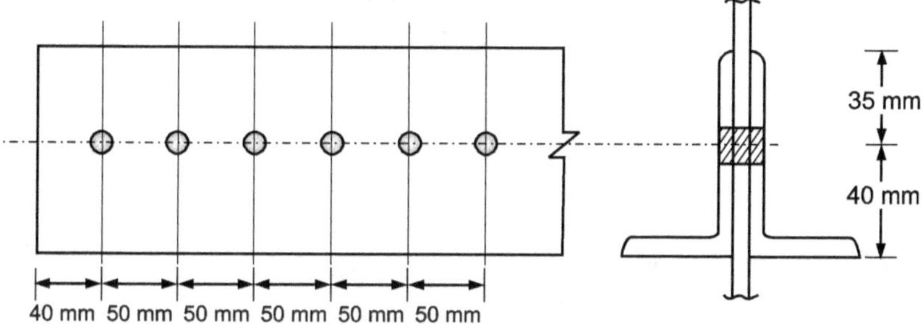

Fig. 4.32

$$A_{vg} = [5 \times 50 + 40] \times 8 = 2320 \text{ mm}^2$$

$$A_{vn} = [5 \times 50 + 40 - 5.5 \times 22]\ 8 = 1352 \text{ mm}^2$$

$$A_{tg} = 35 \times 8 = 280 \text{ mm}^2$$

$$A_{tn} = [40 - 0.5 \times 22]\ 8 = 232 \text{ mm}^2$$

Block shear strength

$$T_{db1} = \frac{A_{vg} \cdot f_y}{\sqrt{3}\ \gamma_{mo}} + \frac{0.9\ A_{tn}\ f_y}{\gamma_{m1}} = \frac{2320 \times 250}{\sqrt{3} \times 1.1} + \frac{0.9 \times 232 \times 410}{1.25}$$

$$T_{db1} = 304.421 + 68.486$$

$$[T_{db1} = 372.907 \text{ kN}]$$

$$T_{db2} = \frac{A_{tg} \cdot f_y}{\gamma_{m0}} + \frac{0.9\ A_{vn} \cdot f_u}{\sqrt{3}\ \gamma_{m_1}}$$

$$T_{db2} = \frac{280 \times 250}{1.1} + \frac{0.90 \times 1352 \times 410}{\sqrt{3} \times 1.25} = 63.636 + 230.426$$

$$T_{db2} = 294.062 \text{ kN}$$

For two Angles Block Shear Strength (T_{db})

$$T_{db} = 294.062 \times 2$$

$$T_{db} = 588.124 \text{ kN}$$

Design Tensile strength for Double Angle

$$\text{Section} = \text{Minimum of } T_{dg},\ T_{dn},\ T_{db} = 505.90 \text{ kN}$$

Ex. 4.19: *A tension member consists of 2 ISA 90 × 90 × 8 mm connected back to back with 10 mm thick gusset plate by 20 mm diameter bolts. Tacking bolts are provided. Calculate design tensile load in any one of the case below if they are placed on*

(i) both side of gusset plate

(ii) same side of gusset plate

Consider A_{nb} = 245 mm^2, Ag of single ISA 90 × 90 × 8 = 1379 mm^2.

Sol.: Diameter of bolt hole, $d_h = 20 + 2 = 22$ mm.

Gross area of two angles, $A_g = 2 \times 1379 = 2758$ mm^2.

Case 1: When angles are placed on both side of gusset plate

1. Design strength governed by gross section yielding

$$T_{dg} = \frac{A_g\, f_y}{\gamma_{m_0}} = \frac{2758 \times 250}{1.10} = 626818 \text{ N}$$

2. Design strength governed by net section rupture

Net area of section,

$$A_n = 2758 - 2\,(22 \times 8) = 2406 \text{ mm}^2$$

Assuming bolts ≥ 4 in connection, $\alpha = 0.8$

Approximate rupture strength.

$$T_{dn} = \alpha \cdot \frac{A_n\, f_u}{\gamma_{m_1}} = 0.8 \times \frac{2406 \times 410}{1.25} = 631334 \text{ N}$$

$\therefore$ Design tensile strength = Minimum of T_{dg} and T_{dn} = 626818 N = 626.82 kN

Design shear stress for a bolt $= \dfrac{f_{ub}}{\sqrt{3}\ \gamma_{mb}} = \dfrac{400}{\sqrt{3} \times 1.25} = 185 \text{ N/mm}^2$

Design shear strength of bolt in double shear

$$= 2 \cdot A_{nb} \times 185 = 2 \times 245 \times 185 = 90650 \text{ N} = 90.65 \text{ kN}$$

$\therefore$ Number of bolts required $= \dfrac{626.82}{90.65} = 6.90 \approx \mathbf{7}$

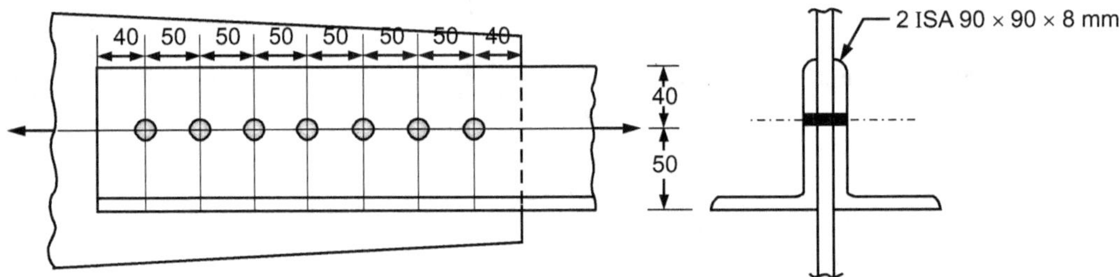

Fig. 4.33

Provide

Minimum pitch, $p = 2.5\, d = 2.5 \times 20 = \mathbf{50\ mm}$

Edge distance, $e \approx 2d = 2 \times 20 = \mathbf{40\ mm}$

3. Design strength governed by block shear

Consider one angle

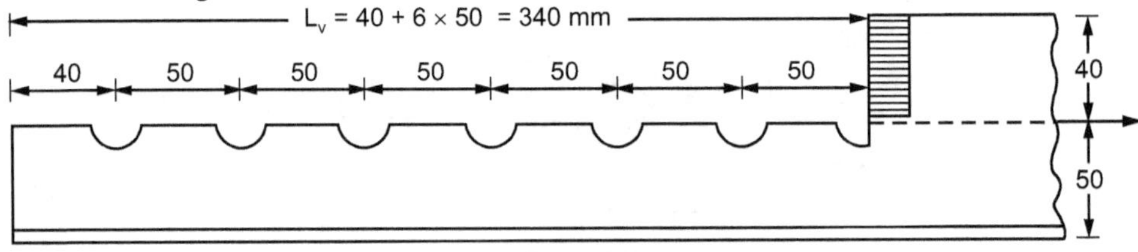

Fig. 4.34

A_{vg} = Minimum gross area in shear along bolt line $= L_v \cdot t$

     $= [340] \times 8 = 2720 \text{ mm}^2$

A_{vn} = Minimum net area in shear from bolt line

$$= \left[L_v - \left(6 + \frac{1}{2}\right) d_h \right] \cdot t = [340 - 6.5 \times 22] \times 8 = 1576 \text{ mm}^2$$

A_{tg} = Minimum gross-area in tension from bolt hole to the toe of angle perpendicular to line of force $= L_t \cdot t$

     $= 40 \times 8 = 320 \text{ mm}^2$

A_{tn} = Minimum net area in tension from bolt hole to the toe of angle perpendicular to the line of force $= [L_t - 0.5\, d_h] \cdot t$

     $= [40 - 0.5 \times 22] \times 8 = 232 \text{ mm}^2$

$$T_{db_1} = \frac{A_{vg}\, f_y}{\sqrt{3}\, \gamma_{m_0}} + \frac{0.9\, A_{tn} \cdot f_u}{\gamma_{m_1}} = \frac{2720 \times 250}{\sqrt{3} \times 1.10} + \frac{0.9 \times 232 \times 410}{1.25} = 425394 \text{ N}$$

or

$$T_{db_2} = \frac{A_{tg}\, f_y}{\gamma_{m_0}} + \frac{0.9\, A_{vn} \cdot f_u}{\sqrt{3}\, \gamma_{m_1}} = \frac{320 \times 250}{1.10} + \frac{0.9 \times 1576 \times 410}{\sqrt{3} \times 1.25} = 341330 \text{ N}$$

Minimum of T_{db_1} and T_{db_2} = 341330 N

For 2 angles T_{db} = 2 × 341330 = 682660 N = 682.66 kN

∴   Design tensile strength for double angle section

= Minimum of T_{dg}, T_{dn}, T_{db} = **626.82 kN**

Case 2: When angles are placed on same side of gusset

1. Design tensile strength governed by gross-section yielding

$$T_{dg} = \frac{A_g\, f_y}{\gamma_{m_0}} = \frac{2758 \times 250}{1.10} = 626818 \text{ N}$$

2. Design tensile strength governed by net section rupture

$$A_n = 2\left[\left(90 - 22 - \frac{8}{2}\right) + \left(90 - \frac{8}{2}\right)\right] \times 8 = 2400 \text{ mm}^2$$

$$\alpha = 0.8 \text{ (Consider minimum 4 bolts)}$$

∴ $$T_{dn} = \alpha \frac{A_n\, f_u}{\gamma_{m_1}} = 0.8 \times \frac{2400 \times 410}{1.25} = 629760 \text{ N}$$

∴   Design tensile strength = Minimum of T_{dg} and T_{dn} = 626818 kN = 626.82 kN

Design shear strength of bolt in single shear

= 185 × A_{nb} = 185 × 245 = 45325 N = 45.33 kN

Number of bolts required = $\dfrac{626.82}{45.33}$ = 13.82 say 14

Provide 14 bolts as shown in Fig. 4.25 with min. pitch.

p = 50 mm and edge distance, e = 40 mm.

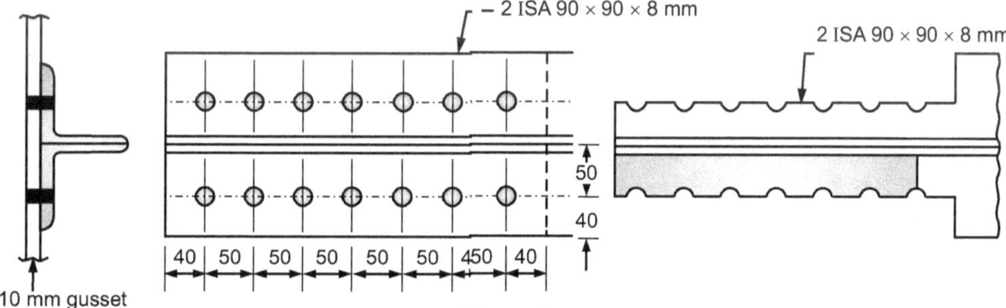

Fig. 4.35

3. Design tensile strength governed by block shear

For single angle 90 × 90 × 8 mm.

$$T_{db} = 341330 \text{ N}$$

(This step is similar to step 3 of case 1) as number of bolts are matching along one angle.

For 2 angles, T_{db} = 2 × 341330 = 682660 N = 682.66 kN

∴   Design tensile strength for double angle section

= Minimum of T_{dg}, T_{dn} and T_{db} = **626.82 kN**

Note: *For both the cases above, the design tensile strength is observed to be same. However, case 2 will be uneconomical due to additional bolting. As per specifications of IS : 800-2007 the effective net area of the sections are independent of whether the sections are tacked or not.*

Important Points

- A tension member is a structural member which transmits a direct axial pull applied at its ends.
- Generally flats, single and double angle sections and built-up sections made of rolled steel sections (L, I, C sections) and/or with plates are used as tension members.

- **Net Sectional Area**
  - (a) For chain bolting, $A_n = (b - n \cdot d_h) \cdot t$
  - (b) For zig-zag/staggered bolting

$$A_n = \left[b - n \cdot d_h + \left(\Sigma \frac{p^2}{4g} \right) \right] \times t$$

- **Design strength of axially loaded tension member should be minimum of following three cases:**
  - (a) Cross-section yielding:

$$T_{dg} = \frac{A_g \cdot f_y}{\gamma_{m_0}}$$

  - (b) Net section rupture:
    - (i) For plate

$$T_{dn} = 0.9 \frac{A_n \cdot f_u}{\gamma_{m_1}}$$

    - (ii) For Angles

$$T_{dn} = 0.9 \frac{A_{nc} \cdot f_u}{\gamma_{mn}} + \beta \frac{A_{go} f_y}{\gamma_{mo}}$$

where, $\beta = 1.4 - 0.076 \times \dfrac{w}{t} \times \dfrac{f_y}{f_u} \times \dfrac{b_s}{L_c} \leq 0.9 \dfrac{f_u}{f_y} \times \dfrac{\gamma_{m_0}}{\gamma_{m_1}} \geq 0.7$

$\quad w = $ Outstand length

$\quad b_s = $ shear lag width

$\qquad = w + g - t \qquad$ for bolted connection

$\qquad = w \qquad\qquad$ for welded connection

$\quad L_c = $ Length of end connection

Approximate rupture strength

$$T_{dn} = \alpha \cdot \frac{A_n \cdot f_u}{\gamma_{m_1}}$$

where, $\alpha = $ Reduction factor = 0.6, 0.7, 0.8 as appropriate

  - (c) Block shear
    - (i) For shear yield and tension fracture

$$T_{db_1} = \frac{A_{vg} \cdot f_y}{\sqrt{3} \cdot \gamma_{m_0}} + 0.9 \frac{A_{tn} \cdot f_u}{\gamma_{m_1}}$$

    - (ii) For shear fracture and tension yield

$$T_{db_2} = \frac{A_{tg} \cdot f_y}{\gamma_{m_0}} + 0.9 \frac{A_{vn} \cdot f_u}{\sqrt{3} \cdot \gamma_{m_1}}$$

Important Points

1. Define tension member.
2. Enlist four types of sections used as a tension member with sketches.
3. Write the stepwise procedure for design of tension members.
4. Enlist two types of sections used as a tension member along with sketches. Also, write function of gusset plate used.
5. Explain gross-section yielding and net section rupture in case of design strength of tension member. Also write two measures are taken to prevent rupture.
6. List types of failures in case of Tension member and describe any one of them with a neat sketch along with labelling.
7. A flat 150 mm × 8 mm is used as a tension member with 20 mm diameter bolts. Find the minimum net area.
8. Design a single angle tension member to carry a pull of 220 kN (service load). It is connected with gusset plate with fillet weld. Design also the welded end connection.

9. Design a tie member of a roof truss subjected to an axial service load of 150 kN. Use single equal angle connected by bolts of 16 mm diameter at ends. Refer Table 4.2.

Table 4.2

	A (mm^2)	Wt. (kg/m)
ISA $80 \times 80 \times 8$	1221	9.6
ISA $80 \times 80 \times 10$	1505	11.8
ISA $90 \times 90 \times 8$	1379	10.8

10. Design a tension member using single equal angle section to carry a force of 150 kN (service load). Assume welded connections. Refer Table 4.3.

Table 4.3

	A in mm^2	Z_{XX} in mm^3
ISA $80 \times 80 \times 10$	1505	15.5×10^3
ISA $90 \times 90 \times 6$	1047	12.2×10^3
ISA $90 \times 90 \times 8$	1379	16.0×10^3
ISA $100 \times 100 \times 6$	1167	15.2×10^3
ISA $90 \times 60 \times 8$	1137	15.1×10^3
ISA $90 \times 60 \times 10$	1401	18.6×10^3
ISA $100 \times 65 \times 6$	955	14.2×10^3

11. Design a tension member to carry an axial service load of 210 kN. Double angle section with gusset plate in between is to be used. 18 mm diameter bolts are used. Available angle sections are given in Table 4.4.

Table 4.4

Angle size	Cross-section area in mm^2
$65 \times 65 \times 6$	744
$70 \times 70 \times 6$	806
$80 \times 80 \times 6$	929
$70 \times 70 \times 8$	1058

12. Using two unequal angles placed back to back and connected with their longer legs on the same side of gusset plate with 16 mm diameter bolts, design a section to transmit a service tensile load of 300 kN. Also calculate number of bolts for connection with gusset plate.

Angle section	Wt/m (N/m)	Cross-sectional area (mm^2)	Thickness (mm)	Z_{xx} (mm^3)
ISA 60×40	58	717	8	6.5×10^3
ISA 65×45	64	817	8	7.7×10^3
ISA 100×75	105	1336	8	19.1×10^3
ISA 125×75	130	1650	10	23.6×10^3
ISA 150×75	137	1748	8	42.0×10^3

13. Calculate the strength of a tension member consisting of 2 ISA $90 \times 60 \times 8$ mm placed back to back on the opposite sides of gusset plate. Tack bolts are provided along the length of member. Use 10 mm thick gusset plate with 200 mm bolts.

[**Ans.:** No. of bolts = 6, T_d = 504.33 kN]

14. Design a suitable single angle section as a tie member to carry a tensile load of 100 kN (service load) to be connected by 16 mm diameter bolts to 10 mm thick gusset plate.

Sr. No.	Available section	Sectional area
(i)	ISA $65 \times 45 \times 8$ mm	817 mm^2
(ii)	ISA $75 \times 50 \times 8$ mm	938 mm^2
(iii)	ISA $75 \times 50 \times 10$ mm	1100 mm^2

15. Calculate the net effective area of one ISA $90 \times 90 \times 8$ mm thickness by means of fillet welds.

16. Design a tie member using single equal angle section to carry a pull of 150 kN. Diameter of bolts for connection is 16 mm.

Size of angles	Wt kg/m.	A (mm^2)	Z_{xx} (mm^3)
ISA $100 \times 75 \times 10$	13.0	1650	26.6×10^3
ISA $125 \times 75 \times 8$	12.1	1538	29.4×10^3
ISA $125 \times 95 \times 6$	10.1	1286	23.1×10^3

17. A tension member consists of 2 ISA $100 \times 100 \times 10$ mm connected back to back at same face of gusset plate. Calculate its net area if 20 mm diameter bolts are used for the connection. Also find design tensile strength.

18. Design a tension member using single equal angle section connected with gusset plate using fillet weld. It carries a service force of 190 kN. Use IS 800 specifications.

	A (mm^2)	Z_{xx} (mm^3)
ISA $80 \times 80 \times 8$	1221	12.6×10^3
ISA $80 \times 80 \times 10$	1505	15.5×10^3
ISA $90 \times 90 \times 8$	1379	16.0×10^3
ISA $90 \times 90 \times 10$	1703	19.8×10^3
ISA $80 \times 50 \times 10$	1202	14.4×10^3
ISA $90 \times 60 \times 6$	865	11.5×10^3
ISA $90 \times 60 \times 8$	1137	15.1×10^3
ISA $100 \times 65 \times 6$	955	14.2×10^3

19. A tension member consists of 2 ISA $90 \times 60 \times 8$ mm connected back to back on opposite side by long leg to 10 mm thick gusset plate by 16 mm diameter bolts. Find design tensile load carrying capacity and number of bolts required. Tack bolts are used. Use IS 800 specifications.

20. The tie of truss carries an axial tension of 225 kN. Design the section of the members and also the connection of the member to 10 mm thick gusset plate. Use 20 mm diameter bolts.

21. Two ISA $100 \times 75 \times 8$ mm placed back to back on same side of gusset plate 10 mm thick to act as a tension member. Calculate the design strength of tension member if

    (i) Tack bolts are provided

    (ii) If tack bolts are not provided.

22. Calculate the strength of a tie member composed of 2 ISA $150 \times 75 \times 8$ mm, when they are placed back to back with their longer legs connected on the same side of the gusset plate by 20 mm diameter bolts. Tacking bolts have been used.

23. The longer leg of a single angle $90 \times 60 \times 10$ mm is connected to the gusset with plate with 3 bolts in a line of 20 mm diameter at a pitch of 60 mm for this tension member. Determine the block shear strength.

24. Determine the design strength of a single angle tension member consisting of single angle 80 mm $\times$ 50 mm $\times$ 8 mm. The shorter leg of the angle is connected to the gusset plate by 5 mm weld as shown in Fig. 1. Gross sectional (Ag) for this angle is 978 mm2.

■■■

Chapter **5**

DESIGN OF COMPRESSION MEMBERS BY L.S.M.

Syllabus

- Compression Members - Effective Length and Effective Sectional Area of Compression Members - Design Stress and Design Strength - Buckling Class of Cross Sections - Imperfection Factor - Stress Reduction Factor - Thickness of Elements - Analysis and Design of Axially Loaded Column. Introduction to Lacing and Battening (No Numerical Problem on Lacing and Battening)

About this Chapter

After reading this chapter students can understand:
- Types of Sections used as Compression Members
- Effective length, Slenderness ratio
- Angle Struts
- Strength and Design of Axially Loaded Compression Members

5.1 DEFINITION

- A compression member is a structural member which is straight and subjected to two equal and opposite compressive forces applied at its ends.
- If the load is concentric with longitudinal axis of the member it is called axially loaded compression member.
- On the other hand if load is eccentric with longitudinal axis of the member it is called eccentric column/compression member.
- The components of a truss carrying compression are popularly known as *struts*.
- The vertical members in a building carrying direct compression are called *columns*, *stanchions* or *posts*.

5.2 TYPES OF SECTIONS

- The different types of sections used as compression members are shown in Fig. 5.1.
- The selection of compression member section depends on the loading and its nature (whether axial or eccentric).
- For compression members of trusses and bracings of plates girders single angle, channels or I-sections may be used.
- For principal rafter and main tie in a roof truss double angle sections are preferred.
- For columns carrying axial or eccentric load I-sections are generally used.
- For heavy compressive loads, combinations of I-sections, channels with or without cover plates may be used.

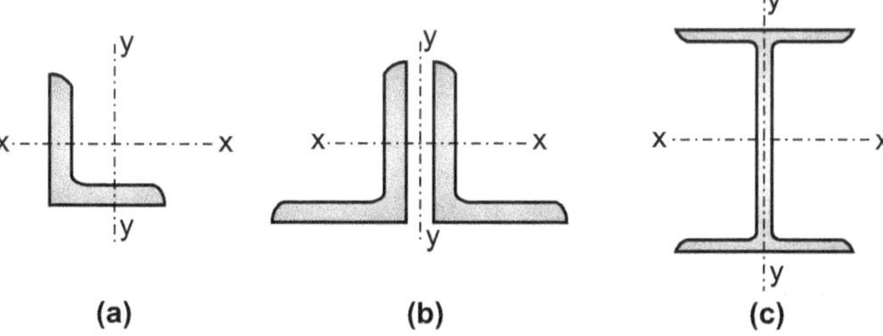

(a)   (b)   (c)

contd. ...

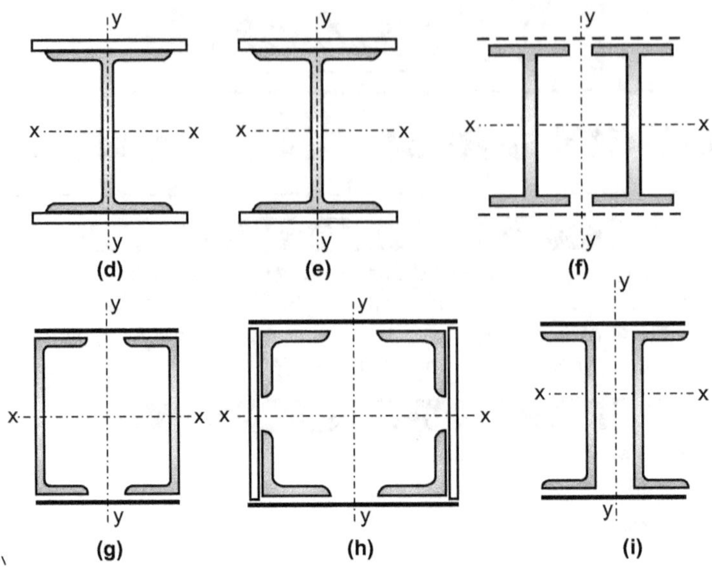

Fig. 5.1: Various Forms of Compression Members

5.3 EFFECTIVE LENGTH

- *It is defined as that length of column for which it acts as if both the ends are hinged.* At these points, the flexure changes its sign or in other words it is the distance between two points of zero moments.
- The effective length *l* is derived from the actual length L for different end conditions. According to IS: 800 – 2007, it is taken as given in Table 5.1.
- The effective length KL is calculated from actual length L of the member, considering rotational and relative translational boundary conditions at the ends.
- The actual length shall be taken as the length from the centre to centre of its intersections with the supporting members in the plane of bucking deformation.

Table 5.1: Effective length of prismatic compression members
(I.S. 800, clause 7.2.2)

Boundary conditions				Schematic Representation	Effective Length
At one end		**At the other end**			
Translation	Rotation	Translation	Rotation		
Restrained	Restrained	Free	Free		2 L
i.e. Restrained against rotation and translation at one end but free at the other end					
Restrained	Free	Free	Restrained		2 L
i.e. Restrained against translation and free rotation at one End but roller supported at the other end					
Restrained	Free	Restrained	Free		1 L
i.e. Restrained against translation and free against rotation at both ends					

contd. ...

Restrained	Restrained	Free	Restrained		1.2 L
i.e. Restrained against rotation and translation at one end and restrained against rotation and free (roller Support)					
Restrained	Restrained	Restrained	Free		0.8 L
i.e. Restrained against rotation and translation at one end and restrained against translation but not against rotation at the other end					
Restrained	Restrained	Restrained	Restrained		0.65 L
i.e. Restrained against rotation and translation at both ends					

5.4 RADIUS OF GYRATION

It is the distance from an axis at which all elemental areas of a given section may be assumed to be concentrated so as not to alter the moment of inertia about the given axis.

Mathematically, it is the property of a section and is equal to

$$r = \sqrt{\dfrac{I}{A}}$$

where, I = Moment of inertia of the section, (mm^4).

           A = Area of section, (mm^2).

5.5 SLENDERNESS RATIO

It is the ratio of effective length to the least radius of gyration.

i.e. $S.R. = \dfrac{kL}{r_{min}}$

where, kL = effective length of the member

         r_{min} = least radius of gyration of the member

5.6 MAXIMUM SLENDERNESS RATIO

- The slenderness ratio should not exceed the values given in Table 5.3.

Notes:

1. L is the unsupported length of compression member.
2. For battered columns, the effective length shall be increased by 10%.

Table 5.2: Maximum Slenderness Ratio (λ) for compression members

Sr. No.	Type of Member	λ
1.	A tension member in which a reversal of direct stress occurs due to loads other than wind or seismic forces.	180
2.	A member carrying compressive loads resulting from dead loads and imposed loads.	180
3.	A member subjected to compressive forces resulting only from combination with wind/earthquake actions, provided the deformation of such members does not adversely effect the stress in any part of the structure.	250
4.	Compression flange of a beam restrained against lateral torsional buckling.	300
5.	A member normally acting as a tie in a roof truss or a bracing system is not considered effective when subjected to possible reversal of stresses resulting from the action of wind or earthquake forces.	350

Table 5.3: Angle Struts (Clause 4.5, IS: 800 – 1984)

Sr. No.	Type	End Connections		Effective length l	Allowa-ble stress	Slender-ness ratio
1.	Single angle discontinuous	(i) One rivet or bolt at each end.		$l = L$	$0.8\,\sigma_{ac}$	$\dfrac{l}{r} \not> 180$
		(ii) Two or more rivets or bolts or welding at each end.		$l = 0.85\,L$	σ_{ac}	–
2.	Double angle discontinuous tacked	(i) Connected on same side of gusset plate (a) One rivet or bolt at each end.		$l = L$	$0.8\,\sigma_{ac}$	$\dfrac{l}{r} \not> 180$
		(b) Two or more rivets or bolts or welding at each end.		$l = 0.85\,L$	σ_{ac}	–
		(ii) Connected on both sides of gusset plate by two or more rivets, bolts or welding.		$l = 0.7$ to $0.85\,L$ depending on rigidity of joint	σ_{ac}	–
3.	Single or double angle continuous	One or more rivet bolt or welding		$l = 0.7$ to $0.85\,L$ depen-ding on end rigidity	σ_{ac}	–

Notes: 1. L = Unsupported length.

         2. σ_{ac} = Permissible axial compressive stress.

5.7 DESIGN COMPRESSIVE STRENGTH

- Common hot rolled and built-up steel members used for carrying axial compression, usually fail by flexural buckling. The buckling strength of these members is affected by residual stresses, initial bow and accidental eccentricities of load. To account for all these factors, the strength of members subjected to axial compression is defined by buckling class a, b, c or d as given in Table 5.3.

- The design compressive strength P_d of a member is given by:

$$P < P_d$$

where, $P_d = A_e\,f_{cd}$

where, A_e = Effective sectional area

            f_{cd} = Design compressive stress, obtained as per equation given below

- The design compressive stress f_{cd}, of axially loaded compression members shall be calculated using the following equation:

$$f_{cd} = \dfrac{\dfrac{f_y}{\gamma_{m_0}}}{\phi + \sqrt{\phi^2 - \lambda^2}} = \dfrac{\chi f_y}{\gamma_{m_0}} \le \dfrac{f_y}{\gamma_{m_0}}$$

where, $\phi = 0.5\left[1 + \alpha\left(\lambda - 0.2\right) + \lambda^2\right]$

λ = Non-dimensional effective slenderness ratio = $\sqrt{\dfrac{f_y}{f_{cc}}} = \sqrt{\dfrac{f_y\left(\frac{kL}{r}\right)^2}{\pi^2 E}}$

f_{cc} = Euler buckling stress = $\dfrac{\pi^2 E}{\left(\frac{kL}{r}\right)^2}$

where, $\dfrac{kL}{r}$ = Effective slenderness ratio or ratio of effective length, kL to appropriate

radius of gyration, r;

α = imperfection factor given in Table 5.3.

χ = Stress reduction factor (see Table 5.4) for different buckling class, slenderness ratio and yield stress

$= \dfrac{1}{\phi + \sqrt{\phi^2 - \lambda^2}}$

γ_{mo} = Partial safety factor for material strength

Note: *Calculated values of design compressive stress, f_{cd} for different buckling classes are given in Table 5.4.*

- The classification of different sections under different buckling class a, b, c or d, is given in Table 5.5.

- The stress reduction factor χ, and the design compressive stress f_{cd}, for different buckling class, yield stress and effective slenderness ratio is given in Table 5.4 for convenience. The curves corresponding to different buckling class are presented in non-dimensional form, in Fig. 5.2.

Table 5.4: Imperfection Factor, α
(Clauses 7.1.1 and 7.1.2.1)

Buckling Class	a	b	c	d
α	0.21	0.34	0.49	0.76

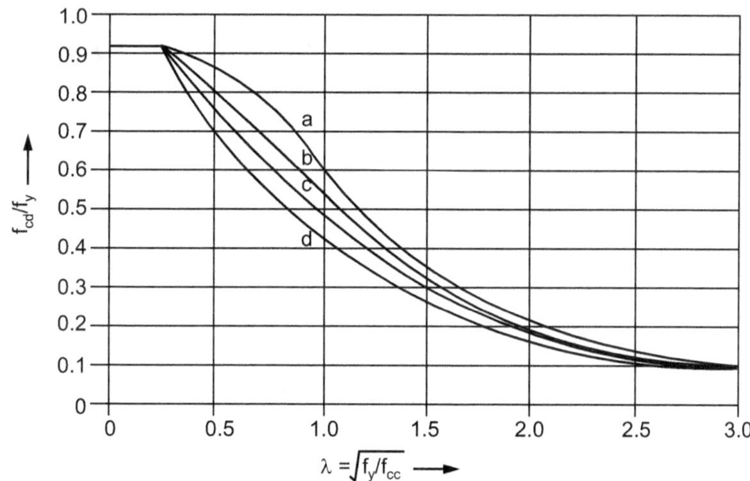

Fig. 5.2: Column Buckling Curves

Table 5.5: Stress Reduction Factor (χ) and Design Compressive Stress (f_{cd}) for different Buckling Curves (f_y = 250 MPa)

KL/r (SR)	Curve a		Curve b		Curve c		Curve d	
	χ	f_{cd}	χ	f_{cd}	χ	f_{cd}	χ	f_{cd}
10	1.000	227	1.000	227	1.000	227	1.000	227
20	0.994	226	0.991	225	0.987	224	0.980	223
30	0.969	220	0.950	216	0.930	211	0.896	204
40	0.939	213	0.906	206	0.870	198	0.815	185
50	0.904	205	0.855	194	0.807	183	0.736	167
60	0.859	195	0.798	181	0.740	168	0.659	150
70	0.803	182	0.732	166	0.670	152	0.587	133
80	0.734	167	0.661	150	0.600	136	0.521	118
90	0.657	149	0.589	134	0.533	121	0.461	105
100	0.579	132	0.520	118	0.471	107	0.408	92.6
110	0.507	115	0.458	104	0.416	94.6	0.361	82.1
120	0.443	101	0.403	91.7	0.368	83.7	0.321	73.0
130	0.388	88.3	0.356	81.0	0.327	74.3	0.287	65.2
140	0.342	77.8	0.316	71.8	0.291	66.2	0.257	58.4
150	0.303	68.9	0.281	64.0	0.261	59.2	0.231	52.6
160	0.270	61.4	0.252	57.3	0.234	53.3	0.209	47.5
170	0.242	55.0	0.227	51.5	0.212	48.1	0.190	43.1
180	0.218	49.5	0.205	46.5	0.192	43.6	0.173	39.3
190	0.197	44.7	0.186	42.2	0.175	39.7	0.158	35.9
200	0.179	40.7	0.169	38.5	0.160	36.3	0.145	33.0
210	0.163	37.1	0.155	35.2	0.146	33.3	0.134	30.4
220	0.149	34.0	0.142	32.3	0.135	30.6	0.123	28.0
230	0.137	31.2	0.131	29.8	0.124	28.3	0.114	26.0
240	0.127	28.8	0.121	27.5	0.115	26.2	0.106	24.1
250	0.117	26.6	0.112	25.5	0.107	24.3	0.099	22.5

Note:

- For steel with $f_y \neq 250$ MPa Tables 8 and 9 of IS: 800 – 2007 may be referred.
- Interpolation formula for intermediate SR value is

$$f_{cd} = f_{cd_1} - \frac{f_{cd_1} - f_{cd_2}}{SR_2 - SR_1}\,(SR - SR_1)$$

5.8 CLASSIFICATION OF DIFFERENT SECTIONS UNDER DIFFERENT BUCKLING CLASS

- This classification is given in following Table 5.6.

Table 5.6: Buckling Curves and Limiting thickness for Cross-Sections

Cross-section	Limits	Buckling about axis	Buckling curve
Rolled I-Sections	$h/b_f > 1.2$: $t_f \leq 40$ mm	x-x	a
		y-y	b
	40 mm $< t_f \leq 100$ mm	x-x	b
		y-y	c
	$h/b_f \leq 1.2$: $t_f \leq 100$ mm	x-x	b
		y-y	c
	$t_f \geq 100$ mm	x-x	d
		y-y	d

contd. ...

Section	Condition	Buckling about axis	Buckling Class
Welded I-Sections:	$t_f \leq 40$ mm	x-x	b
		y-y	c
	$t_f \geq 40$ mm	x-x	c
		y-y	d
Hollow Sections	Hot rolled	Any	a
	Cold formed	Any	b
Welded Box Section	Generally (Except as below)	Any	b
	Thick welds and $b/t_f < 30$	x-x	c
	$b/t_w < 30$	y-y	c
Channel, Angle, T and Solid Sections		Any	c
Built-up Member		Any	c

5.9 ANGLE STRUTS

- Angle struts are generally used in roof trusses. The section may be composed of single angle or double angles. These members may be continuous members (like principal rafter) or discontinuous member (like vertical and diagonal members). The effective length kL of angle strut may be taken as 0.7 to 1.0 times, the distance between centres of connections depending on the degree of end restraints.

- For buckling in the plane perpendicular to the plane of trusses, the effective length kL shall be taken as the distance between the centres of intersection.

- Generally the angle sections fall under buckling class 'c' as per Table 5.6. Hence for angles design compressive stress f_{cd} corresponding to buckling class c shall be interpolated using the values in Table 5.5.

5.9.1 Single Angle Struts

- The compression in single angles may be transferred either concentrically to its centroid through end gusset or eccentrically by connecting one of its legs to a gusset or adjacent member.

    (a) Concentric loading: When a single angle is concentrically loaded in compression the design strength may be determined as in the case of axially loaded columns.

    (b) Eccentric loading (Angle loaded through one leg): A single angle when loaded through one of its leg it is subjected to flexural-torsional buckling. The equivalent slenderness ratio in such a case.

$$\lambda_e = \sqrt{k_1 + k_2\, \lambda_{vv}^2 + k_3\lambda_\phi^2}$$

where, k_1, k_2, k_3 = Constant depending upon the end condition, as given in Table 5.6.

$$\lambda_{vv} = \frac{\left(\dfrac{l}{r_{vv}}\right)}{\varepsilon\sqrt{\dfrac{\pi^2 E}{250}}} \quad \text{and} \quad \lambda_\phi = \frac{(b_1 + b_2)}{\varepsilon\sqrt{\dfrac{\pi^2 E}{250}} \times 2t}$$

where,

l = Centre to centre length of the supporting member

r_v = Radius of gyration about the minor axis

b_1, b_2 = Width of the two legs of the angle

t = Thickness of the leg

ε = Yield stress ratio $\sqrt{\dfrac{250}{f_y}}$

Table 5.7: Constants k_1, k_2 and k_3

No. of Bolts at each end connection	Gusset/Connecting member fixity	k_1	k_2	k_3
≥ 2	Fixed	0.20	0.35	20
	Hinged	0.70	0.60	5
1	Fixed	0.75	0.35	20
	Hinged	1.25	0.50	60

5.9.2 Double Angle Struts

- For double angle discontinuous struts, connected back to back, on opposite sides of the gusset or section, by not less than two bolts or by an equivalent in welding, the load may be regarded as applied axially. The effective length, kL, in the plane of end gusset shall be taken as between 0.7 and 0.85 times the distance between intersections depending on the degree of restraint provided. The effective length kL in the plane perpendicular to that of the end gusset, shall be taken as equal to the distance between centres of intersections. Design strength is calculated as in axially loaded columns.

- Double angle discontinuous struts connected back to back, to one side of a gusset or section by one or more bolts in each angle or by the equivalent in welding, shall be designed as in case of single angle struts.

- Double angle continuous struts shall be designed as axially loaded compression members with effective length as 0.7 L to 1.0 L.

5.10 TACKING BOLTS

- When double angle section is used as a strut, the two angles together must act as one unit so as to avoid each of them buckle separately and fail. Therefore, the angles are connected together at the interval of 32 t or 300 mm whichever is less through out their length. Where the plates are exposed to the weather, the spacing in line shall not be greater than 16 t or 200 mm whichever is less. Where 't' is the thickness of the thinner outside plate. The bolts stitch the sections together and hence are called stitching or tacking bolts.

- In tension members tacking fasteners, with solid distance pieces shall be provided at a spacing in line not exceeding 1000 mm.

- In compression members the spacing shall not exceed 600 mm.

5.11 STANCHIONS AND COLUMNS

- Stanchion and column is a vertical compression member supporting floors or girders in a building. These compression members are subjected to heavy loads. The sections used for stanchions/columns may be simple or built-up. (See Fig. 5.3 and 5.4)

(i) Sketch of stanchion:

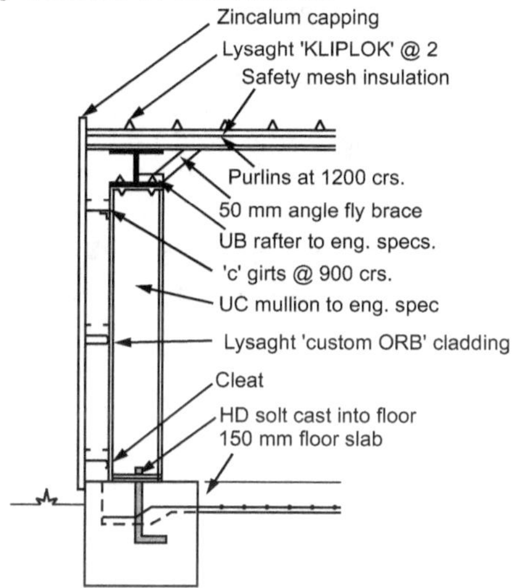

Fig. 5.3

(ii) Sketch of column supporting girder

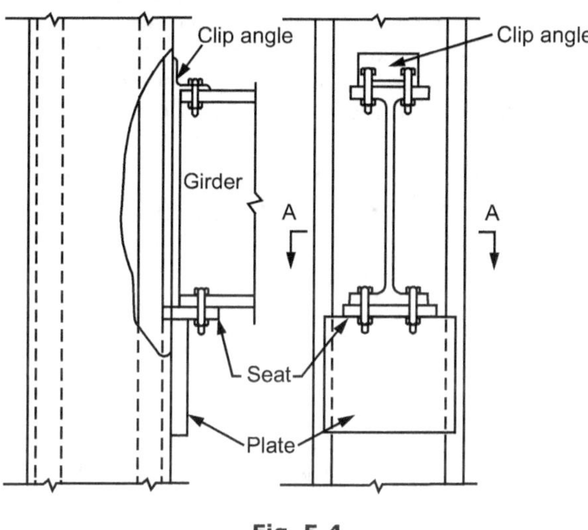

Fig. 5.4

5.11.1 Simple Sections

- The single rolled steel sections like angle (equal and unequal). I section and channel section are frequently used as columns in the buildings. They are used due to ease of availability and savings in the cost of fabrication. Hence, these sections are called simple sections. Some of the simple sections are shown in Fig. 5.3 (A). However, there use is limited for heavy loads because of insufficient strength. In such case built-up sections are used.

5.11.2 Built-up Sections

- In the market the limited number of rolled steel sections are available. The radius of gyration about minor axis is small giving large slenderness ratio with the result design stress gets sufficiently reduced making the section uneconomical. Further the rolled steel sections do not have sufficient strength to resist heavy loads. In such cases the sections are fabricated consisting of two or more individual sections such as angles, channels, I-sections etc., properly connected along their length by lacing or battening so that they act together as one unit. They are called as built-up sections. These are shown in Fig. 5.3 (B) and 4.1 (d) to (i).

- Built-up sections commonly used consist of two channels connected with lacing or-battening to hold the two parts together so as to act as one unit. A pair of channels connected by cover plates on one side and latticing on the other are sometimes used as top chord members on large span bridge trusses. For moderate loads plated I-sections are used.

- All built-up sections should preferably be so proportioned that the slenderness ratio about both principal axes are the same.
- In case if the unsupported length in both direction is the same then the radius of gyration about both the axes should be approximately the same to effect economy. Further, from economy point of view, other considerations such as simplicity of fabrication and ease of construction should also be taken into account.

5.12 LIMITS OF WIDTHS TO THICKNESS RATIOS TO PREVENT LOCAL BUCKLING

- The width to thickness ratio of plates depends on the local buckling of flanges and webs. The buckling of the plate element of the cross-section under compression/ shear may take place before overall buckling or yielding. This phenomenon is called local buckling.
- Local buckling involves distortion of cross section. It has the effect of reducing the overall load carrying capacity and also the rotation capacity of columns and beams due to reduction in stiffness of a locally buckled section.
- To prevent local buckling the width to thickness ratios are limited and are given along with the symbols used in Table 5.6 and Fig. 5.5. These limits depend on the way in which the, outstanding flanges of the sections are supported (either one edge supported as for the flange outstand of an I-section or two edges supported as for the internal flange element of a box section), the bending stress distribution, the type of section and residual stress distribution.

Table 5.8: Limiting Width to Thickness Ratio to Prevent Local Buckling

Compression Element		Ratio	Class 1 Plastic	Class 2 Compact	Class 3 Semi-compact
Outstanding element of compression flange	Rolled section	b/t_f	$9.4\,\varepsilon$	$10.5\,\varepsilon$	$15.7\,\varepsilon$
	Welded section	b/t_f	$8.4\,\varepsilon$	$9.4\,\varepsilon$	$13.6\,\varepsilon$
Internal element of compression flange	Compression due to bending	b/t_f	$29.3\,\varepsilon$	$33.5\,\varepsilon$	$42\,\varepsilon$
	Axial compression	b/t_f	Not applicable		
Web of an I, H or box section	Neutral axis at mid-depth	d/t_w	$84\,\varepsilon$	$105\,\varepsilon$	$126\,\varepsilon$
	Generally — If r_1 is negative	d/t_w	$\dfrac{84\,\varepsilon}{1 + r_1}$ but $\leq 42\,\varepsilon$	$\dfrac{105.0\,\varepsilon}{1 + r_1}$	$\dfrac{126.0\,\varepsilon}{1 + 2r_2}$ but $\leq 42\,\varepsilon$
	Generally — If r_1 is positive	d/t_w		$\dfrac{105.0\,\varepsilon}{1 + 1.5\,r_1}$ but $\leq 42\,\varepsilon$	
	Axial compression	d/t_w	Not applicable		$42\,\varepsilon$
Web if a channel		d/t_w	$42\,\varepsilon$	$42\,\varepsilon$	$42\,\varepsilon$
Angle, compression due to bending (Both criteria should be satisfied)		b/t	$9.4\,\varepsilon$	$10.5\,\varepsilon$	$15.7\,\varepsilon$
		d/t	$9.4\,\varepsilon$	$10.5\,\varepsilon$	$15.7\,\varepsilon$
Single angle, or double angles with the components separated, axial compression (All three criteria should be satisfied)		b/t	Not applicable		$15.7\,\varepsilon$
		d/t			$15.7\,\varepsilon$
		$(b + d)/t$			$25\,\varepsilon$
Outstanding leg of an angle in contact back-to-back in a double angle member.		$\dfrac{d}{t}$	$9.4\,\varepsilon$	$10.5\,\varepsilon$	$15.7\,\varepsilon$
Outstanding leg of an angle with its back in continuous contact with another component		$\dfrac{d}{t}$	$9.4\,\varepsilon$	$10.5\,\varepsilon$	$15.7\,\varepsilon$
Stem of a T-section, rolled or cut from a rolled I or H-section		$\dfrac{D}{t_f}$	$8.4\,\varepsilon$	$9.4\,\varepsilon$	$18.9\,\varepsilon$

Note:

1. $\varepsilon = \sqrt{\dfrac{250}{f_y}}$

2. Elements which exceed semi-compact limits are to be taken as of slender cross-section.

3. Webs shall be checked for shear buckling when d/t > 67 ε, where, b is the width of the element (may be taken as clear distance between lateral supports or between lateral support and free edge, as appropriate), t is the thickness of element, d is the depth of web, D is the outer diameter of the element.

4. Different elements of a cross-section can be different classes. In such cases the section is classified based on the least favourable classification.

5. The stress ratio r_1 and r_2 are defined as:

$$r_1 = \frac{\text{Actual average axial stress (negative if tensile)}}{\text{Design compressive stress of web alone}}$$

$$r_2 = \frac{\text{Actual average axial stress (negative if tensible)}}{\text{Design compressive stress of overall section}}$$

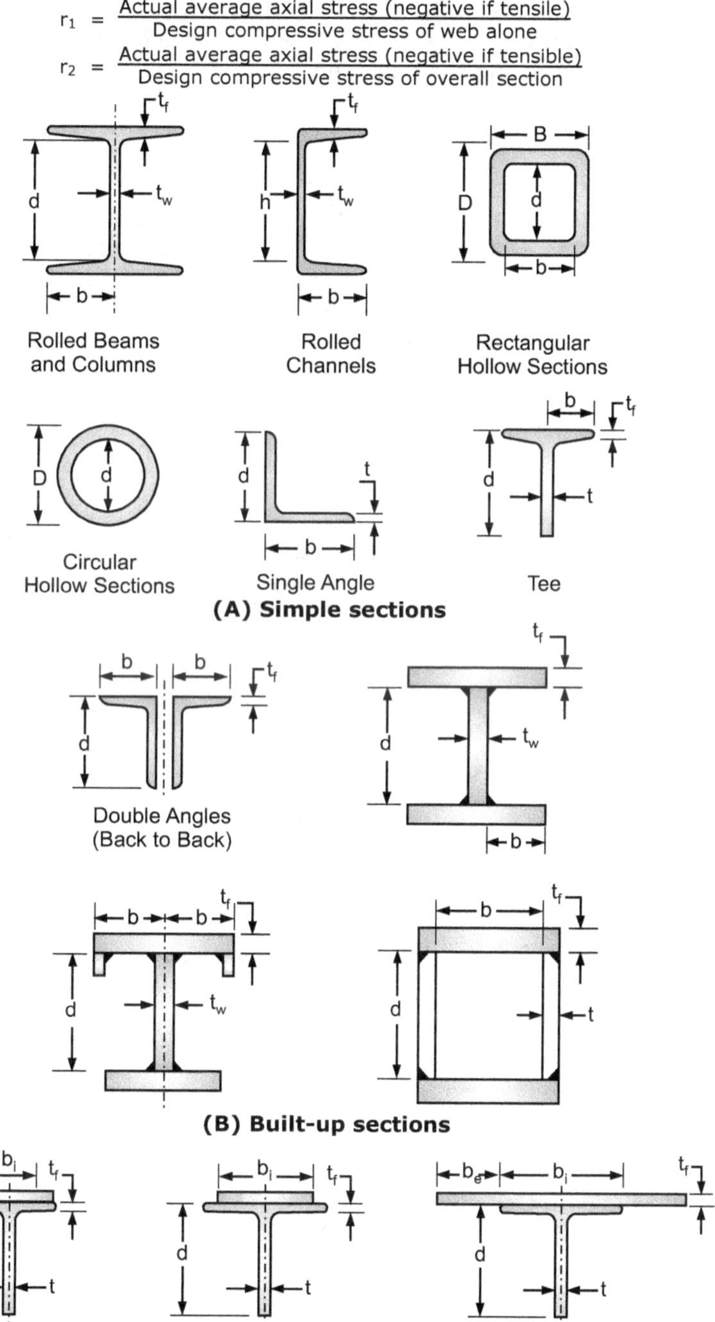

Fig. 5.5: Dimensions of Sections (symbols used in Table 5.6)

5.13 LACING AND BATTENING

- Built-up columns sections consists usually of two or more sections connected together by cover plates or without cover plates. The lateral sides or all the four sides of section may contain perforations.

- It is essential to connect the component parts so that they will act as a single unit, otherwise each element will act individually and the design strength of such a column will be the sum of the design strengths of the individual components, thereby killing the whole purpose of built-up column.

- By providing an arrangement to connect the individual components, full use of the strength of the material is utilized. Connecting the individual components assists in holding the components in their respective positions. The components of a built-up column can be connected by (i) lacing and (ii) battening and accordingly such columns are called laced columns and battened columns respectively.

5.14 LACED COLUMNS

- In this case the components of the composite section are connected by lacing. Lacing consists of connecting the components of the column by a system of generally flat plates. (In some cases angles and channels are also used as lacings). Lacing plates may be 50 mm to 75 mm wide and 8 mm to 10 mm thick. See Fig. 5.6.

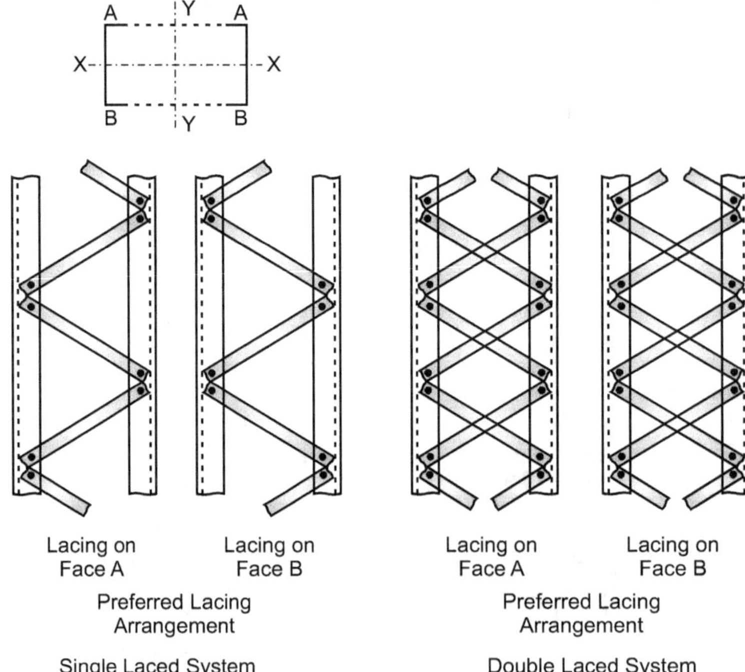

Fig. 5.6: Laced Columns

5.14.1 General Requirements for Lacing as per IS : 800

- Compression members comprising two main components laced and tied, should where practicable have a radius of gyration about the axis perpendicular to the plane of lacing not less than the radius of gyration about the axis parallel to the plane of lacing.

- As far as possible the lacing system shall be uniform throughout the length of the column.

- Single laced systems on opposite faces of the components being laced together shall preferably be in the same direction so that one is the shadow of the other instead of being mutually opposed in direction.

- The effective slenderness ratio $\left(\dfrac{KL}{r}\right)_e$ of laced columns shall be taken as 1.05 times the $\left(\dfrac{KL}{r}\right)_o$, the actual maximum slenderness ratio, in order to account for shear deformation effects.

- Width of lacing bars. In bolted/riveted construction, the minimum width of lacing bars shall be three times the nominal diameter of the end bolt/rivet.

- **Thickness of lacing bars:** The thickness of flat lacing bars shall not be less than one-fortieth of its effective length for single lacing and one-sixtieth of the effective length for double lacings.

- Rolled sections or tubes of equivalent strength may be permitted instead of flats for lacings.

- **Angles of inclination:** Lacing bars, whether in double or single or double system shall be inclined at an angle not less than 40° nor more than 70° to the axis of the built-up member.

- **Spacing:** The maximum spacing of lacing bars, whether connected by bolting, riveting or welding, shall also be such that the maximum slenderness ratio of the components of the main member $\left(\dfrac{a_1}{r_1}\right)$ between consecutive lacing connections is not greater than 50 or 0.7 times the most unfavourable slenderness ratio of the member as a whole, whichever is less, where a_1 is the unsupported length of the individual member between lacing points and r_1 is the minimum radius of gyration of the individual member being laced together.

5.15 BATTENED COLUMNS

- Compression members composed of two main components battened should preferably have the individual members of the same cross-section and symmetrically disposed about their major axis.

- Where practicable, the compression members should have a radius of gyration about the axis perpendicular to the plane of the batten not less than the radius of gyration about the axis parallel to the plane of the batten.

- The battens should be placed opposite to each end of the member and at points where the member is stayed in its length and as far as practicable, be spaced and proportioned uniformly throughout.

- The number of battens shall be such that the member is divided into not less than three bays within its actual length from centre to centre of end connections. Refer Fig. 5.7.

- The effective slenderness ratio $\left(\dfrac{KL}{r}\right)_e$ of battened columns, shall be taken as 1.1 times $\left(\dfrac{KL}{r}\right)_o$, the maximum actual slenderness ratio of the column, to account for shear deformation effects.

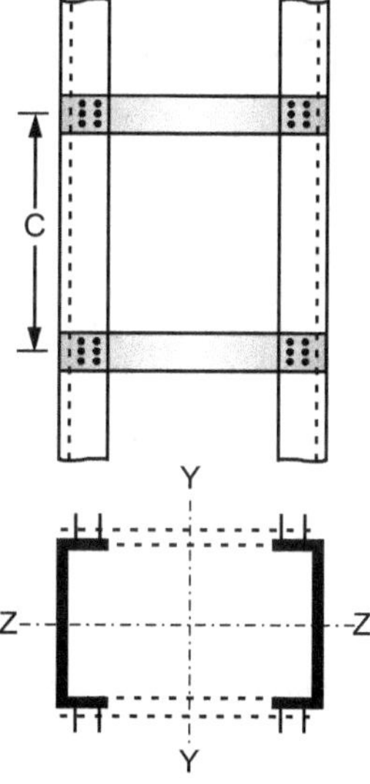

Fig. 5.7: Battened columns

5.16 STEPS FOR DESIGN OF SIMPLE COMPRESSION MEMBERS

- Calculate the ultimate axial load to be resisted. For the case of buildings it consists of the reactions from the beams, (having approximately equal spans and carrying equal loads), connecting the column.
- Assume suitable trial section The general guide lines to obtain the approximate section are as under:

  For single and double angles assume the design stress, f_{cd} = 90 N/mm^2 and for I-section = 150 N/mm^2.
- Arrive at the effective length of the column considering end conditions.
- Calculate the slenderness ratio. Check that they satisfy the maximum limits given in Table 5.2.
- Check the buckling class of the cross-section and select proper buckling class a or b or c or d from Table 5.5. Corresponding to the selected buckling class obtain the design compressive stress f_{cd} appropriately from Table 5.4.
- The allowable compressive load = f_{cd} × gross cross-sectional area.
- Check that the design compressive load is equal to or greater than the applied load to be resisted.
- If the design compressive load of the column exceeds the factored load over the column by more than 5%, the section needs a revision (being over safe and uneconomical). However, many times it may not be possible due to the limitation of sections available.

SOLVED EXAMPLES

Ex. 5.1: *A single angle ISA 90 × 90 × 8 mm is used as a strut in a roof truss. The c/c distance between intersection points at each end is 2.75 m. Determine the design strength of the strut for following cases.*

(i) At each end of connection two bolts are provided.

(ii) At each end of connection only one bolt is provided.

(iii)The ends of the strut connected by welds.

Take f_y = 250 N/mm^2.

Sol.: From steel tables, Area of single ISA 90 × 90 × 8 mm, A = 1379 mm^2, r_{min} = r_{vv} = 17.5 mm.

Case (i): At each end of connection two bolts are provided

Equivalent slenderness ratio

$$\lambda_e = \sqrt{k_1 + k_2 \cdot \lambda_{vv}^2 + k_3\, \lambda_\phi^2}$$

and

$$\varepsilon = \text{Yield stress ratio} = \sqrt{\frac{250}{f_y}} = \sqrt{\frac{250}{250}} = 1$$

$$\lambda_{vv} = \frac{\left(\dfrac{l}{r_{vv}}\right)}{\varepsilon\sqrt{\dfrac{\pi^2 E}{250}}} = \frac{\left(\dfrac{2.75 \times 10^3}{17.5}\right)}{1 \times \sqrt{\dfrac{\pi^2 \times 2 \times 10^5}{250}}} = 1.7685$$

$$\lambda_\phi = \frac{(b_1 + b_2)}{\varepsilon\sqrt{\dfrac{\pi^2 E}{250}} \times 2t} = \frac{(90 + 90)}{1 \times \sqrt{\dfrac{\pi^2 \times 2 \times 10^5}{250}} \times 2 \times 8} = 0.1266$$

(a) When the ends of member are fixed:

For which k_1 = 0.20; k_2 = 0.35; k_3 = 20 (As per Table 5.6)

$$\therefore \qquad \lambda_e = \sqrt{0.20 + 0.35 \times 1.7685^2 + 20 \times 0.1266^2} = 1.271$$

$\therefore$ Design compressive stress,

$$f_{cd} = \frac{\dfrac{f_y}{\gamma_{m0}}}{\phi + \sqrt{\phi^2 - \lambda_e^2}}$$

where, $\quad \phi = 0.5 \left[1 + \alpha (\lambda_e - 0.2) + \lambda_e^2\right]$

λ = Non-dimensional effective slenderness ratio = $\lambda_e = 1.271$

Imperfection factor $\alpha = 0.49 \qquad$ ... (for buckling class 'c' see Table 5.3)

$\therefore \qquad \phi = 0.5 [1 + \alpha (\lambda_e - 0.2) + \lambda_e^2]$

$\qquad = 0.5 [1 + 0.49 (1.271 - 0.2) + 1.271^2] = 1.57$

$\therefore \qquad f_{cd} = \dfrac{\dfrac{250}{1.1}}{1.57 + \sqrt{1.57^2 - 1.271^2}} = 91.20 \text{ N/mm}^2$

$\therefore \qquad$ Design compressive force $= f_{cd} \times A_g$

$\qquad = 91.20 \times 1379 = 125765 \text{ N} = \textbf{125.76 kN}$

(b) When the ends of member are hinged:

For this case $\lambda_{vv} = 1.7685$, $\lambda_\phi = 0.1266$ (already calculated)

$k_1 = 0.7$; $k_2 = 0.6$; $k_3 = 5 \qquad\qquad$ ... (As per Table 5.6)

$\therefore \qquad \lambda_e = \sqrt{0.70 + 0.6 \times 1.7685^2 + 5 \times 0.1266^2} = 1.63$

$\qquad \phi = 0.5 [1 + 0.49 (1.63 - 0.2) + 1.63^2] = 2.179$

$\therefore \qquad f_{cd} = \dfrac{\dfrac{250}{1.1}}{2.179 + \sqrt{2.179^2 - 1.63^2}} = 62.69 \text{ N/mm}^2$

$\therefore \qquad$ Design compressive force $= f_{cd} \times A_g = 62.79 \times 1379$

$\qquad = 86449.5 \text{ N} = \textbf{86.45 kN}$

Case (ii): When one bolt is provided at each end

Again $\lambda_{vv} = 1.7685$ and $\lambda_\phi = 0.1266$

(a) When the ends of member are fixed:

For this condition, $k_1 = 0.75$; $k_2 = 0.35$ and $k_3 = 20$.

$\qquad \lambda_e = \sqrt{0.75 + 0.35 \times 1.7685^2 + 20 \times 0.1266^2} = 1.471$

$\qquad \phi = 0.5 [1 + 0.49 (1.471 - 0.2) + 1.471^2] = 1.893$

$\therefore \qquad f_{cd} = \dfrac{\dfrac{250}{1.1}}{1.893 + \sqrt{1.893^2 - 1.471^2}} = 73.68 \text{ N/mm}^2$

$\therefore \qquad$ Design compressive force $= f_{cd} \times A_g$

$\qquad = 73.68 \times 1379 = 101604.72 \text{ N} = \textbf{101.60 kN}$

(b) When the ends of member are hinged

For this condition, $k_1 = 1.25$; $k_2 = 0.50$; $k_3 = 60$

$\qquad \lambda_e = \sqrt{1.25 + 0.50 \times 1.7685^2 + 60 \times 0.1266^2} = 1.943$

$\qquad \phi = 0.5 [1 + 0.49 (1.943 - 0.2) + 1.943^2] = 2.814$

$\therefore \qquad f_{cd} = \dfrac{\dfrac{250}{1.1}}{2.814 + \sqrt{2.814^2 - 1.943^2}} = 46.86 \text{ N/mm}^2$

$\therefore \qquad$ Design compressive force $= f_{cd} \times A_g$

$\qquad = 46.86 \times 1379 = 64620.72 \text{ N} = \textbf{64.62 kN}$

Case (iii): When the ends of member are welded

For welded member,

$\qquad$ Strength = Strength of member connected at each end with two bolts considering the ends as fixed

$\therefore \qquad$ Design compressive force = **125.76 kN**

Ex. 5.2: *A single angle discontinuous strut is to carry a service load of 100 kN, the length being 3.00 m. Minimum two bolts are to be used for end connections. Use 20 mm diameter bolts of 4.6 grade. Design the strut. Take f_y = 250 N/mm². Check the suitability of following section.*

(i) ISA 100 × 100 × 10; $A = 1903$ mm², $r_{min} = 19.4$ mm

(ii) ISA 110 × 110 × 8; $A = 1708$ mm², $r_{min} = 21.8$ mm

(iii) ISA 110 × 110 × 10; $A = 2106$ mm², $r_{min} = 21.4$ mm

Sol.:   Design compressive load $= \gamma_f \times$ Service load

$$P_u = 1.5 \times 100 = 150 \text{ kN}$$

Assuming $f_{cd} = 90$ N/mm² initially

$$A_{approx.} = \frac{P_u}{f_{cd}} = \frac{150 \times 10^3}{90} = 1667 \text{ mm}^2$$

Referring steel table, select ISA 110 × 110 × 8 mm (Slightly greater area)

For this angle $A = 1708$ mm², $r_{min} = r_{vv} = 21.8$ mm

Design shear strength of 20 mm diameter bolt in single shear

$$= \frac{f_{ub}}{\sqrt{3}\ \gamma_{mb}} A_{nb} = \frac{400}{\sqrt{3} \times 1.25} \times \left(0.78 \times \frac{\pi}{4} \times 20^2\right)$$

$$= 45272 \text{ N} = 45.27 \text{ kN}$$

Design bearing strength $= \dfrac{2.5\ k_b \cdot d \cdot t \cdot f_{ub}}{\gamma_{mb}} = \dfrac{2.5 \times 1 \times 20 \times 8 \times 400}{1.25}$ (Taking $k_b = 1$)

$$= 128000 = 128.00 \text{ kN}$$

∴   Design strength of one bolt $= 45.27$ kN (lesser value)

Number of bolts required at each end $= \dfrac{150}{45.27} = 3.31 \approx 4$

With bolts > 2 and assuming fixity at ends.

$k_1 = 0.2$; $k_2 = 0.35$ and $k_3 = 20$.

Imperfection factor $\alpha = 0.49$ ... (class c buckling)

$$\lambda_{vv} = \frac{\left(\dfrac{l}{r_{vv}}\right)}{\varepsilon\sqrt{\dfrac{\pi^2 E}{250}}} = \frac{\left(\dfrac{3000}{21.8}\right)}{1 \times \sqrt{\dfrac{\pi^2 \times 2 \times 10^5}{250}}} = 1.5487$$

$$\lambda_\phi = \frac{(b_1 + b_2)}{\varepsilon\sqrt{\dfrac{\pi^2 E}{250}} \times 2t} = \frac{(110 + 110)}{1 \times \sqrt{\dfrac{\pi^2 \times 2 \times 10^5}{250}} \times 2 \times 8} = 0.1547$$

$$\lambda_e = \sqrt{k_1 + k_2\,\lambda_{vv}^2 + k_3\,\lambda_\phi^2}$$

$$= \sqrt{0.2 + 0.35 \times 1.5487^2 + 20 \times 0.1547^2} = 1.232$$

$$\phi = 0.5\left[1 + \alpha\,(\lambda_e - 0.2) + \lambda_e^2\right]$$

$$= 0.5\left[1 + 0.49\,(1.232 - 0.2) + 1.232^2\right] = 1.512$$

∴ Design compressive stress,

$$f_{cd} = \frac{\dfrac{f_y}{\gamma_{m0}}}{\phi + \sqrt{\phi^2 - \lambda_e^2}}$$

∴ $$f_{cd} = \frac{\dfrac{250}{1.1}}{1.512 + \sqrt{1.512^2 - 1.232^2}} = 95.15 \text{ N/mm}^2$$

∴   Design compressive load $= f_{cd} \times A_g = 95.15 \times 1708 = 162516$ N

$$= \mathbf{162.51\ kN} > 150 \text{ kN}$$

Hence the design is safe.

Ex. 5.3: *Explain limit of width to thickness ratio of prevent buckling for a single angle struct, the limiting width to thickness ratio for a semi-compact class is 15.7 ϵ. Check whether ISA 90 $\times$ 90 $\times$ 6 mm is of semi-compact class or not f_y = 250 MPa.*

Sol.: For limits of width - Refer Section 4.12.

Compression element	Ratio
Single angle, or double	b/t
Angles with the components	d/t

ISA 90 $\times$ 90 $\times$ 6 mm

A_g = 1047 mm^2, b = 90, d = 90, t = 6 mm

$$\frac{b}{t} = \frac{90}{6} = 15 < 15.7\ \varepsilon$$

$$\frac{d}{t} = \frac{90}{6} = 15 < 15.7\ \varepsilon$$

The given section is in semi-compacted class.

Ex. 5.4: A strut 2.4 m long of a roof truss consists of a single angle 90 $\times$ 90 $\times$ 6 mm. Calculate load carrying capacity if it is connected to 8 mm thick gusset plate by welding.

Assume - Properties of ISA 90 $\times$ 90 $\times$ 6 mm, f_y = 250 N/mm^2

Area = 1047 mm^2, C_{xx} = C_{yy} = 2.42 mm

r_{xx} = r_{yy} = 27.7 mm, r_w = 17.5 mm

KL/r	80	90	100	110	120	130
f_{cd} (N/mm^2)	136	121	107	94.6	83.7	74.4

Sol.: Actual length (l) = 2.4 = 2400 mm

$\therefore$　Effective length (l) = KL = 0.7 $\times$ L　　　　for angle connected by weld

= 0.7 $\times$ 2400

L = 1680 mm

Least radius of gyration for single angle = γ_{min} = γ_{vv}

$\therefore$　r_{min} = r_{vv} = 17.5 mm

Now find slenderness ratio (λ)

$$\lambda = \frac{\text{Effective length}}{\gamma_{min}} = \frac{1680}{17.5}$$

$$\boxed{\lambda = 96}$$

Now find design compressive stress (f_{cd}) for buckling class c from table of slenderness ratio

S Y λ	F_{ccd} (N/mm^2)	By interpolation
90	121	$\dfrac{121 - 107}{100 - 90} \times (100 - 96) + 107$
96	2	
100	107	$\lambda = 112.6$ N/mm^2

Now find design strength of strut (P_d)

$$P_d = A_g \times f_{cd} = 1047 \times 112.6$$

$$\boxed{P_d = 117.89 \times 10^3 \text{ N}}$$

Ex. 5.5: *Design a single unequal angle strut connected by 6 mm fillet weld at the ends with its longer leg on gusset plate of 10 mm thickness. It carries a service load of 50 kN. The centre to centre distance between intersections is 3 m and f_y = 250 N/mm^2. Select the sections from following Table 5.8.*

Table 5.8

Section	Wt. N/m	Cross-sectional Area (mm^2)	r_{max} (mm)	r_{min} (mm)
ISA 90 × 60 × 6	68.0	865	30.7	12.8
ISA 100 × 65 × 6	75.0	955	34.0	13.9
ISA 100 × 75 × 6	80.0	1014	35.0	15.9
ISA 125 × 75 × 6	92.0	1166	42.3	16.2
ISA 125 × 95 × 6	101.0	1292	44.3	20.7

Sol.: Assume f_{cd} = 90 N/mm^2.

Factored load, $P_u = \gamma_f \cdot P = 1.5 \times 50 = 75$ kN

$$A_{approx.} = \frac{P_u}{f_{cd}} = \frac{75000}{90} = 833.33 \ mm^2$$

Referring steel stable, select ISA 100 × 75 × 6 mm having A = 1014 mm^2, $r_{min} = r_{vv}$ = 15.9 mm.

Effective length kL = 1 × 3 = 3 m = 3000 mm

(k = 1 since buckling may take place perpendicular to plane of truss on the outstanding leg).

For a member connected by weld at each end, the ends may be assumed as fixed at both ends. For which k_1 = 0.2, k_2 = 0.35 and k_3 = 20. Imperfection factor, α = 0.49 (class c buckling).

$$\lambda_{vv} = \frac{\dfrac{l}{r_{min}}}{\varepsilon\sqrt{\dfrac{\pi^2 E}{250}}} = \frac{\dfrac{3000}{15.9}}{1 \times \sqrt{\dfrac{\pi^2 \times 2 \times 10^5}{250}}} = 2.123$$

$$\lambda_\phi = \frac{\dfrac{b_1 + b_2}{2t}}{\varepsilon\sqrt{\dfrac{\pi^2 E}{250}}} = \frac{\dfrac{100 + 75}{(2 \times 6)}}{1 \times \sqrt{\dfrac{\pi^2 \times 2 \times 10^5}{250}}} = 0.164$$

$$\lambda_e = \sqrt{k_1 + k_2 \times \lambda_{vv}^2 + k_3\,\lambda_\phi^2}$$

$$= \sqrt{0.2 + 0.35 \times 2.123^2 + 20 \times 0.164^2} = 1.522$$

$$\phi = 0.5\left[1 + \alpha\,(\lambda_e - 0.2) + \lambda_e^2\right]$$

$$= 0.5\left[1 + 0.49\,(1.522 - 0.2) + 1.522^2\right] = 1.982$$

Design compressive stress,

$$f_{cd} = \frac{\dfrac{f_y}{\gamma_{m0}}}{\phi + \sqrt{\phi^2 - \lambda_e^2}}$$

$$\therefore \qquad f_{cd} = \frac{\dfrac{250}{1.1}}{1.982 + \sqrt{1.982^2 - 1.522^2}} = 69.89 \ N/mm^2$$

∴ Design compressive load,

$$P_d = f_{cd} \times A_g$$

$$P_d = 69.89 \times 1014 = 70874 \ N = \mathbf{70.87 \ kN} < 75 \ kN$$

Design is unsafe, hence revise the section,

Revised Design: Referring to steel table select ISA 125 × 75 × 6 mm having A = 1166 mm^2, $r_{min} = r_{vv}$ = 16.2 mm.

Adopting same data in earlier design l = kL = 3000 mm

k_1 = 0.2, k_2 = 0.35, k_3 = 20 and α = 0.49.

$$\lambda_{vv} = \frac{\dfrac{l}{r_{min}}}{\varepsilon\sqrt{\dfrac{\pi^2 E}{250}}} = \frac{\dfrac{3000}{16.2}}{1 \times \sqrt{\dfrac{\pi^2 \times 2 \times 10^5}{250}}} = 2.084$$

$$\lambda_\phi = \frac{b_1 + b_2}{\varepsilon \sqrt{\dfrac{\pi^2 E}{250}} \times 2t} = \frac{125 + 75}{1 \times \sqrt{\dfrac{\pi^2 \times 2 \times 10^5}{250}} \times (2 \times 6)} = 0.1876$$

$$\lambda_e = \sqrt{k_1 + k_2 \lambda_{vv}^2 + k_3 \lambda_\phi^2}$$
$$= \sqrt{0.2 + 0.35 \times 2.084^2 + 20 \times 0.1876^2} = 1.557$$

$$\phi = 0.5 \left[1 + \alpha (\lambda_e - 0.2) + \lambda_e^2\right]$$
$$= 0.5 \left[1 + 0.49 (1.557 - 0.2) + 1.557^2\right] = 2.044$$

∴ Design compression stress,

$$f_{cd} = \frac{\dfrac{f_y}{\gamma_{m0}}}{\phi + \sqrt{\phi^2 - \lambda_e^2}} = \frac{\dfrac{250}{1.10}}{2.044 + \sqrt{2.044^2 - 1.557^2}}$$

$$= 67.47 \text{ N/mm}^2$$

∴

$$p_d = f_{cd} \times A_g = 67.47 \times 1166 = 78670 \text{ N}$$
$$= \textbf{78.67 kN} > 75 \text{ kN}$$

Now design is safe hence ISA 125 × 75 × 6 mm can be adopted as a strut.

Ex. 5.6: *A strut 3.0 m long of a truss consists of 2 ISA 100 × 100 × 10 mm. Calculate the design strength of strut if it is bolted to 12 mm thick gusset plate on either sides by two rivets at each end.*

Take $r_{vv} = 19.4$ mm. Values of f_{cd} are as:

KL/r	130	140	150
f_{cd}(N/mm²)	74.4	66.2	59.2

Properties of ISA 100 × 100 × 10 mm are A = 19.03 cm², $I_{xx} = I_{yy} = 177.00$ cm⁴, $Z_{xx} = Z_{yy} = 24.7$ cm³.

Sol:

$$A = 1903 \text{ mm}^2$$
$$I_{xx} = I_{yy} = 177 \times 10^4 \text{ mm}^4$$
$$Z_{xx} = Z_{yy} = 24.7 \times 10^3 \text{ mm}^3$$

For a section calculate

Step 1:
$$I_{xx} = 2 \times 177 \times 10^4$$
$$= 3.54 \times 10^6 \text{ mm}^4$$

Step 2: ∴
$$r_{xx} = \sqrt{\frac{I_{xx}}{A_g}}$$
$$= \sqrt{\frac{3.54 \times 10^6}{2 \times 1903}} = 30.49 \text{ mm}$$

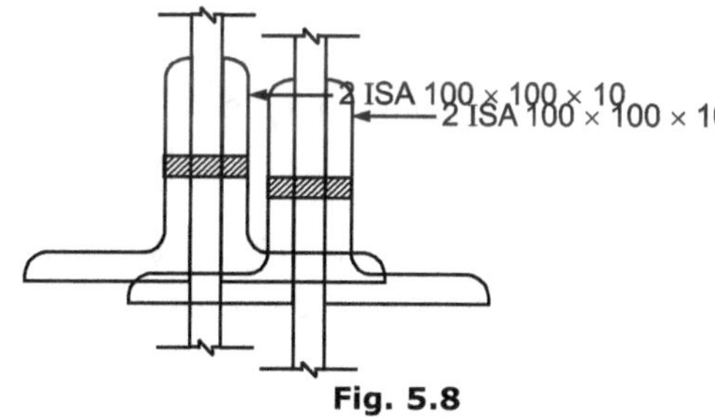

Fig. 5.8

∴ r_{xx} is minimum

Step 3:
$$SR = \frac{K_L}{r_{min}} = \frac{0.85 \times 3000}{30.49} = 80.631344 = 83.63$$

Step 4: (SR)(f_{cd}) by interpolation

	80	136
83.63		
	90	121

$$f_{cd} = 136 - \frac{15 \times 3.63}{10} = 130.55 \frac{N}{mm^2}$$

Step 5: Design compress load = $f_{cd} \times A_g$
$$P_d = f_{cd} \times A_g$$
$$f_{cd} = 130.557$$

Design compressive load $P_d = f_{cd} \times A_g = 130.573.21 \times (2 \times 1903)$

$$\boxed{P_d = 496.878163 \text{ kN}}$$

$$\boxed{P_d = 496.87 \text{ kN}}$$

Ex. 5.7: *In a truss 2 ISA 100 x 100 x 6 mm, 2.80 m long is used as a strut. It is connected to 10 mm thick gusset plate on either sides by two bolts at each end.*

Determine the load carrying capacity of the angle strut:

(i) If connected by bolts

(ii) If connected by weld.

Properties of ISA $100 \times 100 \times 6$ mm:

$\qquad$ A =1167 mm^2, $I_{xx} = I_{yy} = 111.3 \times 10^4$ mm^4

$C_{xx} = C_{yy} = 26.70$ mm

KL /r →	60	70	80	90	100
f_{cd} (N/mm^2) →	163	152	136	121	107

Sol.: **Given:** A = 1167 mm^2, $I_x = I_y = 111.3 \times 10^4$ mm^4

$\qquad$ $C_x = C_y = 26.70$ mm, L = 2.8 m = 2800 m

(i) With bolted connection (two bolts at each end)

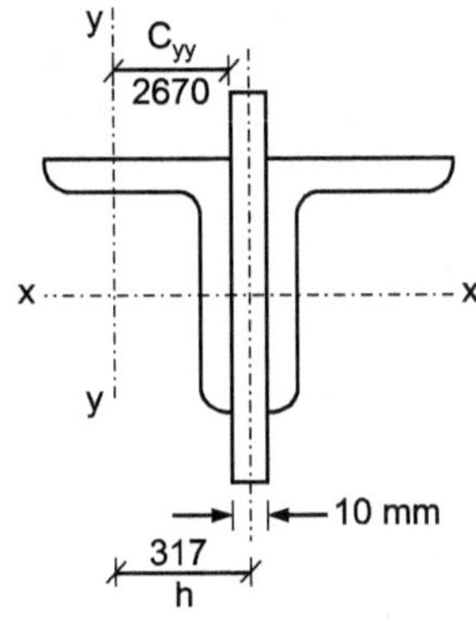

Fig. 5.9

$$I_{xx} \text{ of compound section } = [I_{xx} \text{ of individual}] \times 2 = 111.3 \times 10^4 \times 2$$
$$I_{xx} = 2.226 \times 10^6 \text{ mm}^4$$
$$I_{yy} = 2\,[I_{yy} + Ah^2] = 2\,[111.30 \times 10^4 + 1167 \times 31.7^2]$$
$$I_{yy} = 4.57 \times 10^6 \text{ mm}^4$$

As the angle are equal angles, with calculation we can say that $I_{min} = I_{xx}$

$\therefore \qquad r_{min} = \sqrt{\dfrac{I_{xx} \text{ (compound)}}{2A}} = \sqrt{\dfrac{2.226 \times 10^6}{2 \times 1167}}$

$\therefore \qquad r_{min} = 30.88$ mm

$\qquad$ Slenderness ratio $= \dfrac{kL}{r_{min}} = \dfrac{0.85 \times 2800}{30.88}$

$\qquad \lambda = 77.07$

For $\qquad \lambda = 77.07$ by interpolation f_{cd}

$$f_{cd} = 152 - \left[\frac{152 - 136}{80 - 70} \times (77.07 - 70)\right]$$

$\qquad I_{cd} = 140.70$ N/mm^2

$\because \qquad P_d = A \times f_{cd} = 2 \times 1167 \times 140.70 = 328393.8$ N

$\because \qquad p_d = 328.394$ kN

(b) For welded connection:

$\qquad$ Slenderness ratio $\qquad \lambda = \dfrac{kL}{r_{min}} = \dfrac{0.7 \times L}{30.88} = \dfrac{0.7 \times 2800}{30.88}$

$\qquad \lambda = 63.47$

For λ 63.67 by interpolation f_{cd}

$$f_{cd} = 163 - \left[\frac{163 - 152}{70 - 60} \times (63.47 - 60)\right]$$

$$f_{cd} = 159.18 \text{ N/mm}^2$$

$\therefore \quad P_d = A \times f_{cd} = 2 \times 1167 \times 159.18 = 371526.12 \text{ N} = 371.53 \text{ kN}$

$$\boxed{p_d = 371.53 \text{ kN}}$$

Ex. 5.8: *A discontinuous compression member consists of 2 ISA 90 × 90 × 10 mm connected back to back on opposite sides of 12 mm thick gusset plate and connected by welding. The length of strut is 3 m. It is welded on either side. Calculate design compressive strength of strut. For ISA 90 × 90 × 10, $C_{xx} = C_{yy} = 25.9$ mm, $I_{xx} = I_{yy} = 126.7 \times 10^4$ mm^4, $r_{zz} = 27.3$ mm values of f_{cd} are:*

KL/r	90	100	110	120
f_{ed} (N/mm^2)	121	107	94.6	83.7

Sol.: $r_{xx} = r_x = 27.3$ mm (Due to symmetry @ yy axis)

$$I_{yy} = 2\,[I_y + A \cdot h^2] = 2\left[126.7 \times 10^4 + 1703\left(25.9 + \frac{12}{2}\right)^2\right] = 5999979 \text{ mm}^4$$

$$r_{yy} = \sqrt{\frac{I_{yy}}{A_g}} = \sqrt{\frac{5999979}{2 \times 1703}} = 41.97 \text{ mm}$$

$\therefore \quad r_{min} = r_{xx} = 27.3$ mm

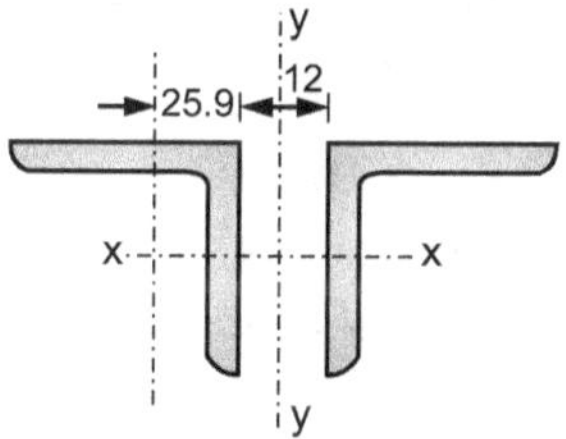

Fig. 5.10

For discontinuous double angle, effective length

$$kL = 0.85\,L = 0.85 \times 3 = 2.55 \text{ m} = 2550 \text{ mm}$$

$\therefore \quad$ S.R. $= \dfrac{kL}{r_{min}} = \dfrac{2550}{27.3} = 93.40$

From table 4.4, for buckling class

$\dfrac{kL}{r}$ (SR)	f_{cd}
90	121
100	107

Design compressive stress f_{cd} is interpolated as:

$$f_{cd} = f_{cd_1} - \frac{f_{cd_1} - f_{cd_2}}{SR_2 - SR_1}(SR - SR_1) = 121 - \frac{121 - 107}{100 - 90}(93.40 - 90)$$

$$= 116.24 \text{ N/mm}^2$$

$\therefore \quad$ Design compressive strength $P_d = f_{cd} \times A_g = 116.24 \times (2 \times 1703)$

$$= 395913 \text{ N} = \textbf{395.91 kN}$$

Ex. 5.9: *A discontinuous compression member consists of 2 ISA 90 × 90 × 10 mm connected back to back on opposite sides of 10 mm thick gusset plate. Tacking rivets are provided along the length along with one bolt at each end. Determine the design compressive strength of the member. The centre to centre distance of connection is 2.8 m. For single ISA 90 × 90 × 10 mm, A = 1703 mm^2, $\gamma_x = 27.3$ mm, $C_x = C_y = 25.9$ mm, $I_x = I_y = 12.67 \times 10^5$ mm^4.*

KL/γ	80	90	100	110	120	130
f_{cd} (MPa)	136	121	107	94.6	83.7	74.4

Sol.: Given: ISA 90 × 90 × 10 mm, A = 1703 mm^2, r_{xx} = 27.03 mm, L = 2800 mm, $c_x = c_y$ = 25.9 mm, $I_{xx} = I_{yy}$ = 12.67 × 10^5 mm^4.

1. $\qquad\qquad r_{min}$ = 27.03 min

2. $\qquad\qquad L_{eff}$ = 0.85 × 2800 = 2380

3. $\qquad\qquad \lambda = \dfrac{L_{eff}}{r_{min}} = \dfrac{2380}{27.03} = 87.17$

4. $\qquad\qquad \lambda_1 = 80 \quad f_{cd_1} = 136$

    $\qquad\qquad \lambda_2 = 90 \quad f_{cd_2} = 121$

To calculate the f_{cd} for λ by interpolation.

5. $\qquad\qquad f_{cd} = f_{cd_1} - \dfrac{f_{cd_1} - f_{cd_2}}{\lambda_2 - \lambda_1}(\lambda - \lambda_1)$

    $\qquad\qquad = 136 - \dfrac{136 - 121}{90 - 80}(87.17 - 80) = 125.24 \text{ N/mm}^2$

6. Design strength $\qquad P_d = f_{cd} \times A_g$

    $\qquad\qquad P_d = f_{cd} \times A_g = 125.24 \times 2A = 125.24 \times 1703 = 426.56 \text{ kN}$

Ex. 5.10: *A discontinuous double angle strut is composed of two ISA 90 × 60 × 8 mm connected to a gusset plate of 10 mm thickness. The length of the strut from centre to centre of fastenings is 3.00 m. Calculate its design strength when*

(A) the longer legs of angles are connected on either side of the gusset plate.

(B) when the longer legs of the angles are connected on the same side of the gusset plate.

Sol.: (A) When the longer legs are connected on either side of gusset

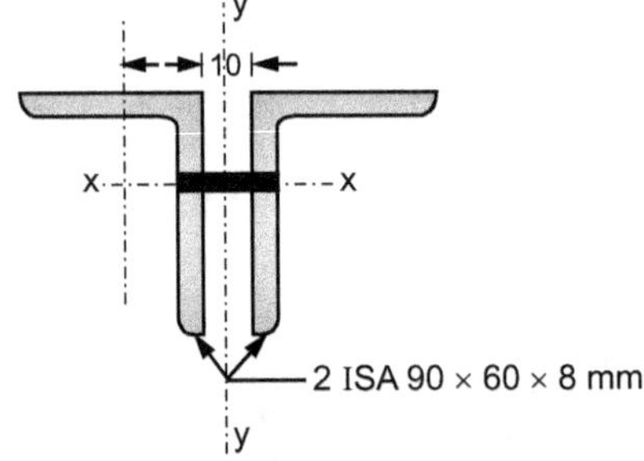

Fig. 5.11

From steel table properties of single ISA 90 × 60 × 8 mm are:

A = 1137 mm^2, r_x = 28.4 mm, I_x = 91.5 × 10^4, C_x = 29.6 mm, I_y = 32.4 × 10^4, C_y = 14.8 mm.

$\qquad\qquad r_{xx} = r_x$ = 28.4 mm

$\qquad\qquad I_{yy} = 2\,[I_y + A \cdot h^2]$

$\qquad\qquad\qquad = 2\left[32.4 \times 10^4 + 1137\left(14.8 + \dfrac{10}{2}\right)^2\right] = 1539499 \text{ mm}^4$

$\qquad\qquad r_{yy} = \sqrt{\dfrac{I_{yy}}{A_g}} = \sqrt{\dfrac{1539499}{2 \times 1137}} = 26.02 \text{ mm}$

$\qquad\qquad \text{S.R.} = \dfrac{kL}{r_{min}} = \dfrac{0.85 \times 3000}{26.02} = 98$

For buckling class c, from Table 5.4

$SR = \dfrac{kL}{r}$	f_{cd}
90	121
100	107

Design compressive stress f_{cd} is interpolated as:

$$f_{cd} = f_{cd_1} - \frac{f_{cd_1} - f_{cd_2}}{SR_2 - SR_1} (SR - SR_1)$$

$$= 121 - \frac{121 - 107}{100 - 90} (98 - 90) = 109.8 \text{ N/mm}^2$$

$\therefore$ Design compressive load $P_d = f_{cd} \times A_g = 109.8 \times (2 \times 1137)$

$$= 249685 \text{ N} = \textbf{249.68 kN}$$

(B) When the longer legs are connected on same side of gusset

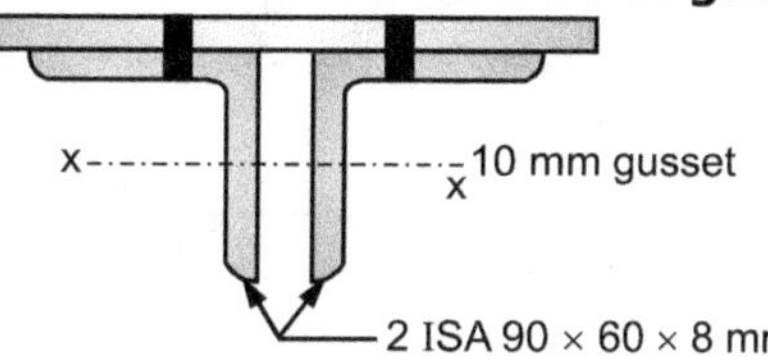

Fig. 5.12

Here x direction of single angle becomes y and y becomes x.

$$r_{xx} = r_y = 16.9 \text{ mm}$$
$$I_{yy} = 2 [I_y + A \cdot h^2]$$
$$= 2 [91.5 \times 10^4 + 1137 (29.6)^2] = 3822388 \text{ mm}^4$$

$\therefore$
$$r_{yy} = \sqrt{\frac{I_{yy}}{A_g}} = \sqrt{\frac{3822388}{2 \times 1137}} = 41 \text{ mm}$$

$$S.R. = \frac{kL}{r_{min}} = \frac{0.85 \times 3000}{16.9} = 150.89$$

For buckling class c and above value of SR.

$SR = \dfrac{kL}{r}$	f_{cd}
150	59.2
160	53.3

$$f_{cd} = f_{cd_1} - \frac{f_{cd_1} - f_{cd_2}}{SR_2 - SR_1} (SR - SR_1)$$

$$= 59.2 - \frac{59.2 - 53.3}{160 - 150} (150.89 - 150) = 58.67 \text{ N/mm}^2$$

$\therefore$ Design compressive load $P_d = f_{cd} \times A_g$

$\therefore$
$$P_d = 58.67 \times (2 \times 1137)$$

$$= 133415 \text{ N} = \textbf{133.41 kN}$$

On comparing cases (A) and (B) above it is concluded that the longer legs of double angles shall be connected on either side of gusset plate as it gives more design strength.

Important Points

- A compression member is a structural member which is straight and subjected to two equal and opposite compressive forces applied at its ends.
- The members of a roof truss are generally composed of single or double angle sections and assumed to be acted upon by axial compression.
- The vertical members in a building carrying direct compression are called columns, stanchions or posts.
- The types of section used as compression member are single/double angles, channels or I-sections with or without cover plates.
- Effective length is defined as that length of column for which it acts as if both the ends are hinged.
- The least radius of gyration is considered to calculate effective slenderness ratio and is given by

$$r_{min} = \sqrt{\frac{I_{min}}{A_g}}$$

- The effective slenderness ratio is given as

$$\text{S.R.} = \frac{KL}{r_{min}}$$

  where, value of K depends on rotational and translational boundary conditions at the ends.

- The design compressive strength f_{cd} depends on buckling class of section and grade of steel (f_y) and effective slenderness ratio.

- For single angle loaded through one leg

$$\lambda_e = \sqrt{k_1 + k_2\,\lambda_{vv}^2 + k_3\,\lambda_\phi^2}$$

  where, k_1, k_2, k_3 = Constants depending upon end condition as given in Table 5.6.

$$\lambda_{vv} = \frac{\dfrac{l}{r_{vv}}}{\varepsilon\sqrt{\dfrac{\pi^2 E}{250}}} \quad \text{and} \quad \lambda_\phi = \frac{b_1 + b_2}{\varepsilon\sqrt{\dfrac{\pi^2 E}{250}} \times (2t)}$$

  For Fe 410, $f_y = 250$.

$$\therefore \quad \varepsilon = \sqrt{\frac{250}{f_y}} = \sqrt{\frac{250}{250}} = 1$$

- Generally angle sections fall under buckling class c for which imperfection factor $\alpha = 0.49$.

- Design compressive stress,

$$f_{cd} = \frac{\dfrac{f_y}{\gamma_{m0}}}{\phi + \sqrt{\phi^2 - \lambda_e^2}}$$

  where, $\phi = 0.5\left[1 + \alpha\,(\lambda_e - 0.2) + \lambda_e^2\right]$

  Design strength $P_d = f_{cd} \times A_e$ or $f_{cd} \times A_g$

- Welded end connections are equivalent to bolts > 2 and fixed ends.
- Stanchions or columns may be made of simple or built-up sections.
- To prevent local buckling of the sections limiting width to thickness ratio shall be followed as per Table 5.6.
- It is essential in case of built-up sections that their components shall be laced or battened so that they will act as a single unit.

Practice Questions

1. Define the effective length and slenderness ratio for the column.
2. State the effective length for column having one end fixed and other free.
3. State maximum slenderness ratio for a member carrying compressive loads resulting from D.L. and L.L.
4. State effective length for a column 3 m long having both ends fixed.
5. Define slenderness ratio and radius of gyration. Why is the radius of gyration taken minimum?
6. State any two end conditions of column and their equivalent length.
7. Sketch the factors on which permissible compressive stress depends.
8. When lacing is provided for column? Give the function of lacing.
9. Why lacings are used? How much load is taken by lacings?
10. State step-by-step procedure of designing a built-up column. Also state the effective length of columns in terms of actual length for any four end conditions. Sketch the buckled shape of columns for such end conditions.
11. Sketch the four end conditions of column showing their effective length.
12. Draw and label any four form of built-up compression.
13. State the function of lacing and battening. Draw neat sketches of single lacing and battening.
14. Draw neat sketches of single and double lacing system. State its purpose.
15. Write any four generation requirements for lacing.

16. Define effective length of column. Also state the effective length of column having translation and rotation restrained at one end and translation free but rotation restrained at other end along with sketch showing end conditions and effective length.

17. What is local buckling in case of compression member? What is its effect? What is to be done to prevent it?

18. A double angle discontinuous strut is made of ISA 125 × 95 × 10 mm with long legs connected back to back on both the sides of gusset plate 10 mm thick with two bolts. The length of the strut between centre to centre of intersections is 4 m. Determine the design load carrying capacity of the section.

19. In Ex. 12 above, if double discontinuous strut is connected to one side of a gusset, determine design load carrying capacity.

20. Calculate the design compressive load carried by a double angle discontinuous strut composed of 2 ISA 80 × 50 × 8 mm placed back to back and connected on both sides of the gusset plate 8 mm thick by two bolts. The actual length of the strut is 3 m.

    For ISA 80 × 50 × 8 mm,
    $$A = 978 \text{ mm}^2$$
    $$C_x = 27.3 \text{ mm}$$
    $$C_y = 12.4 \text{ mm}$$
    $$I_x = 0.38 \times 10^4 \text{ mm}^4$$
    $$I_y = 61.9 \times 10^4 \text{ mm}^4$$

21. A single angle discontinuous strut ISA 75 × 75 × 8 mm of a roof truss is 1.2 m long. It is connected by two bolts at each end. Determine the design load this strut can carry.

Table 5.9

Slenderness ratio	40	50	60	70	80	90	100	110
f_{cd} (MPa)	198	183	168	152	136	121	107	94.6

Table 5.10

Section	Wt/m. (N/m)	Cross-sectional area (mm²)	$r_{max.}$	$r_{min.}$	C_x (mm)
75 × 75 × 8	89	1140	28.80	14.50	21.40

[Ans. P_{cd} = 127.32 kN]

22. A compression member is build using two channel sections ISLC 350@ 389 N/m placed back-to-back. The unsupported length is 8 m. It is effectively held in position at both ends and restrained against rotation at one end. Determine the design compressive load the member can carry. Also determine the distance between these two channels for $I_{xx} = I_{yy}$. For ISLC 350 @389 N/m,
    $$A = 49.50 \text{ cm}^2$$
    $$b_f = 100 \text{ mm}$$
    $$C_{yy} = 2.42 \text{ cm}$$
    $$I_{xx} = 9330 \text{ cm}^4$$
    $$I_{yy} = 396 \text{ cm}^4$$

23. An ISA 200 × 100 × 10 mm has been used as a discontinuous angle strut. Length of the strut upto centres of the holes at the ends is 3.21 m. Calculate the design load bearing capacity of the strut in following case.

    Case: There are two bolts at each end.

    For ISA 200 × 100 × 10 mm → A = 2921 mm², $r_{min.}$ = 21.70 mm.

Slenderness ratio, λ	10	20	30	40	100	110	120	130	140	150	160
f_{cd} (MPa)	227	224	211	198	107	94.6	83.7	74.3	66.2	59.2	53.3

24. A column 5 m long is to support a service load of 4.5 MN. The ends of the column are effectively held in position and direction. Design the column in rolled steel beam sections and 18 mm plates are only available. Take yield stress in steel as 260 MPa.

Table 5.11

Section	Wt/m. (N/m)	Cross-sectional area (mm^2)	r_{xx} (mm)	r_{yy} (mm)
ISLB 550	863	10997	219.9	34.8
ISLB 600	995	12669	239.8	37.9
ISMB 500	869	11100	202.0	35.2
ISWB 450	794	10115	186.3	41.1
ISWB 500	952	12122	207.7	49.6

Table 5.12

Slenderness ratio, (λ)	10	20	30	40	100	110	120	130	140	150	160
f_{cd} (MPa)	227	224	211	198	107	94.6	83.7	74.3	66.2	59.2	53.3

25. A double angle discontinuous strut 2 ISA 100 × 100 × 8 mm is connected to both sides of a gusset plate of 10 mm thickness by two bolts. The length of the strut is 4 m. Determine the design load carrying capacity of the strut.

For one ISA,

$$A \ = \ 1539 \text{ mm}^2$$

$$C_{xx} = C_{yy} \ = \ 27.1 \text{ mm}$$

$$I_{xx} = I_{yy} \ = \ 1.45 \times 10^6 \text{ mm}^4$$

Tacking bolts are used.

Slenderness ratio	30	40	50	60	70	80	90	100	110	120
f_{cd} (N/mm^2)	211	198	183	168	152	136	121	107	94.6	83.7

26. A strut in a roof truss 2.2 m long is of 65 × 65 × 6 mm, double angle connected to a 12 mm thick gusset plate on same side of it. It is connected by two bolts at each end. Calculate the design load the strut can carry. For ISA 65 × 65 × 6 mm,

$$A \ = \ 744 \text{ mm}^2$$

$$I_{xx} \ = \ I_{yy} = 29.1 \times 10^4 \text{ mm}^4$$

$$C_{xx} \ = \ C_{yy} = 18.1 \text{ mm}$$

27. A column consists of 2 ISMC 200 placed back to back at clear spacing of 150 mm. Calculate its design load carrying capacity if it's actual length is 5 m with one end fixed and other hinge.

The properties of ISMC 200 are:

$$A \ = \ 2821 \text{ mm}^2$$

$$I_{xx} \ = \ 18.193 \times 10^6 \text{ mm}^4$$

$$I_{yy} \ = \ 1.403 \times 10^6 \text{ mm}^4$$

$$C_{yy} \ = \ 21.7 \text{ mm}$$

Slenderness ratio	10	20	30	40	50	60	70	80
f_{cd} (MPa)	227	224	211	198	183	168	152	136

28. Design a single angle strut to carry a service load of 100 kN. The effective length of member is 2.70 m. The ends of strut connected by weld.

29. Design a single angle strut for a roof truss connected by two bolts in one leg subjected to an axial service load of 130 kN. The centre to centre distance between intersections is 3.0 m. Check the section for wind forces and outstanding leg. Permissible stresses (f_{cd}) and slenderness ratio (λ) for f_y = 250 MPa are:

λ (S.R.)	90	100	110	120	130	140	150	160
f_{cd} (MPa)	121	107	94.6	83.7	174.3	66.2	59.2	53.3

Table 5.13

Angle	Wt. N/m	C/s area mm^2	r_{max} mm	r_{min} mm
ISA 100 × 100 × 8	121	1539	38.8	19.5
ISA 110 × 110 × 8	134	1708	42.8	21.8
ISA 110 × 110 × 10	166	2112	42.5	21.6
ISA 130 × 130 × 8	159	2028	51.0	25.9

30. Two ISMC 300 placed back to back at 100 mm to act as a column. The length of column is 5 m effectively held in position and restrained against rotation at both ends. Calculate its design load carrying capacity. Refer Table 5.17.

Table 5.14: Properties of ISMC - 300

$$A = 4564 \text{ mm}^2$$
$$r_{xx} = 118.1 \text{ mm}$$
$$r_{yy} = 26.1 \text{ mm}$$
$$C_{yy} = 23.6 \text{ mm}$$
$$I_{xx} = 6362.6 \times 10^4 \text{ mm}^4$$
$$I_{yy} = 310.8 \times 10^4 \text{ mm}^4$$

31. Design a column section to support a service load of 1000 kN. The section consists of four equal angles, the overall dimensions of the section being 240 mm × 240 mm. The column has an effective length of 4 m. Use f_y 250 steel. Refer Table 5.18.

Table 5.15

Angle	Area (mm^2)	I_{xx} (mm)	C_{xx} (mm)
100 × 100 × 10	1903	177 × 10^4	28.4
110 × 110 × 8	1708	196 × 10^4	30
90 × 90 × 8	1379	104.2 × 10^4	25.1

32. A column consists of four angles 100 mm × 100 mm × 10 mm forming square of 300 mm × 300 mm. Find its design strength if effective length is 5.0 m.

For ISA 100 × 100 × 100

$$A = 1903 \text{ mm}^2$$
$$r_{min} = 19.4 \text{ mm}^2$$

33. A column consists of ISLB - 300 and two cover plates 200 × 10 mm connected to each flange. The effective length of column is 6.0 m. Find the design strength of column.

For ISLB 300

$$A = 4808 \text{ mm}^2$$
$$I_{xx} = 7332.9 \times 10^4 \text{ mm}^4$$
$$I_{yy} = 376.2 \times 10^4 \text{ mm}^2$$

■■■

DESIGN OF COLUMN BASES BY L.S.M.

- Slab Base and Gusseted Base - Code Provisions (IS:800-2007) - Minimum Thickness and Effective Area of Base Plate - Design of Slab Base for Axially Loaded Columns using Bolts/Riveted/Welds. Introduction to Gusseted Base (No Numerical Problems on Gusseted Base).

About this Chapter

After reading this chapter students can understand:

- Types of Column Bases
- Slab Base
- Gusseted Base
- Design of Slab Base and Gusseted Base

6.1 INTRODUCTION

- Beams transfer the load to the column and columns transfer its load to the foundation bed evenly over a large area, so that stresses induced in foundation beds are within permissible limit.
- The column bases are used to rest the columns on the concrete bed of the foundation.
- The column load should be applied on sufficient area of concrete foundation so that bearing strength of concrete is not exceeded.
- A steel base plate is therefore provided to distribute the column load on sufficient area of concrete foundation.
- Column bases rest on the concrete slab, which further transfer the load on the soil.
- The area of concrete slab is so designed, that the stresses induced in the soil should not exceed the bearing capacity of the soil.
- The column bases and the holding down bolts connecting them to the concrete foundation, are to be designed so that stability, stiffness and the strength of the foundation is achieved.

6.2 TYPES OF COLUMN BASES

- Three types of column bases are usually used.
  1. Slab base (for light columns loads say < 2060 kN)
  2. Gusseted base (for large column loads or when column is subjected to BM along with axial load).
  3. Grillage foundation (for foundation on soils having very low bearing capacity).

6.3 SLAB BASE

- It consists of a base plate underneath a column end which is machined so as to have a complete bearing on the plate.
- The column is properly secured to the base plate by means of fastenings as shown in Fig. 6.2.
- Fastenings are simply used to secure it with the base plate and secondly to resist all moments and forces due to transit, unloading and erection.
- Those are not designed to resist the direct compression in the column.

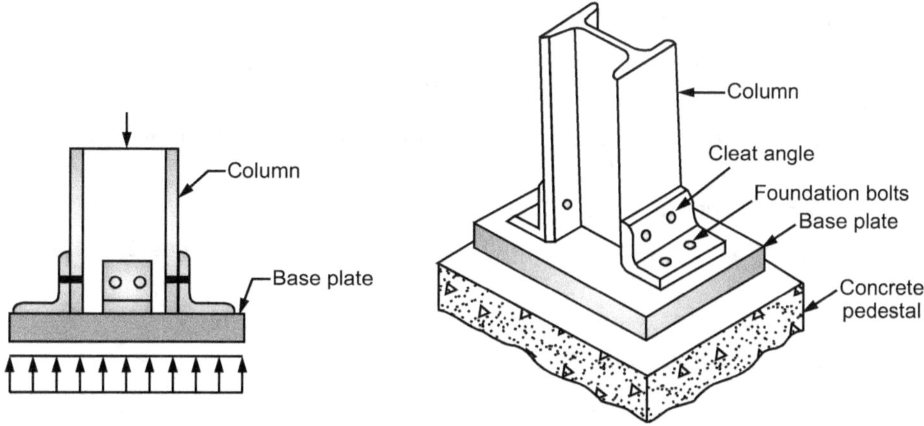

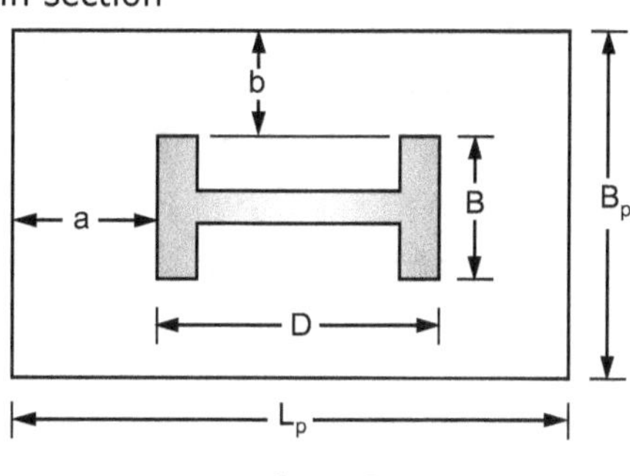

Fig. 6.1: Assumed Pressure Distribution   **Fig. 6.2: Slab Base**

Under Base Plate

6.4 DESIGN OF A SLAB BASE AND CONCRETE BLOCK

- Following steps are to be followed when axial load to which the column is subjected is known.

1. Calculate the bearing area (A) of the base plate:

$$\text{Bearing area} = \frac{\text{Factored load on the column}}{\text{Design bearing strength of concrete}}$$

$$A = \frac{P_u}{0.6\, f_{ck}}$$

where, f_{ck} = Characteristic strength of concrete

2. Select the size of base plate:

Let L_p and B_p be the sizes of base plate

$$D = \text{Length of longer side of column section}$$

$$B = \text{Width of shorter side of column section}$$

For equal projections (a and b),

$$L_p = \frac{D - B}{2} + \sqrt{\left(\frac{D - B}{2}\right)^2 + A}$$

$$\left(\begin{array}{c}\text{Round-off the value}\\ \text{in multiples of 10 on higher side}\end{array}\right)$$

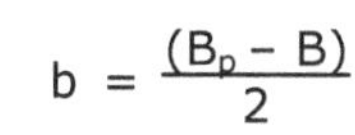

$$\therefore \quad B_p = \frac{A}{L_p}$$

Larger projection, $a = \dfrac{(L_p - D)}{2}$

Fig. 6.3

Smaller projection, $b = \dfrac{(B_p - B)}{2}$

$\therefore$ Area of base plate provided,

$$A_p = L_p \times B_p = (D + 2a) \times (B + 2b) \geq A$$

3. Calculate the ultimate pressure from below on the slab base:

$$w = \frac{P_u}{L_p \times B_p}$$

4. Calculate thickness of base plate:

$$t_s = \sqrt{\frac{2.5\, w\, (a^2 - 0.3\, b^2)\, \gamma_{mo}}{f_y}} \; > \; t_f$$

where, w = Uniform pressure from below on the slab under the factored axial compressive force

        t_f = Flange thickness of compression member

Secure the column section with base plate by providing cleat angles 2 ISA 100 × 100 × 10 mm.

Provide 4-20 mm diameter holding down bolts one at each corner.

5. Calculate the size of concrete block or pedestal

$$A_f = \frac{P_u \cdot \gamma_{mo}}{SBC \cdot \gamma_f} \qquad \begin{cases} \gamma_{mo} = 1.1 \\ \gamma_f = 1.5 \end{cases}$$

where, A_f = Area of foundation block

    SBC = Safe bearing capacity

      P_u = Factored axial load

For equal projection,

$$L_f = \frac{L_p - B_p}{2} + \sqrt{\left(\frac{L_p - B_p}{2}\right)^2 + A_f}$$

$$B_f = \frac{A_f}{L_f} \quad \text{(Consider } L_p, B_p \text{ in m)}$$

SOLVED EXAMPLES

Ex. 6.1: *A column ISMB 300 carries an axial load of 1200 kN. Design a slab base and concrete pedestal for the column. The SBC of the soil is 180 kN/m² and M20 grade concrete is used for concrete pedestal.*

Sol.: Size of ISMB 300, h = 300 mm, b_f = 140 mm, t_f = 13.1 mm.

∴   Longer side, D = 300 mm

     Shorter side, B = 140 mm

            f_{ck} = 20 N/mm²

       SBC = 180 kN/m²

Factored load, $P_u = \gamma_f \cdot P = 1.5 \times 1200 = 1800$ kN

1. Bearing area of base plate

$$A = \frac{P_u}{0.6\, f_{ck}} = \frac{1800 \times 10^3}{0.6 \times 20} = 1{,}50{,}000 \text{ mm}^2$$

2. Size of base plate:

For equal projections (a and b)

$$L_p = \frac{D - B}{2} + \sqrt{\left(\frac{D - B}{2}\right)^2 + A} = \frac{300 - 140}{2} + \sqrt{\left(\frac{300 - 140}{2}\right)^2 + 150000}$$

$$L_p = 475.47 \text{ mm} \quad \text{say } 480 \text{ mm}$$

$$B_p = \frac{A}{L_p} = \frac{150000}{480} = 312.5 \text{ mm say } 320 \text{ mm}$$

Larger projection, $a = \dfrac{(L_p - D)}{2} = \dfrac{480 - 300}{2} = 90$ mm

Smaller projection, $b = \dfrac{(B_p - B)}{2} = \dfrac{320 - 140}{2} = 90$ mm

∴   Area of base plate provided,

$$A_p = L_p \times B_p$$

$$= 480 \times 320 = 153600 \text{ mm}$$

3. Ultimate pressure from below on the slab base:

$$w = \frac{P_u}{L_p \times B_p}$$

$$= \frac{1800 \times 10^3}{480 \times 320}$$

$$= 11.72 \text{ N/mm}^2$$

4. Thickness of base plate:

$$t_s = \sqrt{\frac{2.5\, w\, (a^2 - 0.3\, b^2)\, \gamma_{mo}}{f_y}}$$

$$= \sqrt{\frac{2.5 \times 11.72\, (90^2 - 0.3 \times 90^2) \times 1.1}{250}}$$

$$= 27.03 \text{ mm say 28 mm}$$

$$> t_f\ (= 13.1 \text{ mm})$$

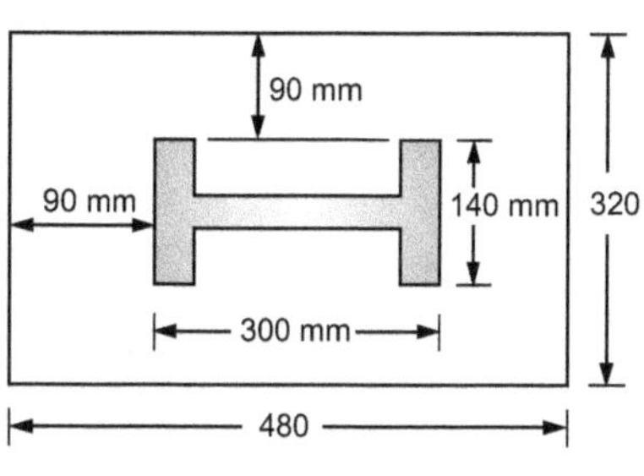

Fig. 6.4

$$\left(\begin{array}{l} \text{See table B-4 in appendix B} \\ \text{for available thickness of plate} \end{array} \right)$$

∴ Provide size of base plate as 480 × 320 × 28 mm. Secure the column section with base plate by providing cleat angles 2 ISA 100 × 100 × 10 mm along with 4-20 mm diameter holding down bolts, one at each corner.

5. Size of concrete block:

$$A_f = \frac{P_u \cdot \gamma_{mo}}{SBC \cdot \gamma_f} = \frac{1800 \times 1.1}{180 \times 1.5} = 7.33 \text{ m}^2$$

For equal projections,

$$L_f = \frac{L_p - B_p}{2} + \sqrt{\left(\frac{L_p - B_p}{2}\right)^2 + A_f}$$

$$= \sqrt{\frac{2.5 \times 11.72\, (90^2 - 0.3 \times 90^2) \times 1.1}{250}}$$

$$= 27.03 \text{ mm} = \frac{0.48 - 0.32}{2} + \sqrt{\left(\frac{0.48 - 0.32}{2}\right)^2 + 7.33}$$

$$= 2.788 \text{ m}$$

$$\therefore\quad B_f = \frac{A_f}{L_f} = \frac{7.33}{2.80} = 2.62 \text{ say 2.65 m}$$

Provide M 20 concrete pedestal of size 2.80 m × 2.65 m.

Actual projections:

(i) $\quad \dfrac{L_f - L_p}{2} = \dfrac{2800 - 480}{2}$

$$= 1160 \text{ mm}$$

(ii) $\quad \dfrac{B_f - B_p}{2} = \dfrac{2650 - 320}{2}$

$$= 1165 \text{ mm}$$

∴ Considering 45° angle of dispersion, provide depth of concrete block D_f = 1.165 m say 1.20 m.

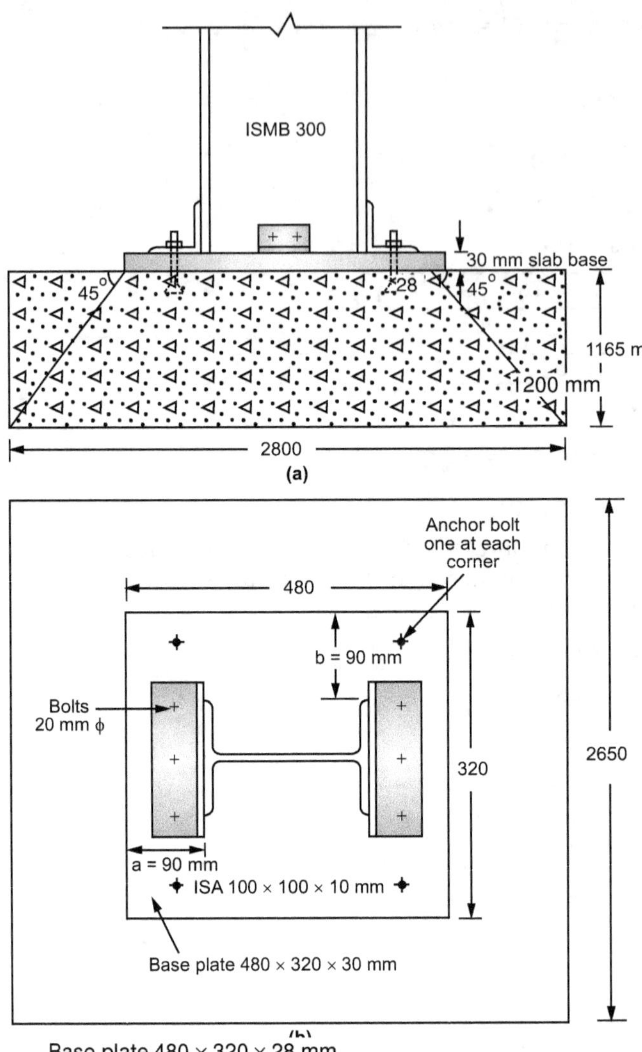

Base plate 480 × 320 × 28 mm

Fig. 6.5

Ex. 6.2: *Design a square slab base and concrete block for column section of SC 250 with two cover plates 300 mm × 25 mm carrying an axial load of 3000 kN. The SBC of soil is 300 kPa and grade of concrete is M15. Draw neat sketch showing plan and elevation for designed slab base with suitable cleat angles.*

Sol.: Arrange the given column section as shown in Fig. 6.6.

$h = 300$ mm, $b_f = 250$, $t_f = 17.0$ mm, $f_{ck} = 15$ N/mm^2,

SBC $= 300$ kPa $= 300$ kN/m^2,

Factored load $P_u = P \times \gamma_f = 3000 \times 1.5 = 4500$ kN,

$D = 300$ mm, $B = 250 + 2 \times 25 = 300$ mm.

1. Bearing area of base plate:

$$A = \frac{P_u}{0.6\, f_{ck}}$$

$$= \frac{4500 \times 10^3}{0.6 \times 15}$$

$$= 5,00,000 \text{ mm}^2$$

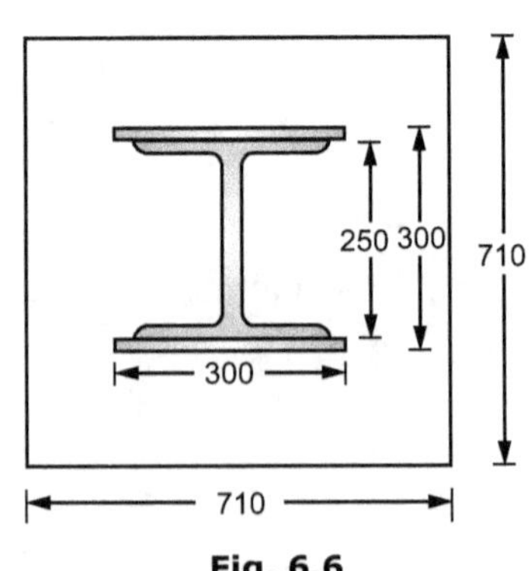

Fig. 6.6

2. Size of base plate:

As both the dimensions of built-up column D and B are equal, projections on both sides will be equal.

$$\therefore \qquad L_p = B_p = \sqrt{A} = \sqrt{500000} = 707 \text{ mm say } 710 \text{ mm}$$

Larger projection = Smaller projection

$$= \frac{L_p - D}{2}$$

$$= \frac{710 - 300}{2}$$

$$= 205 \text{ mm}$$

$\therefore$ Area of base plate provided,

$$A_p = L_p \times B_p = 710 \times 710 = 504100 \text{ mm}^2 > A$$

3. Ultimate pressure from below on the slab base:

$$w = \frac{P_u}{A_p} = \frac{4500 \times 10^3}{504100} = 8.93 \text{ N/mm}^2$$

4. Thickness of base plate:

$$t_s = \sqrt{\frac{2.5 \, w \, (a^2 - 0.3 \, b^2) \, \gamma_{mo}}{f_y}}$$

$$= \sqrt{\frac{2.5 \times 8.93 \, (205^2 - 0.3 \times 205^2) \times 1.1}{250}}$$

$$= 53.75 \text{ mm say } 56 \text{ mm (See table B-4 in Appendix B)}$$

$$> t_f \, (= 17.0 \text{ mm})$$

Provide square size base plate 710 × 710 × 56 mm. Secure the column section with base plate by providing cleat angles 2 ISA 100 × 100 × 10 mm alongwith 4-20 mm diameter holding down bolts, one at each corner.

5. Size of concrete block

$$A_f = \frac{P_u \cdot \gamma_{mo}}{SBC \times \gamma_f}$$

$$= \frac{4500 \times 1.1}{300 \times 1.5}$$

$$= 11 \text{ m}^2$$

$$L_f = B_f = \sqrt{A_f} = \sqrt{11} = 3.31 \text{ m say } 3.40 \text{ m.}$$

Considering 45° angle of dispersion, depth of concrete block

$$D_f = \frac{L_f - L}{2}$$

$$= \frac{3.4 - 0.71}{2}$$

$$= 1.345 \text{ m} \qquad \text{say } 1.35 \text{ m}$$

$\therefore$ Providing M15 grade concrete pedestal of size 3.40 m × 3.40 m × 1.35 m as shown in Fig. 6.7.

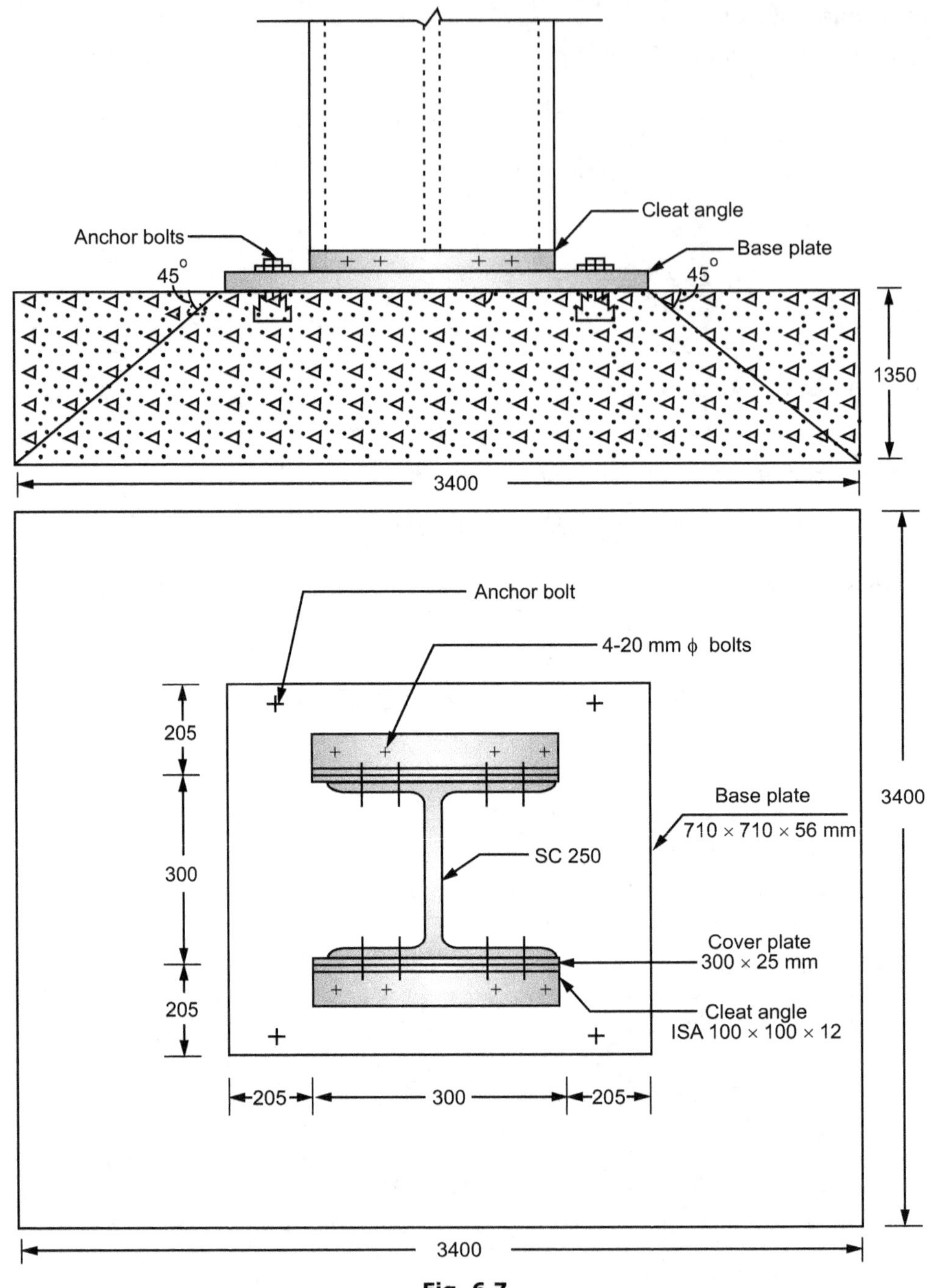

Fig. 6.7

Ex. 6.3: *A column ISMB 300 carries an axial load of 1.58 MN. Design a slab base and concrete pedestal for the column. Take the SBC of soil is 200 KPa and M20 grade of concrete is used for concrete pedestal. For ISMB 300, consider b_f = 140 mm and t_f = 13.1 mm. Take f_y = 250 MPa and γ_{mo} = 1.10.*

Sol.:

$$\text{Axial load} = 1.58 \text{ MN} = 1580 \text{ kN}$$

$$\text{Ultimate load} = 1.5 \times 1580 = 2370 \text{ kN}$$

$$\text{Bearing area of base plate } A = \frac{P_u}{0.6 \, f_{ck}} = \frac{P_u}{0.6 \, f_{ck}}$$

$$A = \frac{2370 \times 10^3}{0.6 \times 20} = 197.500 \text{ mm}^2$$

$$B \times D = 197500 \qquad \dots \text{(i)}$$

$$a = b$$

$$\frac{B - 140}{2} = \frac{D - 300}{2}$$

$$B = D - 300 + 140$$

$$B = D - 160 \qquad \dots \text{(ii)}$$

$$(D - 160)\, D = 197500$$

$$D^2 - 160\, D - 197500 = 0$$

$$\therefore \qquad D = 531.5 \text{ mn}$$

$$\text{Adopt } D = 540 \text{ mm}$$

$$\therefore \qquad B = 380 \text{ mm}$$

$$\text{and } a = b = 120 \text{ mm}$$

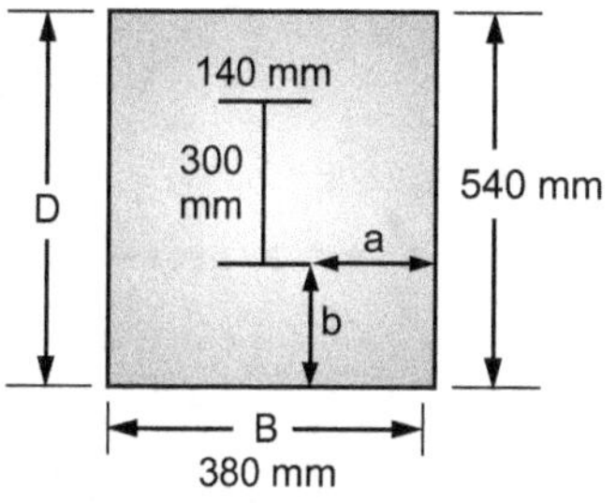

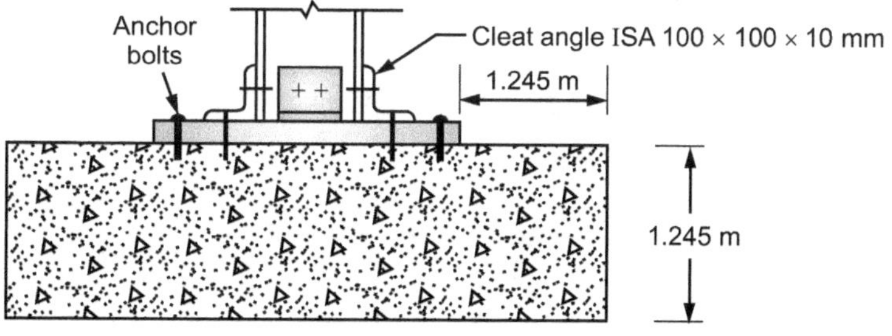

Fig. 6.8

$$w = \frac{2370 \times 10^3}{540 \times 330}$$

$$= 11.55 \text{ N/mm}^2$$

$$t = \sqrt{\frac{2.5\, w\, (a^1 \times 0.3\, b^2)}{f_y}}$$

Thickness of plate

$$t = \sqrt{\frac{2.5 \times 11.55\, (120^2 - 0.3 \times 120^2) \times 1.1}{250}}$$

$$t = 35.78 \text{ mm} \cong 36 \text{ mm} < t_f$$

Provide base plate of size 540 mm × 380 × 36 mm $\therefore$ O.K.

$$\text{Area of concrete block } A = \frac{P_u \cdot \gamma_{mo}}{586\, \gamma_f}$$

$$A = \frac{2370 \times 10^3 \times 1.1}{200 \times 10^3 \times 1.5} = 8.69 \text{ m}^2$$

For equal projections

$$B \times D = 8.69 \text{ m}^2$$

$$= 8.69 \times 10^6 \text{ mm}^2$$

$$\frac{B - 440}{2} = \frac{D - 600}{2}$$

$$\therefore \qquad B = D - 160$$

$$\therefore \qquad D^2 - 160\, D - 8.69 \times 10^6 = 0$$

$$D = 3028.9 \text{ mm}$$

$$\text{Adopt } D = 3030 \text{ mm} = 3.03 \text{ m}$$

$$B = 2870 \text{ mm} = 2.87 \text{ m}$$

$$\text{Actual projections } a = b = 1245 \text{ mm}$$

Considering 45° angle of dispersion of load

$$\text{The depth of concrete black} = 1.245 \text{ m}$$

Fig. 6.9

Ex. 6.4: *Design a slab base for a column ISHB 350 @ 724 N/m to carry factored axial compressive load of 1500 kN. The base rests on concrete pedestal M20. For ISHB 350 @ 724 N/m - b_f = 250 mm, t_f = 11.6 mm, f_u = 410 MPa, γ_{m0} = 1.10.*

Sol.: P_u = 1500 kN, f_{ck} = 20 N/mm^2

Assume f_y = 250 MPa (as f_u = 410 Mpa) and SBC = 150 kN/m^2.

For ISHB 350, D = 350 mm, B = b_f = 250 mm,

(i) Bearing area of base plate:

$$A = \frac{P_u}{0.6\,f_{ck}} = \frac{1500 \times 10^3}{0.6 \times 20} = 125000 \text{ mm}^2$$

(ii) Size of base plate:

For equal projections (a = b)

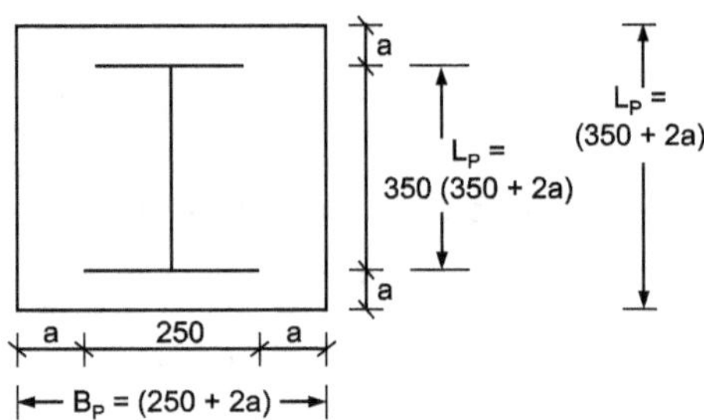

Fig. 6.10

$$L_p = \frac{D - B}{2} + \sqrt{\left(\frac{D - B}{2}\right)^2 + A}$$

$$= \frac{350 - 250}{2} + \sqrt{\left(\frac{350 - 250}{2}\right)^2 + 125000}$$

$$= 407.07 \text{ say } 410 \text{ mm}$$

$$B_p = \frac{A}{L_p} = \frac{125000}{410} = 304.87 \text{ say } 310 \text{ mm}$$

∴ Larger projection,

$$a = \frac{L_p - D}{2} = \frac{410 - 350}{2} = 30 \text{ mm}$$

Smaller projection,

$$b = \frac{B_p - B}{2} = \frac{310 - 250}{2} = 30 \text{ mm}$$

∴ Area of base plate provided

$$A_p = L_p \times B_p = 410 \times 310 = 127100 \text{ mm}^2$$

(iii) Ultimate pressure from below on the slab base

$$w = \frac{P_u}{L_p \times B_p} = \frac{1500 \times 10^3}{410 \times 310} = 11.80 \text{ N/mm}^2$$

(iv) Thickness of base plate

$$t_s = \sqrt{\frac{2.5\,w\,(a^2 - 0.3\,b^2)\,\gamma_{mo}}{f_y}} = \sqrt{\frac{2.5 \times 11.80\,(30^2 - 0.3 \times 30^2) \times 1.1}{250}}$$

$$= 9.04 \text{ mm say } 12 \text{ mm}$$

(10 mm thickness could have been taken, but as t_f = 11.6 mm we have to adopt greater thickness. Provide 410 × 310 × 12 mm base plate and secure the column section with 2 ISA 100 × 100 × 100 mm with 4-20 mm φ bolts one at each corner).

(v) Size of concrete block:

$$A_f = \frac{P_u \cdot \gamma_{mo}}{SBC \cdot \gamma_f} = \frac{1500 \times 1.1}{150 \times 1.5} = 7.33 \text{ m}^2$$

For equal projections

$$L_f = \frac{L_p - B_p}{2} + \sqrt{\left(\frac{L_p - B_p}{2}\right)^2 + A_f} = \frac{0.41 - 0.31}{2} + \sqrt{\left(\frac{0.41 - 0.31}{2}\right)^2 + 7.33}$$

$$= 2.758 \text{ m say } 2.80 \text{ m}$$

$$B_f = \frac{A_f}{L_f} = \frac{7.33}{2.80} = 2.62 \text{ say } 2.65 \text{ m}$$

∴ Provide M20 concrete pedestal of size 2.80 m × 2.65 m

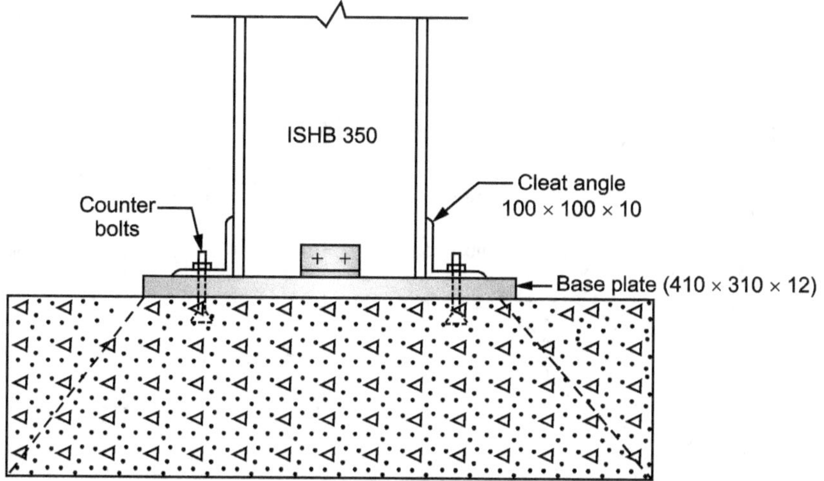

Fig. 6.11

6.5 GUSSETED BASE

- When the load on column is large or subjected to moment, gusseted base is provided.

- A gusseted base consists of a base plate connected to the column through gusset plates.

- The thickness of base plate in this case will be less than thickness of the slab base for the same axial load as the bearing area of the column on base plate increases by the gusset plates.

- As per IS 800 : 2007 gusseted base, consists of a base plate, gusset plates, angle cleats, stiffness fastenings etc. in combination with the bearing area of the shaft should be sufficient to take the loads, bending moment and reaction to the base plate without exceeding the specified design stresses.

- All bearing surfaces are machined to ensure perfect contact where the ends of the column shaft and the gusset plates are not faced for complete bearing. The fastening shall be sufficient to transmit the forces to which the base is subjected.

- A typical view of gusseted base is shown in Fig. 6.12.

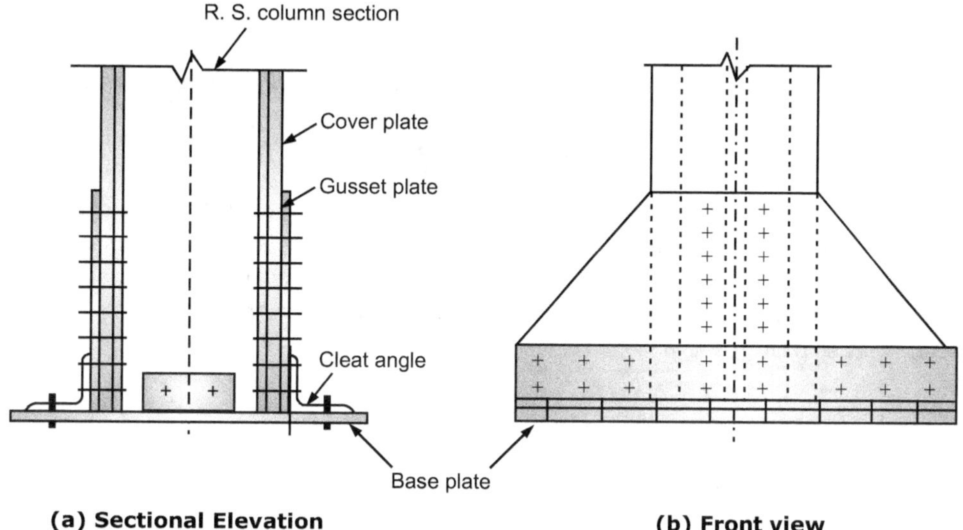

(a) Sectional Elevation   **(b) Front view**

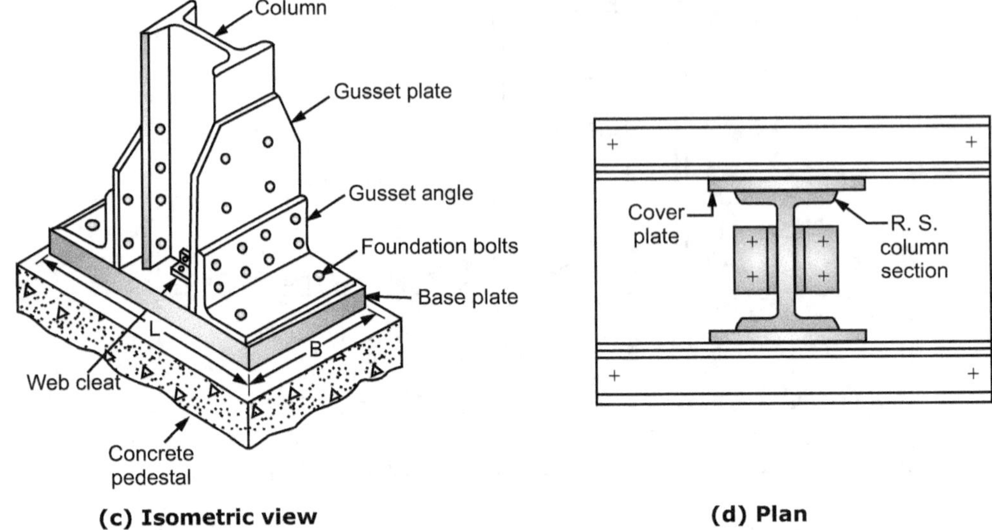

(c) Isometric view   **(d) Plan**

Fig. 6.12: Bolted Gusset Base

6.6 DESIGN OF A GUSSETED BASE AND CONCRETE BLOCK

- Following design steps are to be followed :

1. Calculate the area (A) of base plate.

$$A = \frac{\text{Axial load on column}}{\text{Bearing strength of plain concrete}} = \frac{P_u}{0.6\, f_{ck}}$$

2. Assume thickness of gusset plate $\ngtr 16$ mm. The connecting angle should have thickness equal to that of gusset and its vertical leg $\ngtr 100$ mm, so that it can be accommodate two bolts in vertical line.

3. Calculate the width of gusset plate

= Depth of column (+ cover plates, if provided) + 2 × thickness of gusset plate + 2 × length of connecting leg of angle with base slab + overhang

4. Calculate length of base plate $= \dfrac{A}{\text{Width of base plate}}$

5. Equating BM @ face of gusset angle due to intensity of pressure below the base plate to its moment of resistance, the thickness of base plate can be calculated.

Important Points

- The column bases are used to rest the columns on the concrete bed of the foundation even over a large area.
- The area of concrete slab is so designed, that the design stresses induced in the soil should not exceed the bearing capacity of soil.
- The types of column bases are : slab base, gusseted base and grillage foundation.
- The thickness of base plate of slab base is

$$t_s = \sqrt{\frac{2.5\, w\, (a^2 - 0.3\, b^2)\, \gamma_{mo}}{f_y}}$$

where, t_s = thickness of base plate

           w = The pressure or loading on underside of base

           a = Greater projection of the plate beyond column section

           b = Lesser projection of the plate beyond column section

        γ_{mo} = Partial factor of safety = 1.1

           f_y = Yield stresses of steel (generally = 250 MPa)

Practice Questions

1. Differentiate between slab base and gusseted base on any four parameters.

2. What is the basic concept to decide the plan area of slab base and concrete block below it. State the function of cleat angle.

3. Write the steps to calculate the thickness of base plate used in slab base. Why anchor bolts are used in slab base?

4. Define Gusseted base. Also draw its neat labelled sketch showing all details.

5. State the necessity of column bases. Also state the function of cleat angle and anchor bolts in slab base.

6. Draw plan of gusseted base showing all components.

7. A column ISHB 250 @ 51 kg/m, carries an axial service load of 600 kN. Design a slab base for the column. The grade of concrete is M15. Flange width of ISHB 250 is 250 mm. SBC = 200 kN/m^2.

8. A column section HB 200 @ 373 N/m carries an axial service load of 2000 kN. Determine the area and thickness of slab base for the column. The grade of concrete is M10. Take width of flange = 200 mm.

9. Design a suitable slab base for a column having a section ISHB 300 @ 58.8 kg/m and carrying an axial load of 900 kN. The bearing capacity of the soil is 250 kN/m^2. The grade concrete is M15. Concrete block need not to be designed.

10. Design a slab base for the column ISHB 250 @ 510 N/m, carrying an axial load of 600 kN. Flange width of the column is 250 mm. The grade of concrete is 15 M15. Take SBC of soil = 150 kN/m^2.

11. Design a slab base and concrete pedestal for a column ISLB 350 used to carry axial load of 900 kN. The grade of concrete is M15 and SBC of soil 200 kN/m^2. Width of flange of ISLB 350 is 165 mm. Assume additional data if required.

■■■

Chapter 7

DESIGN OF FLEXURAL MEMBERS FOR
BM AND SF BY L.S.M.

Syllabus

- General - Effective Span of Beams, Design Strength of Bending, (Flexure), Limiting Deflection of Beams – Design of Laterally supported Simple Beams for Bending Moment and Shear Force using Single / Double Rolled Steel Sections (Symmetrical Cross-sections only) - Problems. Names of various Components of Plate Girder and their Functions with Usual IS Recommendations.

About this Chapter

After reading this chapter students can understand:
- Different types of Beams and different sections used for it.
- Design of laterally supported Beams.
- Types of Sections used as Plate Girders.
- Component Parts of Plate Girder and its functions.

7.1 INTRODUCTION

- A beam is defined as a structural member usually straight and horizontal in position and subjected to transverse loads normal to its longitudinal axis.

7.2 TYPES OF BEAMS

- Beams are mainly of two types:
  1. Simple beams
  2. Built-up or Compound beams
- Depending on the function of beams, they are given alternative names as mentioned below:
  - *Joist* is a beam supporting floor construction but not other beams.
  - *Floor Beam* is a major beam usually supporting the joists. This may also be a transverse beam supporting a bridge floor.
  - *Girder* is a floor beam used in buildings. This in general is any major beam in a structure.
  - *Lintel* is a beam spanning door opening, window opening and supporting the wall upto the floor above.
  - *Purlin* is roof beam supported by roof trusses.
  - *Rafter* is roof beam supported by the purlins.
  - *Spandrel beam* is a beam at the outermost wall of building supporting the floor and the wall upto the floor.
  - *Stringer* is a longitudinal beam used in bridge floors and supported by floor beams. This term is also applied to a beam supporting stair steps.

7.3 DIFFERENT STEEL SECTIONS USED AS A BEAM

- Rolled I-sections with or without cover plates are usually used for floor beams. I-sections are also called as rolled steel joist (R.S.J.).
- Channel, tee- and angle-sections are usually used for beams in roof trusses as purlins and common rafters.

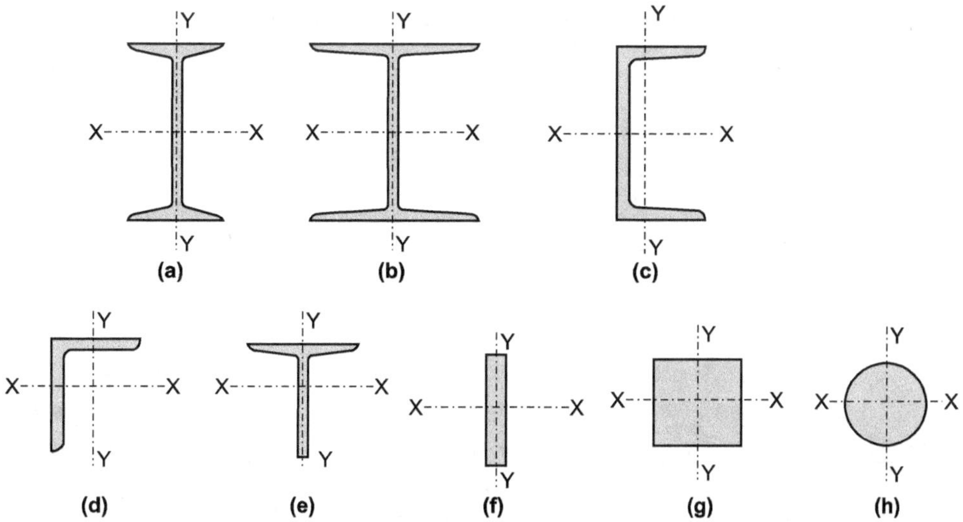

Fig. 7.1: Rolled steel Sections used as Beams

7.4 PERMISSIBLE BENDING STRESS

- When I or channel sections are used as breams, there is a problem of lateral buckling since the moment of inertia of these sections about minor axis is lessor than M.I. about major axis as shown in Fig. 7.2 (a). It may be noted that the upper portion of the beam is subjected to compression and the lower portion to tension.

- Since the upper portion of the beam is subjected to compression, it tries to buckle. As the upper portion is not independent and is integral with the lower portion, it cannot buckle freely. That is, the upper flange tries to buckle laterally whereas the lower flange has no tendency to buckle.

- In the process, the beam is subjected to torsion in addition to lateral buckling and bending in the vertical plane. This type of buckling is known as lateral torsional buckling. (Fig. 7.2 (b)).

- This may be prevented by providing continuous lateral support to the compression flange in which case the beams are known as laterally supported beams.

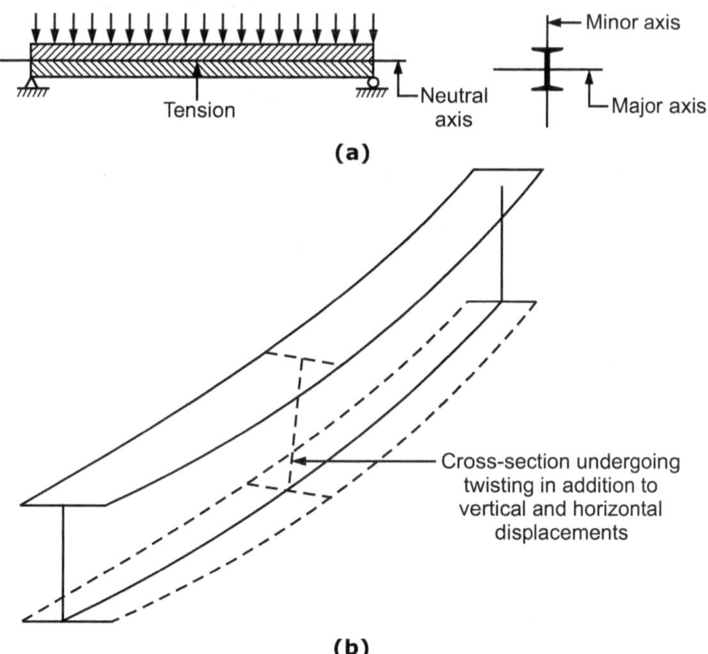

Fig. 7.2: Lateral torsional buckling of simply supported beam

7.5 EFFECTIVE LENGTH FOR LATERAL TORSIONAL

- For simply supported beams and girders of span L where no resistant to the compression flange is provided but each end of the beam is restrained against torsion, the effective L_{LT} for the lateral torsional buckling is given in table 7.1.
- Restraint against torsion at the ends may be provided by
  - (a) web or flange cleats, or
  - (b) bearing stiffeners acting in conjunction with bearing of the beam, or
  - (c) lateral end frames or external supports providing lateral restraint to the compression flanges at the ends, or
  - (d) their being built into walls.

Table 7.1: Effective length of Simply Supported Beams, L_{LT}

Sr. No.	Conditions of Restraint at Supports		Loading condition	
	Torsional Restraint	Warping Restraint	Normal	Destabilizing
(1)	(2)	(3)	(4)	(5)
1.	Fully restrained	Both flanges fully restrained	0.70 L	0.85 L
2.	Fully restrained	Compression flange fully restrained	0.75 L	0.90 L
3.	Fully restrained	Both flanges fully restrained	0.80 L	0.95 L
4.	Fully restrained	Compression flange partially restrained	0.85 L	1.00 L
5.	Fully restrained	Warping not restrained in both flanges	1.00 L	1.20 L
6.	Partially restrained by bottom flange support connection	Warping not restrained in both flanges	1.0 L + 2D	1.2 L + 2D
7.	Partially restrained by bottom flange bearing connection	Warping not restrained in both flanges	1.2 L + 2D	1.4 L + 2D

Notes:
1. Torsional restrained prevents rotation about the longitudinal axis.
2. Warping restraint prevents rotation of the flange in its plane.
3. D is the overall depth of the bream.

7.6 EFFECTIVE SPAN OF LATERALLY RESTRAINED BEAMS

- The effective span of a laterally restrained beam shall be taken as the length of the beam between the centres of the supports or the length between assumed points of application of reactions.

7.7 MAXIMUM AND ALLOWABLE DEFLECTION

- The maximum deflections for following standard cases are given below:
  - (i) Simply supported beam with udl over span

$$\delta_{max} = \frac{5}{384} \frac{wL^4}{EI}$$

  - (ii) Simply supported beam with central point load

$$\delta_{max} = \frac{PL^3}{48EI}$$

  - (iii) Cantilever with udl over entire span

$$\delta_{max} = \frac{wL^4}{8EI}$$

(iv)   Cantilever with point load at free end

$$\delta_{max} = \frac{PL^3}{3EI}$$

The allowable deflection as per IS: 800 - 1984 is $\frac{1}{325}$ of span of the beam.

i.e. $\delta_{allowable} = \frac{L}{325}$

7.8 PLASTIC ANALYSIS CONSIDERATIONS

- When the strain in extreme fibers $\varepsilon_{max} < \varepsilon_y$, i.e. $f_{max} < f_y$, the behaviour of beam is elastic when $\varepsilon_{max} = \varepsilon_y$ i.e. $f_{max} = f_y$ the yielding of extreme fibres takes place, whereas the inner fibres of the beam remain elastic. When $\varepsilon_{max} >> \varepsilon_y$, yielding takes place through the entire depth of the beam and the cross-section becomes plastic.

- This behaviour of beam is explained in details in Sections 9.3 to 9.8 of Chapter 9. Hence before going to following sections, the reader must refer the above sections so as to study idealized stress-strain curve, requirements of steel for plastic analysis and its assumptions, formation of the plastic hinges, plastic moment of resistance, plastic section modulus etc.

7.9 CLASSIFICATION OF BEAM CROSS-SECTIONS

- It is well known that a thin projecting flange of an I-section may buckle due compressive bending stress. The web of an I-section is also liable to buckle under compressive bending stress and shear stress. In order to prevent such local buckling it is essential that outstand/thickness ratios of flanges and depth/thickness ratios of webs must be limited.

- Hence the sectional dimensions of plates constituents should satisfy that following conditions:

  (i) When designs are made by plastic analysis, the members should be able to form plastic hinges with sufficient rotation capacity (ductility) without local buckling, so as to permit redistribution of bending moments needed before reaching a collapse mechanism.

  (ii) When designs are made by elastic method, the members must be able to reach the yield stress under compression without buckling.

- Based on the above beam sections are classified as follows in accordance with their behaviour in bending as per IS: 800 – 2007.

  (a) Class 1 (Plastic): Cross-sections which can develop plastic hinges and have the rotation capacity required for failure of the structure by formation of the plastic mechanism. The width to thickness ratio of plate elements shall be less than that specified under class 1 (Plastic) in the table 4.7.

  (b) Class 2 (Compact): Cross-sections which can develop plastic moment of resistance, but have inadequate plastic hinge rotation capacity for formation of plastic mechanism, due to local buckling. The width to thickness ratio of plate elements shall be less than the specified under class 2 (Compact), but greater than that specified under 1 (Plastic) in the table 4.7.

  (c) Class 3 (Semi-compact): Cross-sections in which the extreme fibre in compression can reach yield stress but cannot develop the plastic moment of resistance, due to local buckling. The Width to thickness ratio of plate elements shall be less than that specified under class 3 (Semi-compact) but greater than the specified under class 2 (compact) in the table 4.7.

  (d) Class 4 (Slender): Cross-sections in which the elements buckle locally even before reaching yield stress. The width of thickness ratio of plate elements shall be greater than that specified under class 3 (semi-compact) in table 4.7. In such cases, the effective sections shall be calculated either by the following provision of IS 801 to account for the post local buckling strength or by deducting with of the compression plate element in excess of semi-compact section limit.

7.10 LATERALLY SUPPORTED BEAMS

- *A beam is defined as a structural member usually straight and horizontal in position and subjected to transverse loads normal to its longitudinal axis.*
- A laterally supported beam is a beam whose compression flange is restrained from buckling. This is a beam whose compression flange is laterally supported in various ways. The compression flange may be connected to the concrete floor either by its embedment or by shear connectors. The compression flange may also be restrained by its connection to cross-beams or bracings. Fig. 7.3 shows some methods of providing lateral support to the compression flange.

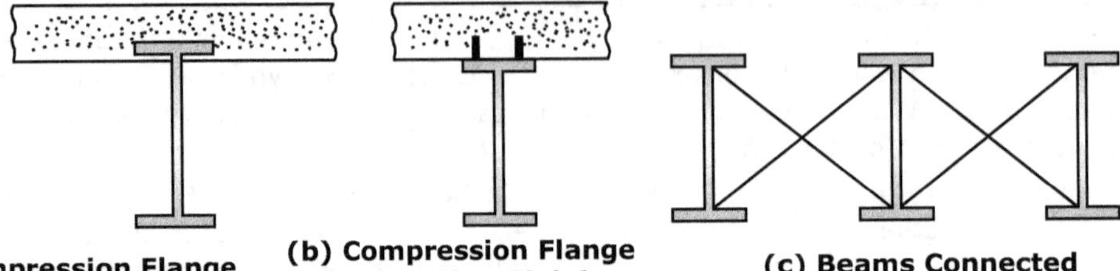

(a) Compression Flange Embedded in Slab   **(b) Compression Flange Connected to Slab by Shear Connectors**   **(c) Beams Connected by Braces**

Fig. 7.3: Laterally Supported Beams

7.11 DESIGN STEPS FOR LATERALLY SUPPORTED BEAMS

(i)   Calculate the factored load, maximum bending moment and shear force for given loading.

(ii)   Obtain the plastic modulus of section required

$$Z_{p\ reqd.} = M_d \times \frac{\gamma_{m0}}{f_y}$$

(iii)   For I-section

$$Z_{e\ reqd.} = \frac{Z_{p\ reqd}}{1.14}$$

Select suitable I-section from steel table (Annexure) such that

$$Z_{e\ reqd.} < Z_{e\ available.}$$

(iv)   Classify the rolled steel section for $f_y = 250$, $\varepsilon = 1$.

For $\frac{b_h}{t_f} = 9.4 \rightarrow$ plastic

$\frac{b_h}{t_f} = 10.5 \rightarrow$ compact

$\frac{b_h}{t_f} = 15.7 \rightarrow$ semi-compact

$\frac{b_h}{t_f} > 15.7 \rightarrow$ slender

Also for $\frac{d}{t_w} = 84 \rightarrow$ plastic

$\frac{d}{t_w} = 105 \rightarrow$ compact

$\frac{d}{t_w} = 126 \rightarrow$ semi-compact

(v)   Calculate design shear for web plastic shear resistance

$$V_{dr} = \frac{f_y \times t_w \times h}{\gamma_{mo}\ \sqrt{3}}$$

Check that $V_{dr} > V_d$ and also $V_d < 0.6\ V_{dr}$

(vi)   Check for web buckling

If $\frac{d}{t_w} < 67\ \varepsilon$, web stiffeners are required (if asked to check web buckling)

(vii) Calculate moment resisted by section

$$M_r = \frac{\beta_b \cdot Z_p \cdot f_y}{\gamma_{mo}}$$

$\beta_b = 1$ for plastic and compact sections

For semi-compact section

$$M_r = \frac{Z_e \cdot f_y}{\gamma_{mo}}$$

(As along as shape factor is < 1.2, this check is not necessary for I-section)

(viii) Check for deflection

$$\delta_{allowable} = \frac{L}{300} \qquad \left(\begin{array}{c} \text{as per table 1.8 for vertical deflection} \\ \text{in other buildings for floors and roofs sections} \\ \text{not susceptible to cracking} \end{array} \right)$$

$$\delta_{max} = \frac{5}{384} \frac{wL^4}{EI} \qquad \text{(for service ud} l \text{ over entire span of S.S. beam)}$$

$$\delta_{max} < \delta_{allowable} \qquad (\therefore \text{ O.K.})$$

7.11.1 Laterally Restrained Beams

- Beams are laterally restrained because:
  - (a) Thin projecting flange is susceptible to buckling under compression.
  - (b) In laterally supported beam, flange is restrained from buckling.
- We can support compression flange in many ways as follows:
  - (i) Connection of compression flange to the floor.
  - (ii) Connection of compression flange to the floor with the help of shear connectors.
  - (iii) Use of cross-beam by braces for connection of compression flange.
- Differentiate between Laterally supported and unsupported beam with neat sketch showing all details.

Laterally Supported Beam	Laterally Unsupported Beam
In laterally supported beam, compression flanges are embedded in concrete.	In laterally unsupported beam, compression flanges are not embedded in concrete.
Compression flange of Beam is restrained against rotation.	Compression flange of Beam is free for rotation.
Lateral deflection of compression flange is not occur.	Lateral deflection of compression flange is occur.
Laterally supported (it means compression flange is restrained) **Fig. 7.4 (a)**	Laterally unsupported **Fig. 7.4 (b)**

- **Sketches of bolted connections in case of:**
  - (i) Beam to Beam connection when flanges are at same level.
  - (ii) Beam to Beam connection when flanges are not at same level.
  - (iii) Beam to column connection.

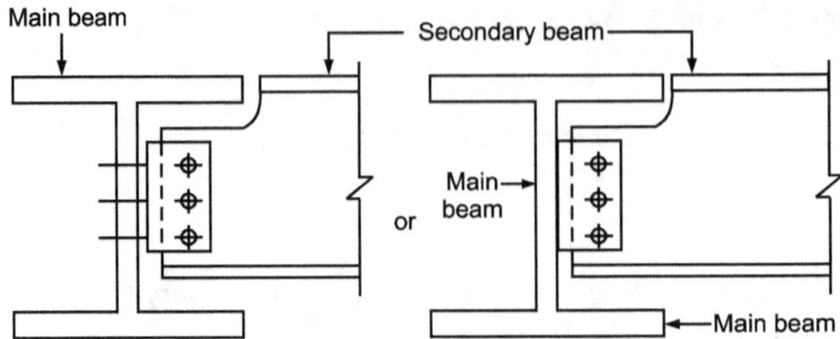

(a) Bolted-Bolted connection using cleat angle

(b) Welded-Bolted connection using plate

Fig. 7.5: Beam to Beam connection when flanges at same level

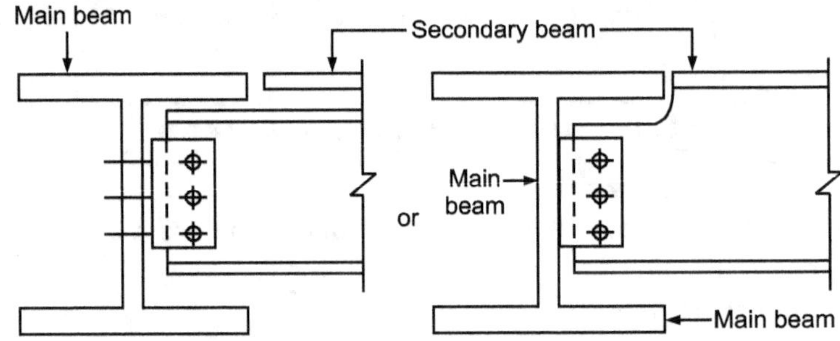

(a) Bolted-Bolted connection using cleat angle

(b) Welded-Bolted connection using plate

Fig. 7.6: Beam to Beam connection when flanges are not a same level

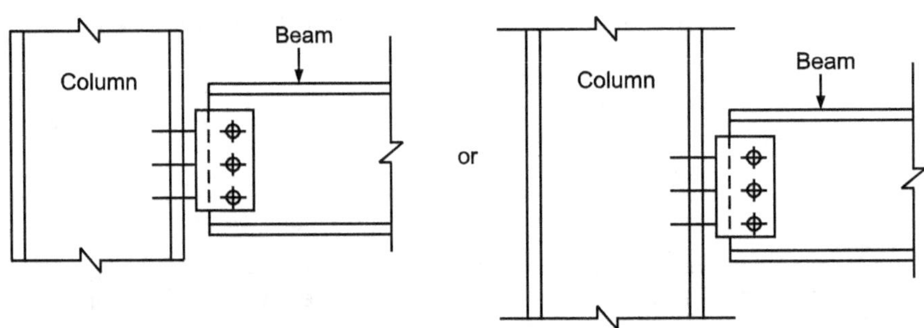

(a) Bolted-Bolted connection using cleat angle

(b) Welded-Bolted connection using plate

Fig. 7.7: Beam to Column connection

Solved Examples

Ex. 7.1: *An ISMB 250 is used as simply supported beam for 3 m span to carry 20 kN/m load. Take f_y = 250 MPa, check the section for shear only, t_w = 6.4 mm.*

Sol.: Check for shear

Given: ISMB 250, h = 250 mm, t_w = 6.4 mm.

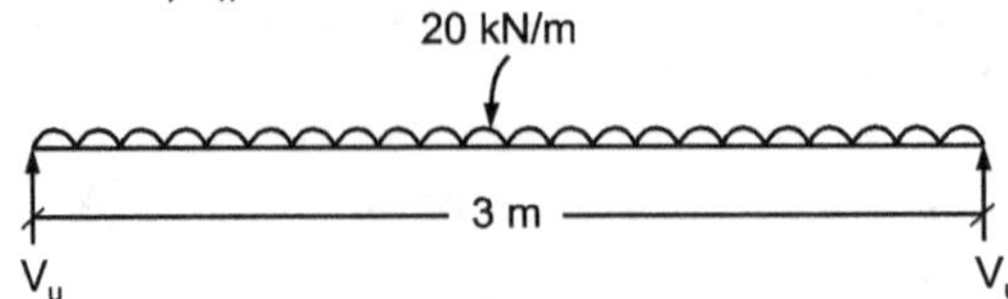

Fig. 7.8

Assuming UDL carried by beam with self weight = 20 kN/m

Factored shear force developed max at support equal to support reactors

$$\therefore \qquad V_u = \frac{1.5\,(w \times l)}{2}$$

$$V_u = \frac{1.5 \times 20 \times 3}{2}$$

$$\therefore \qquad V_u = 45 \text{ kN}$$

Design shear capacity resisted by web of given beam section,

$$V_d = \frac{f_y \times h \times t_w}{\gamma_{m_0} \times \sqrt{3}}$$

$$\therefore \qquad V_d = \frac{250 \times 250 \times 6.4}{1.1 \times \sqrt{3}}$$

$$V_d = 209.95 \times 10^3 \text{ N}$$

$$\therefore \qquad V_d = 209.95 \text{ kN} \nless V_u$$

$\therefore$ Design shear capacity is very large than factored max. shear developed hence beam is safe in shear.

Ex. 7.2: *An ISMB 400 @ 6043 N/m is used as a simply supported beam for 3 m span. The compression flange of beam is laterally supported through out the span. Determine design flexural strength of member. Also calculate working u.d.l. the beam can carry per m span.*

Take Z_p = 1176.18 $\times 10^3$ mm^3, γ_{m_0} = 1.1, β_b = 1, f_y = 250 MPa.

Sol.: Given: ISMB 400 wt. 6043 N/m (which must be 604.3 N/m)

L = 3000 mm, Z_p = 1176.18 $\times 10^3$ mm^3, γ_{m_0} = 1.1, β_b = 1, f_y = 250 N/mm^2.

Assuming UDL = w kN/m

1. To calculate the design flexural strength M_d =

$$M_d = \frac{\beta_b \times Z_p \times f_y}{\gamma_{m_0}} = \frac{1 \times 1176.18 \times 10^3}{1.10} = 267.27 \times 10^6 \text{ kN-m}$$

2.

$$M_u = \frac{wl^2}{8} = \frac{w \times 3^2}{8} = 1.125 \text{ kN/m}$$

$$M_d = M_u$$

$$267.27 \times 10^6 = 1.125 \text{ W}$$

$$w = 237.57 \text{ kN/m}$$

Ex. 7.3: *An ISMB 350 @ 514 N/m is used as a simply supported beam for 5 m span. The compression flange of beam is laterally supported through out span. Determine design bending strength of beam. Also calculate working UDL the beam can carry per m span. Check the member for deflection.*

Take   – Zp = 889.6 x 10^3 mm^3, γ_{mo} = 1.10

β_b = 1, f_y = 250 MPa, I_{xx} = 13630.3 $\times 10^4$ mm^4

E = 2 $\times 10^5$ N/mm^2

Sol.: From steel table explore the properties of ISMB 350@ 514 N/m

h = 350 mm, b_f = 140 mm, t_f = 14.2, t_w = 8.1 mm

Z_e = 778.9 $\times 10^3$ mm^3

Z_p = 889.57 $\times 10^3$ mm^3

r_1 = 14 mm

Classification of section:

$$d = h - 2\,(t_f + r_1) = 350 - 2\,(14.2 + 14) = 293.6 \text{ mm}$$

$$\frac{b_h}{t_f} = \frac{(140/2)}{14.2} = 4.93$$

$$\frac{d}{t_w} = \frac{293.6}{8.1} = 36.25$$

As $\frac{b_h}{t_f}$ < 9.4 and $\frac{d}{t_w}$ < 84

$\therefore$ Section classification: Plastic

$\because$ **Design bending strength**

$$M_d = \frac{\beta_b \times Z_p \times f_y}{\gamma_{mo}}$$

$$= \frac{1 \times 889.57 \times 10^3 \times 250}{1.1} \quad [\beta_b = 1 \text{ for plastic}]$$

$$= 202.175 \times 10^6 \text{ N-mm}$$

$$\boxed{M_d = 202.175 \text{ kN·m}}$$

To find out u.d.l.:

$$M_d = \frac{wl^2}{8}$$

$$202.175 = \frac{w \times 5^2}{8}$$

$$w_f = 64.696 \text{ kN/m}$$

$\therefore$ Working u.d.l. $\quad w = \dfrac{64.696}{1.5}$

$$\boxed{w = 43.13 \text{ kN/m}}$$

Check for deflection:

$$\delta_{allowable} = \frac{L}{300} = \frac{5000}{300} = 16.67 \text{ mm}$$

$$\delta_{max} = \frac{5}{384} \frac{wL^4}{EI} = \frac{5}{384} \times \frac{64.696 \times 5000^4}{2 \times 10^5 \times 13630.3 \times 10^4}$$

$$= 19.3 \text{ mm}$$

$$\boxed{\delta_{max} > \delta_{allowable}} \quad \text{Hence O.K.}$$

Ex. 7.4: *A simply supported beam 5.2 m long carries UDL of 50 kN/m. The beam is laterally supported. Design the section and check for deflection only.*

Take $\gamma_{m0} = 1.1$, $\beta_b = 1.0$, $f_y = 250$ MPa and available section properties as:

Section	b_f (mm)	t_f (mm)	t_w (mm)	R_1 (mm)	h_2 (mm)	Z_{xx} (mm³)
ISWB 350	200	11.4	8.0	12.0	27.25	887×10^3
ISWB 400	200	13.0	8.6	13.0	29.75	1171.3×10^3
ISWB 450	200	15.4	9.2	15.0	33.00	1558.1×10^3

Sol.: L = 5.2 m, udl = 50 kN/m, $\gamma_{m0} = 1.1$

Step I: Loads and factored BMs

Total $\quad\quad\quad\quad\quad\quad$ udl = 50 kN/m

$$w_d = 50 \times 1.5 = 75 \text{ kN/m}$$

$$SF = \gamma_d = \frac{W_e}{2} = \frac{75 \times 5.2}{2} = 195 \text{ kN}$$

$$M_d = \frac{W_d L_e^2}{8} = \frac{75 \times 5.2^2}{8} = 253.50 \text{ kNm}$$

Step II: Elastic modulus of section required

$$Z_{e \text{ reqd}} = \frac{Z_{e \text{ reqd}}}{1.14}$$

Plastic modulus of section

$$Z_{e \text{ reqd}} = \frac{M_d \cdot \gamma_{m0}}{f_y} = \frac{253.50 \times 10^6 \times 1.1}{250} = 1113.20 \times 10^3 \text{ mm}^3$$

$$Z_{e \text{ reqd}} = \frac{1.1132 \times 10^6}{1.14} = 976.49 \times 10^3 \text{ mm}^3$$

Step III: Try ISWB 400

$$h = 400 \text{ mm}$$
$$b_f = 200 \text{ mm}$$
$$t_f = 13.0 \quad t_w = 8.6 \text{ mm} \quad R = 13 \text{ mm}$$
$$Z_{xx} = 1171.30 \times 10^3 \quad I_{xx} = 23426.70 \text{ mm}^4 \times 10^4$$

Classification of beam section

$$d = h - 2\,(t_f + \gamma_1) = 400 - 2\,(13 + 13) = 348 \text{ mm}$$
$$\frac{b_h}{t_f} = \left(\frac{200/2}{13}\right) = 7.69 < 9.4$$
$$\frac{d}{t_w} = \frac{348}{8.6} = 40.46 < 67$$

Section classification is plastic:

Step IV: Check for deflection

$$\delta_{allowable} = \frac{L}{300} = \frac{5200}{300} = 17.33$$

$$\delta_{max} = \frac{5}{384} \times \frac{W_l{}^4}{EI} = \frac{5}{384} \times \frac{75 \times 5200^4}{2 \times 10^5 \times 23426.70 \times 10^4} = 15.23 \text{ mm}$$

$$\therefore \qquad \delta_{max} < \delta_{allowable}$$

Ex. 7.5: *A simply supported beam of span 8 m supports a reinforced concrete slab. The compression flange of beam is restrained due to its connection with the slab. The c/c spacing of beam is 4 m and thickness of R.C.C. slab is 140 mm. The superimposed load is 4 kN/m² and floor finish is 1.5 kN/m². Design the section of intermediate beam if Fe 410 steel section is used.*

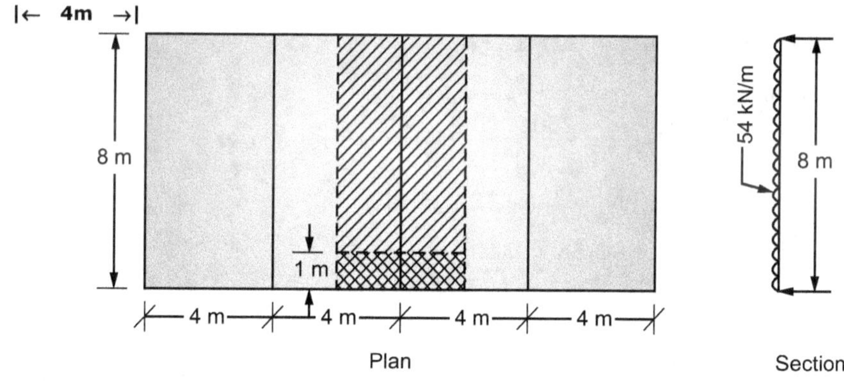

Fig. 7.9

Sol.: (i) Loads and Factored BM

Consider 1 m length of beam which supports 4 m width of slab

Self wt. of slab	$= 1 \times 4 \times 0.14 \times 25$	$=$	14.0 kN/m
Superimposed load	$= 1 \times 4 \times 4$	$=$	16.0 kN/m
Floor finish	$= 1 \times 4 \times 1.5$	$=$	6 kN/m
	w $= 36.0$ kN/m		

Factored u.d.l. $w_d = w \times \gamma_f = 36 \times 1.5 = 54.0$ kN/m

Factored BM, $M_d = \dfrac{w_d \cdot l_e^2}{8} = \dfrac{54 \times 8^2}{8} = 432$ kNm

Factored   S.F., $V_d = \dfrac{w_d \cdot l_e}{2} = \dfrac{54 \times 8}{2} = 216$ kN

(ii) Plastic modulus of section required

$$Z_{p\,reqd.} = \frac{M_d \cdot \gamma_{mo}}{f_y} = \frac{432 \times 10^6 \times 1.1}{250} = 1900.80 \times 10^3 \text{ mm}^3$$

(iii) Elastic modulus of section required

$$Z_{e\ reqd} = \frac{Z_{p\ reqd}}{1.14} = \frac{1900800}{1.14} = 1667.37 \times 10^3 \ mm^3$$

Try ISMB 500 @87.1 kg/m = 87.1 × 9.81 N/m = 854 N/m (from steel table)

Properties of ISMB 500 are:

h = 500 mm, b_f = 180 mm, t_f = 17.2 mm, t_w = 10.2 mm

r_1 = 17.0 mm, Z_{xx} = 1810 × 10^3 mm^3, I_{xx} = 452.18 × 10^6 mm^4

z_p = 2074.67 × 10^3 mm^3

(iv) Classification of section

d = h - 2 $(t_f + r_1)$ = 500 – 2 (17.2 + 17) = 431.6 mm

$$\frac{b_h}{t_f} = \frac{\frac{180}{2}}{17.2} = 5.23 < 9.4$$

$$\frac{d}{t_w} = \frac{431.6}{10.2} = 42.31 < 84$$

As $\dfrac{b_h}{t_f}$ < 9.4 and $\dfrac{d}{t_w}$ < 84

Section classification: plastic

Include self weight in w_d

$\therefore$ Actual $w_d = \left(36 + \dfrac{854}{1000}\right) \times 1.5 = 55.28$ kN/m

Actual $M_d = 55.28 \times \dfrac{8^2}{8} = 442.24$ kN/m

Actual $V_d = 55.28 \times \dfrac{8}{2} = 221.12$ kN

(v) Check for shear

$$V_{dr} = \frac{f_y \times t_w \times h}{\gamma_{mo}\ \sqrt{3}} = \frac{250 \times 10.2 \times 500}{1.1 \times \sqrt{3}} = 669201 \ N$$

$$= 669.20 \ kN > V_d \ (= 221.12 \ kN)$$

Also $\dfrac{V_d}{V_{dr}} = \dfrac{221.12}{669.2} = 0.33 < 0.6$

$\therefore$ Check for shear is satisfied.

(vi) Check for flexure

$$M_r = \frac{\beta_b \cdot Z_p \cdot f_y}{\gamma_{mo}} = \frac{1 \times 2074.67 \times 10^3 \times 250}{1.10} \qquad (\beta_b = 1 \text{ for plastic section})$$

$$= 465.38 \times 10^6 \ Nmm = 465.38 \ kNm > M_d \ (= 442.24 \ kNm)$$

$\therefore$ Flexure Check is O.K.

(vii) Check for deflection

$$\delta_{allowable} = \frac{L}{300} = \frac{8000}{300} = 26.67 \ mm$$

$$\delta_{max} = \frac{5}{384}\frac{wL^4}{EI} = \frac{5}{384} \times \frac{36.871 \times (8000)^4}{2 \times 10^5 \times 452.18 \times 10^6}$$

$$= 21.74 \ mm$$

As $\delta_{max} < \delta_{allowable}$ $\therefore$ Deflection check is satisfied.

Ex. 7.6: *Check whether ISMB 250 is suitable or not as a simply supported beam over an effective span of 6.0 m. It carries u.d.l. of 15 kN/m (including self weight). Properties of ISMB 250 are:*

$b_f = 125$ mm, $t_f = 12.5$ mm, $t_w = 6.9$ mm, $I_{xx} = 5131.6 \times 10^4$ mm^4,

$z_{xx} = 410 \times 10^3$ mm^3, $r_1 = 13.0$ mm, $Z_p = 465.71 \times 10^3$ mm^3.

Sol.: (i) Loads and factored BMS

$$w = 15 \text{ kN/m}$$

Factored u.l.d., $w_d = 15 \times 1.5 = 22.5 \text{ kN/m}$

Factored BM, $M_d = \dfrac{w_d \cdot l_e^2}{8} = \dfrac{22.5 \times 6^2}{8} = 101.25 \text{ kNm}$

Factored S.F., $V_d = \dfrac{w_d \cdot l_e}{2} = \dfrac{22.5 \times 6}{2} = 67.5 \text{ kN}$

(ii) Plastic modulus of section required

$$Z_{p \text{ reqd.}} = \frac{M_d \cdot \gamma_{mo}}{f_y} = \frac{101.25 \times 10^6 \times 1.1}{250} = 445.5 \times 10^3 \text{ mm}^3$$

$$Z_{p \text{ reqd.}} < Z_{p \text{ avail.}} (= 465.71 \times 10^3 \text{ mm}^3)$$

(iii) Classification of beam section

$$d = h - 2\,(t_f + r_1) = 250 - 2\,(12.5 + 13) = 199 \text{ mm}$$

$$\frac{b_h}{t_f} = \frac{\dfrac{125}{2}}{12.5} = 5.0 < 9.4$$

$$\frac{d}{t_w} = \frac{199}{6.9} = 28.84 < 67$$

As $\dfrac{b_h}{t_f} < 9.4$ and $\dfrac{d}{t_w} < 67$ $\therefore$ Section classification is plastic.

(v) Check for shear

$$V_{dr} = \frac{f_y \times t_w \times h}{\gamma_{mo} \sqrt{3}} = \frac{250 \times 6.9 \times 250}{1.1 \times \sqrt{3}} = 226348 \text{ N}$$

$$= 226.35 \text{ kN} > V_d (= 67.5 \text{ kN})$$

Also $\dfrac{V_d}{V_{dr}} = \dfrac{67.5}{226.35} = 0.298 < 0.6$

$\therefore$ Check for shear is satisfied.

(vi) Check for flexure

$$M_r = \frac{\beta_b \cdot Z_p \cdot f_y}{\gamma_{mo}} = \frac{(1) \times 465.71 \times 10^3 \times 250}{1.1}$$

$$= 105.84 \times 10^6 \text{ N-mm} = 105.84 \text{ kNm} > M_d (= 101.25 \text{ kNm})$$

$\therefore$ Flexure check is O.K.

(vii) Check for deflection

$$\delta_{allowable} = \frac{L}{300} = \frac{6000}{300} = 20 \text{ mm}$$

$$\delta_{max} = \frac{5}{384} \frac{wL^4}{EI} = \frac{5}{384} \times \frac{15 \times 6000^4}{2 \times 10^5 \times 5131.6 \times 10^4}$$

$$= 24.66 \text{ mm}$$

As $\delta_{max} > \delta_{allowable}$ $\therefore$ Deflection check is not o.k.

Hence ISMB 250 is not a suitable section for given loading and span.

7.12 PLATE GIRDERS

- Plate girders are built-up beams comprising of plate sections for web and flanges when welded connections are used, and plate sections for web and angle sections with or without cover plates for flanges when bolted connections are used.

- Plate girders are flexural members. Their bending resistance can be increased by increasing the distance between the flanges.

- This in turn also increases the shear resistance as the web area increases. Plate girders are primarily provided in bridges.

- Plate girder is economical when the span is very large, say beyond 20 m.
- Plate girders are also very common in buildings where heavy concentrated loads (from upper storey columns etc.) act on a long span beam.
- For example, a dining hall floor beam of a restaurant requiring a clear space all through.

7.12.1 Types of Sections used for Plate Girders

- The typical plate girder arrangements are shown in Fig. 7.10. Figs. 7.10 (a) and (b) show bolted plate girders whereas (c) and (d) are the welded plate girders.

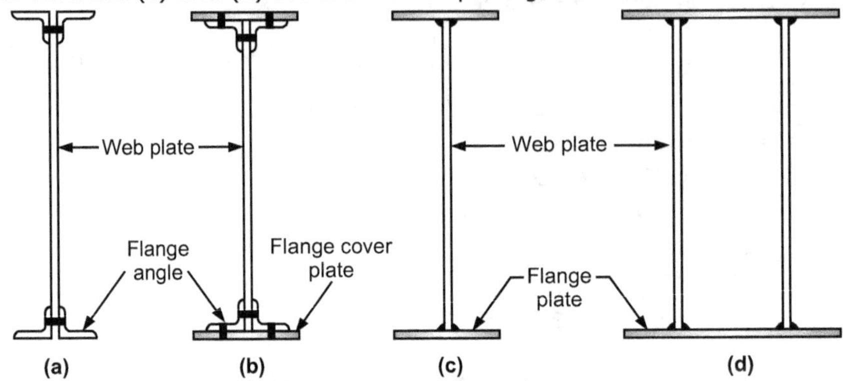

Fig. 7.10: Plate Girder Arrangements

7.12.2 Components of Plate Girder

- The various components of a bolted plate girder are shown in Fig. 7.11 and are as listed below:
  1. Web plate.
  2. Flange angles with or without flange cover plates.
  3. Bearing stiffeners.
  4. Vertical stiffeners.
  5. Horizontal stiffeners.
  6. Web splices.
  7. Flange splices.

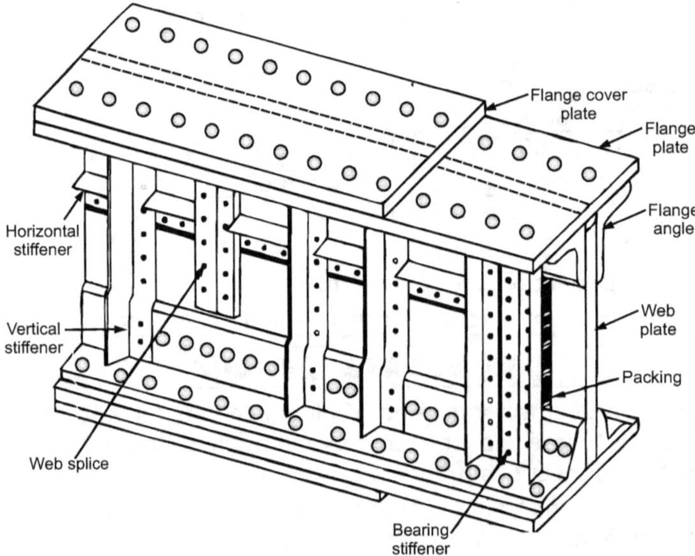

Fig. 7.11: Components of Bolted Plate Girder

7.12.3 Functions of Components of Plate Girder

1. **Web plate:** The web of a plate girder resists the entire shear force. The web plate sometimes buckles due to excess load. To avoid web buckling, vertical and horizontal stiffeners are provided. The factored design shear force $V \le V_d$ where, V_d = design strength

$$= \frac{A_v \cdot f_{yw}}{\gamma_{mo} \sqrt{3}}$$

   where, A_v = $h \cdot t_w$ for hot rolled sections

   f_{yw} = $d \cdot t_w$ for welded sections

   t_w = Thickness of web

   γ_{mo} = 1.1

2. **Flange angles with or without cover plates:** Flanges are essentially consisting of flange angles. It is designed for resisting the maximum bending moment over the beam. It increases as M.I. increases. Hence to increase M.I. additional flange cover plates can be provided.

3. **Bearing stiffeners:** Bearing stiffeners are provided at the points of concentrated loads and at supports. The ends of the load bearing stiffeners should be milled to fit tightly and provide a uniform bearing upon the loaded flange.

4. **Vertical stiffeners:** Where vertical stiffeners are required, they should be provided with the length of the girder at distance not greater than 1.5 d_1 and not less than 0.33 d_1, where d_1 is the clear distance between the flange angles when no horizontal stiffeners are provided. When horizontal stiffeners are provided, d_1 is clear distance between horizontal stiffener and tension flange ignoring fillets.

5. **Horizontal stiffeners:** The horizontal stiffeners are extended between the vertical stiffeners but need not be continuous over them.

6. **Web splices:** Splices in web plates are provided when the span of girder is more than the available size of the plate section. Web splices may also be required to facilitate the fabrication work. Splices in web plates are designed to resist the shear and moment at the spliced section.

7. **Flange splices:** Flanges should not be spliced at the points of maximum bending moment. Area of splice plates should be 5% more than the area of flange element spliced. The c.g. of the splice plates and flange element spliced should coincide as far as possible.

Important Points

- A beam is defined as a structural member usually straight and horizontal in position and subjected to transverse loads normal to its longitudinal axis.
- The types of beams are: Simple beams and Built-up or compound beams.
- Further the sections are classified as plastic, compact, semi-plastic and slender.
- For laterally restrained beams, the bending stress in compression may be taken, same as that for bending stress in tension.
- The maximum deflection for simply supported beam with u.d.l. over entire span, $\delta_{max} = \frac{5}{384} \frac{wL^4}{EI}$ and with central point load $\delta_{max} = \frac{PL^3}{48\,EI}$.
- The allowable deflection is $\frac{L}{300}$.
- Plate girders are built-up beams comprising of plate sections for web and flanges when welded connections are used and plate sections for web and angle sections with or without cover plates for flanges when bolted connections are used.
- Plate girders are used where heavy concentrated loads act on long span beam, say beyond 20 m.
- The components of bolted plate girder are: web plate, flange angles with or without flange cover plates, bearing stiffeners, vertical stiffeners, horizontal stiffeners, web splices and flange spices.

Practice Questions

1. For a beam ISWB - 600 section is insufficient. Suggest a suitable remedy.

2. For a steel beam consisting of R.S.J. –

   flanges are designed for ... and

   the web is designed for ...

   Fill the gaps and rewrite the sentence.

   [Ans. Bending moment, shear force]

3. State the design steps of rolled steel beam when it is laterally restrained.

4. What is plate girder? Write the functions of web plate and bearing stiffeners.

5. Define a plate grider. Draw the neat sketch of plate girder and label any four parts.

6. Define laterally supported beam along with suitable sketch. State any three methods of providing lateral supports to the beam.

7. State four classification of cross-sections of beam based on moment-rotation behaviour as per IS 800-2007.

8. An ISMB 300 is used as a beam over a simply supported span of 6 m (effective). It is subjected to u.d.l. 20 kN/m (inclusive of self weight) over entire span. Check the safety of beam for bending, shearing and deflection.

   Properties of ISMB 300 @ 452 N/m are b_f = 140 mm, t_f = 13.1 mm, t_w = 7.7 mm, I_{xx} = 8985.7 × 10⁴ mm⁴, Z_{xx} = 599 × 10³ mm³, r_1 = 14.0 mm, Z_p = 651.74 × 10³ mm³.

9. An ISLB 300 is to be used as a beam to carry a u.d.l. of 22 kN/m over a span of 5.4 m. Check the adequacy of the is section if A = 4808 mm².

   b_f = 150 mm, I_{xx} = 7332.9 × 10⁴ mm⁴, Z_{xx} = 488.9 × 10³ mm³, t_f = 9.4 mm, t_w = 6.7 mm, Z_p = 554.32 × 10³ mm³, r_1 = 15.0 mm.

10. A beam of 7.5 m effective span carries a u.d.l. of 19.40 kN/m inclusive of its own weight. Design the beam using IS 800 specifications. The compression flange is held against lateral displacement. Consider f_y as 250 MPa.

11. Design a simply supported beam to carry a u.d.l. of 44 kN/m. The effective span of the beam is 8 m. The effective length of compression flange of the beam is also 8 m. The ends of beam are not free to rotate at the bearings. The R.S. section WB 600 @133.7 kg/m is available for the beam. For WB 600 @ 133.7 kg/m (= 1312 N/m).

    b_f = 250 mm, Z_{xx} = 3.54 × 10⁶ mm³, Z_p = 3987 × 10³ mm³, I_{xx} = 10.62 × 10⁸ mm⁴

    Thickness of web, t_w = 11.2 mm, Thickness of flange, t_f = 21.3 mm

    Root radius, r_1 = 17.0 mm

12. A simply supported beam carries u.d.l. of 45 kN/m inclusive of self weight over an effective span of 7 m. The beam is laterally supported throughout. Check whether ISLB 450 will be sufficient or not. Also check the same section for bending, deflection and shear. Take the value of E = 200 GPa.

    Properties of ISLB 450 @ 65.3 kg/m:

    A = 8314 mm², Width of flange, b_f = 170 mm, Z_{xx} = 1223.8 × 10³ mm³, Z_p = 1401.35 × 10³ mm³, t_f = 13.4 mm, t_w = 8.6 mm, Root radius, r_1 = 16.0 mm.

13. Design a suitable ISLB section for a simply supported beam of an effective span 5.0 m subjected to an u.d.l. of 30 kN/m excluding self weight over entire span. The beam is effectively restrained for a lateral buckling along its span. Check the section for shear and deflection. E = 2 × 10⁵ MPa. Refer table below for properties of available rolled steel beam.

Designation	Wt.	A	b	t_f	t_w	I_{xx}	Z_p	Z_{xx}	Root Radius
	N/m	mm²	mm	mm	mm	mm⁴	mm³	mm³	r_1 mm
ISLB 300	377	4808	150	9.4	6.7	7332 × 10⁴	554.32 × 10³	488.9 × 10³	15.0
ISLB 325	431	5490	165	9.8	7.0	9874.6 × 10⁴	687.76 × 10³	607.7 × 10³	16.0
ISLB 350	495	6301	165	11.4	7.4	13158.36 × 10⁴	851.11 × 10³	751.9 × 10³	16.0

14. A simply supported beam of effective span 6 m carrying u.d.*l.* 35 kN/m including self weight and point load 40 kN at centre. Check the beam of ISMB section having following properties for bending, shear and deflection.

    ISMB 450 @ 710.2 N/m

    $Z_{xx} = 1350.7 \times 10^3 \, mm^3$, $Z_p = 1533.35 \times 10^3 \, mm^3$, $I_{xx} = 30390.8 \times 10^4 \, mm^4$

    $t_w = 9.4 \, mm$, $t_f = 17.4 \, mm$, $h = 450 \, mm$, $r_1 = 15.0 \, mm$

15. A floor measuring 8×6 m consists of 10 cm thick RCC slab supported on RSJB, spacing 2 m c/c. Floor finish is 5 cm thick I.P. stone and live load 5000 N/m^2. Design B1 to satisfy bending, shear and deflection criteria and draw sketch showing connection of beam B_1 with main beam.

    Try ISMB - 350 @ 514 N/m, $A = 66.71 \, cm^2$, $t_w = 8.1 \, mm$, $I_{xx} = 13630.3 \, cm^4$.

    $Z_{xx} = 779 \times 10^3 \, mm^3$, $Z_p = 889.6 \times 10^3 \, mm^3$, $t_f = 14.2 \, mm$, $R_1 = 14.0 \, mm$

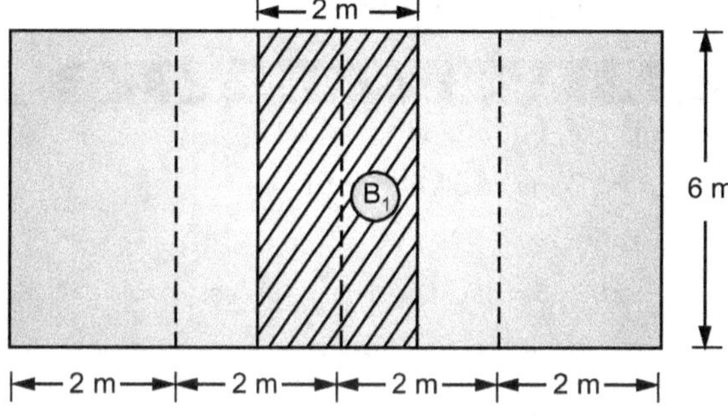

Fig. 7.12

16. An ISMB - 400 is used for a simply supported beam over effective span of 6 m. Calculate maximum UDL it can carry safely. ISMB - 400, $I_{xx} = 20458.4 \times 10^4 \, mm^4$, $t_w = 8.9 \, mm$, $t_f = 16.0 \, mm$, $Z_{xx} = 1020 \times 10^3 \, mm^3$, $R_1 = 14.0$, $Z_p = 1176.18 \times 10^3 \, mm^3$.

■■■

Chapter 8

STEEL ROOF TRUSS

Syllabus

- Types of Steel Roof Truss and its Selection Criteria. Calculation of Panel Point Load for Dead Load; Live Load and Wind Load as per I.S. 875-1987 Analysis and Design of Steel Roof Truss. Design of Angle Purlin as per I.S. Arrangement of Members at Supports.

About this Chapter

After reading this chapter students can understand:

- Analyse and Design component Parts of Steel Roof Truss.
- Types of Trusses used in Industries.
- Calculate Dead load, Live Load and Wind Load on Trusses.
- Design of Roof Truss Members and its connections.

8.1 INTRODUCTION

- *A truss is defined as a framed structure composed of members connected to each other at their ends, and forming triangles which lie in the same plane.*
- The members are subjected to direct (axial) stresses, as the truss is usually loaded at the point of intersection of its members only.

Advantages/Uses of Steel Roof Trusses:

1. At the places of high rainfall to avoid the roof drainage problems;
2. Where roofs have to support an additional load due to snowfall;
3. For very large span, where use of beams will make the construction most uneconomical.
4. For roofs of multistory buildings, industrial buildings, auditorium, cinema halls, malls, commercial complexes, stadium etc.
5. In the form of bracings in horizontal and vertical planes in industrial buildings to resist lateral loads and wind loads.

8.2 COMPONENT PARTS OF A STEEL ROOF TRUSS

- The various component parts of a steel roof truss have been shown in Fig. 8.1 below:

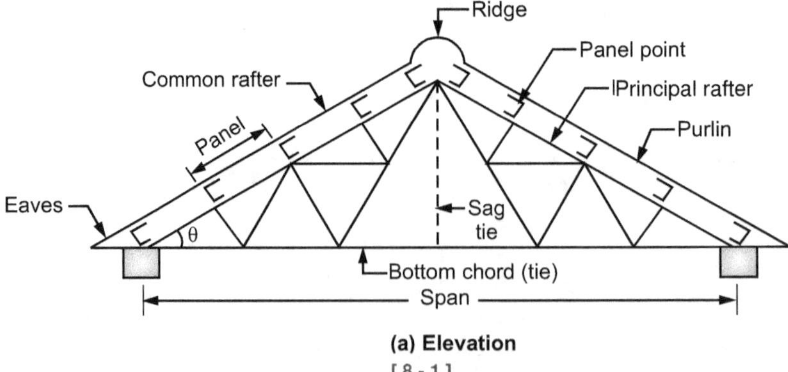

(a) Elevation

[8 - 1]

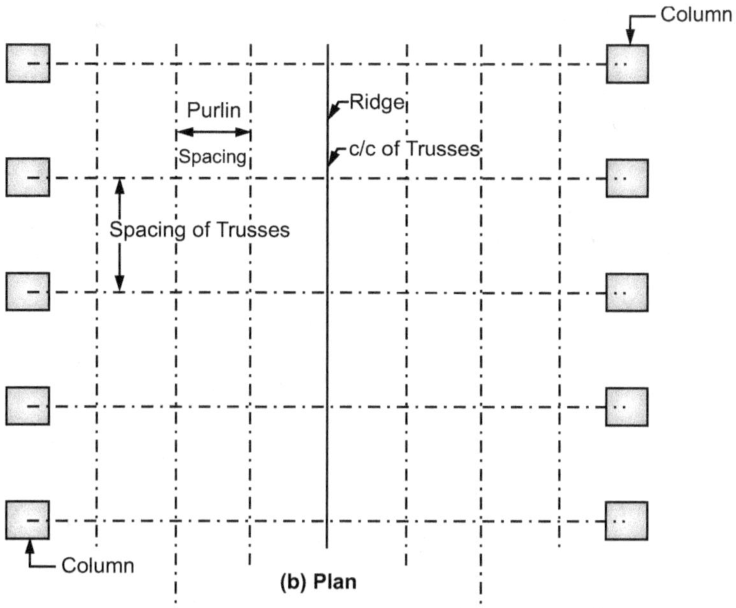

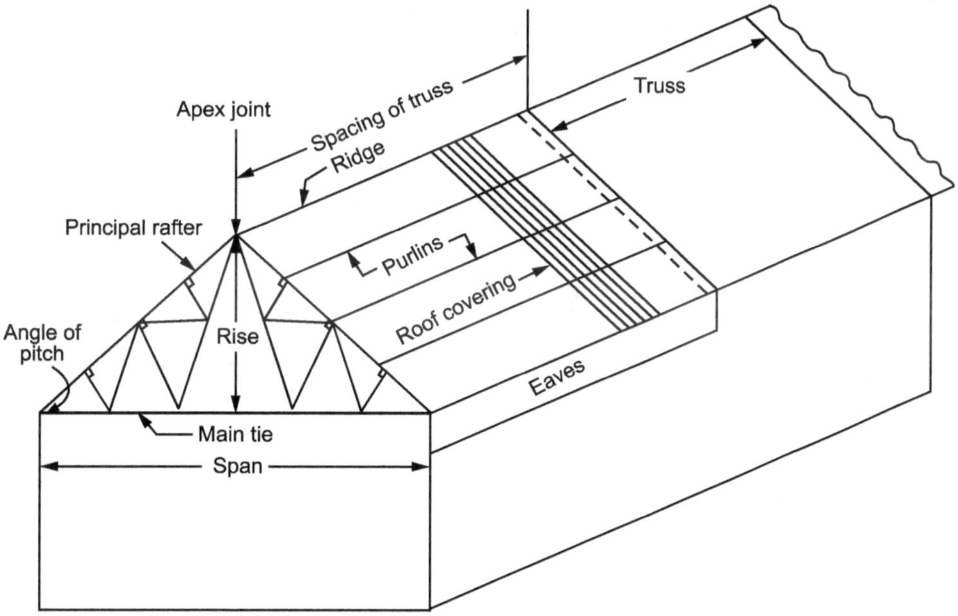

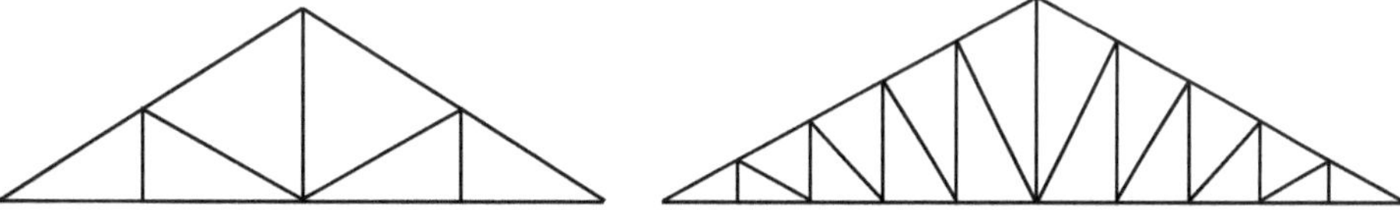

(c) Oblique view

Fig. 8.1: Details of Roof Truss

8.3 DIFFERENT TYPES OF TRUSSES

- Different types of steel roof trusses suitable for different spans are shown in Fig. 8.2.

(a) Howe Truss - With 4 and 8 Panels, Spans 6 m to 24 m

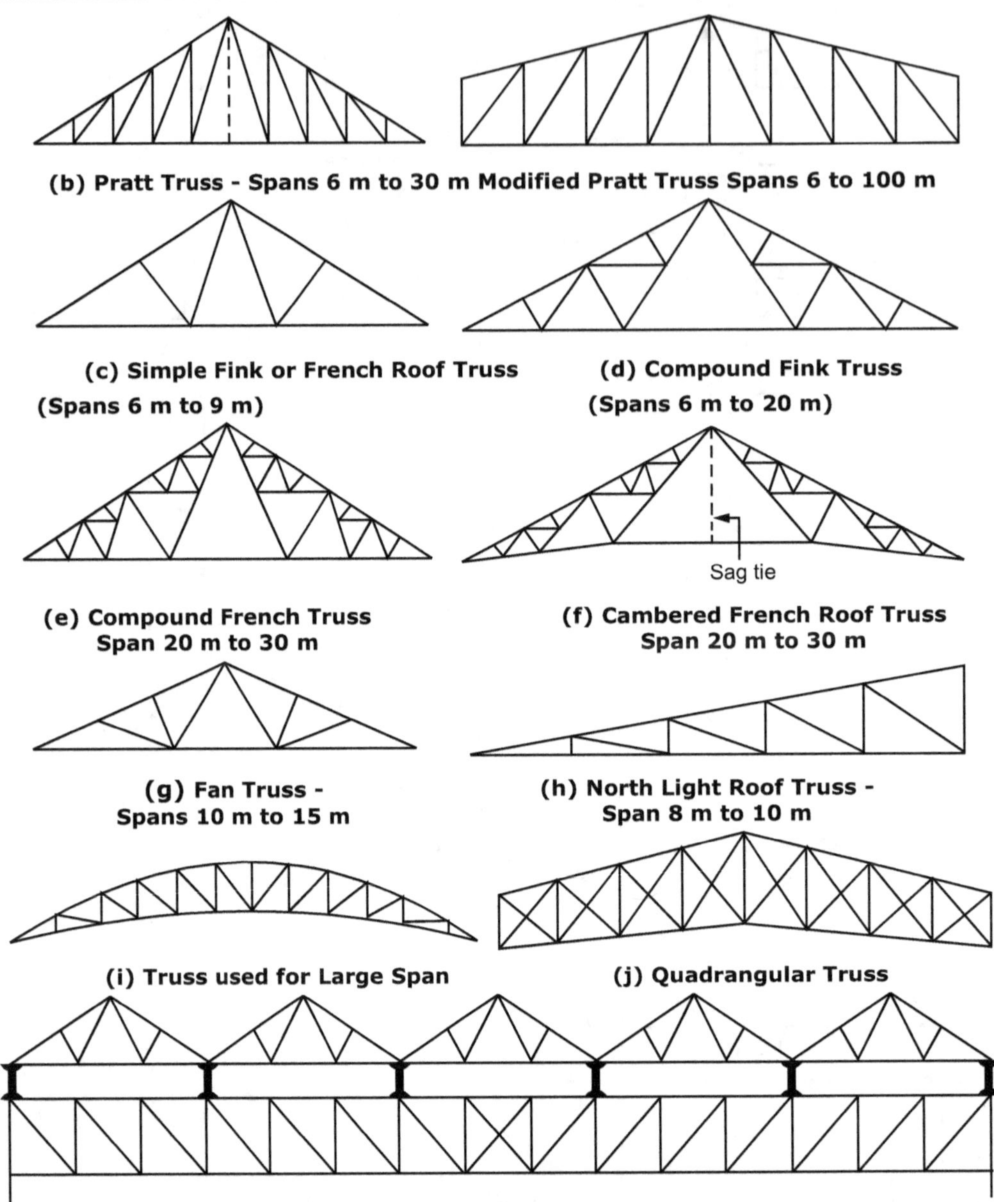

Fig. 8.2: Various Types of Roof Trusses

8.4 SELECTION CRITERIA OF THE TYPE OF TRUSS

- The type of roof truss to be provided depends mainly upon the pitch of the truss, fink trusses are provided for large pitch.
- Pratt and Howe trusses are provided for medium pitch, whereas Warren trusses are provided for small pitch.
- A skylight can be fitted on them for daylight.
- When the layout of the industrial building is such that more daylight is required, a North light truss is most suitable, as the natural light can be obtained from its geometry.
- Steep pitched roofs are avoided as these will be subjected to greater wind pressures and also its members will be long, as uneconomical.

- Some of the other factors which may affect the selection of a particular type of roof truss are:

    (a) Roof Coverings: The pitch of the truss depends upon the roofing material. The minimum recommended rise of the trusses with G.I. sheets is 1 in 6 and with A.C. sheets is 1 in 12.

    (b) Fabrication and Transportation: Normally the trusses are fabricated in the workshop and are transported to the site for erection. From transportation point of view, the depth of the truss becomes a controlling factor as it will not be feasible to transport a very deep truss.

    (c) Aesthetic: From aesthetic point of view, the architect/engineer may give a very flat or very deep truss.

    (d) Climate: The climate of a particular area plays an important role to the selection of the truss. Drainage of water, ice and snow retention etc. will have to be given due consideration.

8.5 TERMS USED IN ROOF TRUSSES

1. **Pitch of roof truss:** It is defined as the ratio between the rise and span of a truss. The value of pitch will depend upon the climatic condition and the nature of load, the truss has to sustain. Generally the value of the pitch range from $\frac{1}{3}$ to $\frac{1}{5}$. The minimum pitch of truss for roof covering of G.I. sheets is $\frac{1}{6}$ and that for asbestos sheets is $\frac{1}{12}$.

2. **Slope:** It is the ratio of rise to half span.

3. **Spacing of trusses:** It is defined as the distance between centre to centre of the trusses. The economical spacing of roof trusses work out to be $\frac{1}{3}$ to $\frac{1}{5}$ of span.

4. **Principal Rafter:** It is the top chord member of the truss and is usually in compression.

5. **Main Tie:** It is the bottom chord member of the truss and is usually in tension.

6. **Purlins:** These are flexural members subjected to transverse loads. It spans between two adjacent trusses and are normally provided at panel points.

7. **Ridge Line:** It is a line joining the apices of the trusses used in the construction of a roof truss.

8. **Sag tie:** A member joining peak of the truss and middle tie member is called sag tie. It helps in decreasing the deflection of maintenance.

9. **Eaves:** These are bottom ends of a sloping roof.

10. **Roof covering:** Corrugated G.I. sheets or A.C. sheets are commonly used for roof covering glass, fibre and slates are also used.

Corrugated G.I. sheets:

- The number of corrugations are 8, 10 or 11 per sheet, the pitch of corrugation is 75 mm and the depth of corrugation is 18 mm.

- These are classified into 1, 2, 3 and 4 classes, according to the zinc coating on the surface.

- Now-a-days colour coated, semi-corrugated G.I. sheets are also available.

- The available sizes and weight of G.I. sheets are as follows:

    Length = 1.2 to 4.8 m increasing by 0.15 m

    Width = 0.75, 0.9, 1.05 m and 1.20 m

Table 8.1: Spacing of Purlins for CGI Sheets

Thickness	Weight, (N/m^2)	Imposed load, (N/m^2)				
		500	750	1000	1250	1500
1.6 (16G)	176	240 cm	210 cm	195 cm	185 cm	180 cm
1.25 (18G)	142	225 cm	200 cm	180 cm	170 cm	165 cm
1.0 (20G)	112.7	210 cm	185 cm	170 cm	155 cm	150 cm
0.8 (22G)	93	195 cm	170 cm	155 cm	150 cm	140 cm
0.63 (24G)	73.4	180 cm	165 cm	150 cm	135 cm	125 cm

- The sheets are tied to purlins by 8 mm J or L type hook bolts with GI nuts plus GI and bituminous felt washers at a maximum pitch of 350 mm.

- In order to make joints water proof side laps of 1.5 to 2 corrugations are usually given.

Corrugated Asbestos Cement Sheets:

- These sheets are manufactured in corrugated and semi-corrugated shapes in thickness of 6 mm or 7 mm.

Table 8.2: (as per IS: 459 - 1962)

	Pitch (mm)	Depth of corrugation (mm)	Width (mm)
For A.C. corrugated sheets	146	48	1050
For semi-corrugated A.C. sheets	338	45	1100

- The lengths of the AC sheets are 1.50, 1.80, 2.10, 2.40, 2.70 and 3.0 m.

- The maximum spacing of purlins is:

- For 6 mm sheet – 1.4 m.

- For 7 mm sheet – 1.6 m.

- The weight of sheets varies from 120 – 160 N/m^2.

- In arriving at the dead load per square metre, the additional area due to side and end lapping and the larger sheet area on the inclined plane are considered.

- Hence the loads per square metre of sheet may be increased by 30 – 40% to get the load per square meter of plan area.

8.6 LOADS ON ROOF TRUSS

- The following loads act on roofs:

    1. **Dead Load:** The following loads per square metre of **plan area** may be used in design.

        (i)   Roof covering

             Corrugated G.I. sheets   = 100 to 150 N/m^2

             Corrugated A.C. sheets   = 120 to 160 N/m^2[*]

        (ii)   Purlins   = 80 to 120 N/m^2[**]

---

[*] Weight of roof covering shall be divided by cos θ because these are provided on inclined principal rafter. However, if in the problem it is mentioned to consider plan area for roof covering then it need to be divided by cos θ.

[**] When weight of purlin is given per m length, then weight = Wt./m × No. of purlins × Spacing of trusses

(iii) Truss

This is estimated from equation

$$W = \left(\frac{L}{3} + 5\right) \times 10$$

where, W = Weight of truss in N/m^2, and

            L = Span of truss in m.

(iv)   Wind bracing = 12 to 13 N/m^2

In appendix A unit weight to some of the common building materials is given in Table A-1 for the assessment of dead loads.

2. **Live or imposed load:** Live or imposed loads assumed in a design should be the greatest loads that might occur during its use, but shall not be less than the minimum loads specified in Table 8.1 and 8.2.

- **Imposed Load on Roofs:**

    (A) Roof membrane, sheets or purlins which directly support the roof covering shall be designed to carry maximum of the following two:

1.
Table 8.3: Imposed Loads on Roofs (Abstracts)

Type of roof	UDL per m^2 of plan area (N/m^2)	Minimum load
(a) Flat sloping or curved roof with slopes upto 10°		
(i) access provided	1500	3750 N u.d.l. over any span of 1 m width of roof slab and 9000 N u.d.*l.* over the span of any beam or truses or wall.
(ii) access not provided	750	Half of above load given in (i).
(b) Sloping roofs with slopes > 10°	750 – (θ - 10) 20	Subjected to a minimum of 400 N/m^2.

2. **Incidental concentrated load of 900 N over a length of 12.5 cm or 12.5 cm^2 in case of roof covering.**

    (B) Trusses, beams, rafters etc. at slope more than 10°: $\frac{2}{3}$ of the load in (A) above.

3. **Snow load:** If the structure is situated in an area where the roof is subjected to snow, the load considered for design should be maximum of the live or snow load. The load due to snow depends upon the pitch of the roof, shape of the roof and roofing material. Snow load may be taken as 2.5 N/m^2 for every mm depth of snow. When the roof slope is greater than 50°, the snow load may be neglected.

4. **Wind load:** Winds of speed over 80 km/hr are said to be very strong winds and are usually associated with cyclonic storms dust storms, thunder storms or active monsoons. Sometimes North-East India faces hurricanes of very high velocities for short durations. Wind causes pressure or suction normal to the surface of a structures. Sudden changes of wind speed cause forces on the trusses which was supporting dead load/live load in its normal course.

The most critical load on steel roof trusses and/or industrial buildings is the wind load. For calculation of wind load on structures IS: 875 – 1987 relates the intensity of wind pressure to the basic maximum wind speed (V_b), over a short interval of 3 seconds, with a 50 years return period, for different zones of the country. The wind pressure intensity at any height of a structure depends upon the velocity and density of air, shape and height of the structure, topography of the surrounding ground surface and the angle of wind attack.

8.7 WIND LOADS AS PER IS: 875 - 1987

- Following important modifications have been made in revised IS: 875.

1. Earlier wind pressure maps have been replaced by a single wind map. This map gives a basic maximum wind speed 'V_b' in m/sec at any location in India. (See Fig. 8.3). The wind speeds have been worked out on the basis of 50 years of observations of wind data. In this map, the whole country is divided into 6 zones depending upon the average wind pressure in that area. There are zones of wind speed 55, 50, 47, 44, 39 and 33 m/sec.

   Table 8.4 gives the values of basic wind speed at 10 m height for some important cities in India.

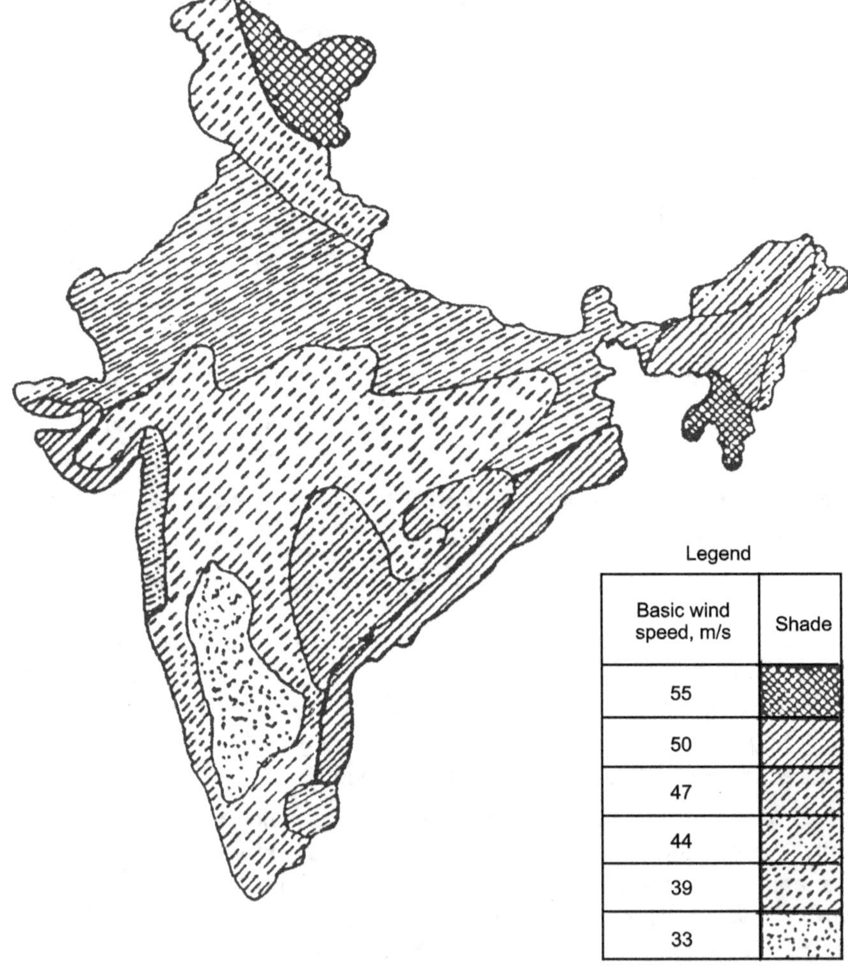

Fig. 8.3: Wind Map of India

Table 8.4: Basic Wind Speed at 10 m Height for Some Important Cities/Towns

City/Town	Basic wind speed, (m/s)	City, Town	Basic wind speed, (m/s)	City/Town	Basic wind speed, (m/s)
Agra	47	Ahmedabad	39	Ajmer	47
Almora	47	Amritsar	47	Asansol	47
Aurangabad	39	Bahraich	47	Bangalore	33
Barauni	47	Bareilly	47	Bhatinda	47
Bhilai	39	Bhopal	39	Bhubaneshwar	50
Bhuj	50	Bikaner	47	Bokaro	47
Chandigarh	47	Coimbatore	39	Cuttack	50
Darbhanga	55	Darjeeling	47	Calcutta	50
Delhi	47	Durgapur	47	Calicut	39
Gauhati	50	Gaya	39	Dehra Dun	47
Hyderabad	44	Imphal	47	Gantok	47
Jaipur	47	Jamshedpur	47	Gorakhpur	47
Jodhpur	47	Kanpur	47	Jabalpur	47
Kurnol	39	Lakshadweep	39	Jhansi	47
Ludhiana	47	Madras	50	Kohima	44
Mandi	39	Mangalore	39	Lucknow	47
Mumbai	44	Madurai	39	Madurai	39
Mysore	33	Nagpur	44	Moradabad	47
Nasik	39	Nellore	50	Nainital	47
Patiala	47	Patna	47	Panjim	39
Port Blair	44	Pune	39	Pondicherry	50
Rajkot	39	Ranchi	39	Raipur	39
Rourkela	39	Simla	39	Roorkee	39
Surat	44	Tiruchirrappalli	47	Srinagar	39
Udaipur	47	Vadodara	44	Trivandrum	39
Vijayawada	50	Visakhapatnam	50	Varanasi	47

2. The basic wind pressure is to be modified by multiplying it with modification factors. These factors are based on effects of terrain, local topography, size of structure etc.
3. Based on characteristics of ground surface irregularities, terrain is classified into four categories.
4. Force and pressure co-efficients have been included for a large range of clad and unclad buildings.
5. The pressure co-efficients for external and internal wind loads have been revised.
6. Some requirements regarding study of dynamic effects in flexible slender structures are included.

8.8 DESIGN WIND SPEED (V_z)

1. Basic wind speed 'V_b' is taken from the map of India given in IS: 875 – 1987 (See Fig. 8.3) and Table 8.4.
2. The basic wind speed is modified (V_z) by taking constants k_1, k_2 and k_3.

    where, V_z = Design wind speed (m/s) at any height z above ground.

    k_1 = Probability factor or risk co-efficient.

    k_2 = Terrain, height, structure, size factor.

    k_3 = Topography factor.

such that $$V_z = V_b \times k_1 \times k_2 \times k_3$$

8.8.1 Risk Co-efficient (k_1)

- Table 8.5 gives the values of Risk Co-efficient (k_1) for basic wind speed (m/s) for terrain category 2 as applicable at 10 m above ground level, based on 50 years life.

Table 8.5: Risk Co-efficient k_1

Class of structure	Probable k_1 for V_b in m/s						
	Life	33	39	44	47	50	55
General buildings and structures.	50 yrs	1.0	1.0	1.0	1.0	1.0	1.0
Temporary sheds, structures built temporarily during construction, boundary walls.	5 yrs	0.82	0.76	0.73	0.71	0.7	0.67
Buildings and structures with low degree of hazard to life and property such as isolated towers in wooded areas, non-residential farm buildings etc.	25 yrs	0.94	0.92	0.91	0.90	0.90	0.89
Important buildings and structures, hospitals, power plants such as communication on buildings/towers etc.	100 yrs	1.05	1.06	1.07	1.07	1.08	1.08

8.8.2 Terrain, Height and Structure Size Factor (k_2)

- Table 8.6 gives the terrain, height and structure size factor (k_2) which is divided into different categories depending upon ground surface roughness.

Table 8.6: k_2 factor

Height	Terrain Category											
	1			2			3			4		
	Class			Class			Class			Class		
(m)	A	B	C	A	B	C	A	B	C	A	B	C
10	1.05	1.03	0.99	1.00	0.98	0.93	0.91	0.88	0.82	0.80	0.76	0.67
15	1.09	1.07	1.03	1.05	1.02	0.97	0.97	0.94	0.87	0.80	0.76	0.67
20	1.12	1.10	1.06	1.07	1.05	1.00	1.01	0.98	0.91	0.80	0.76	0.67
30	1.15	1.13	1.09	1.12	1.10	1.04	1.06	1.03	0.96	0.97	0.93	0.83
50	1.20	1.18	1.14	1.17	1.15	1.10	1.12	1.09	1.02	1.10	1.05	0.95
100	1.26	1.24	1.20	1.24	1.22	1.17	1.20	1.17	1.10	1.20	1.15	1.05

Note:

1. It is permissible to assume constant wind speed between two heights for simplicity.
2. Terrain Categories:
   (a) Category 1 – Exposed open terrain with few or no obstructions and in which the average height of any object surrounding the structure is less than 1.5 m. It includes open sea coasts and flat treeless plains.
   (b) Category 2 – Open terrain with well scattered obstructions having heights between 1.5 to 10 m. It includes airfields, open park lands, open land near sea coasts, town outskirts etc.
   (c) Category 3 – Terrain with numerous closely spaced obstructions having heights upto 10 m. It includes well wooded areas, town and industrial areas etc.
   (d) Category 4 – Terrain with numerous closely spaced large and high obstructions. It includes large city centers and well developed industrial complexes.

3. Building/Structure Classes:

- Class A – Structures and/or their components such as cladding, roofing etc. having greatest horizontal or vertical dimension less than 20 m.
- Class B – Structures and/or their components such as cladding, roofing etc. having greatest horizontal or vertical dimensions between 20 to 50 m.
- Class C – Structures and/or their components such as cladding, roofing etc. having greatest horizontal or vertical dimension greater than 50 m.

8.8.3 Topography Factor (k_3)

- The effect of topography features is to accelerate wind near the summits of hills etc.

k_3 = Topography factor.

k_3 = 1 for upwind slope $\theta < 3°$ i.e. for level ground.

k_3 = 1 to 1.36 for upwind slope $\theta > 3°$ i.e. for sloping grounds.

For a hill or ridge, $k_3 = 1 + C \cdot s$.

where, $C = 1.2 \left(\dfrac{z}{L}\right)$ for upwind slope $3° - 17°$.

$C = 0.36$ for upwind slope $> 17°$.

z = Height of crest of hill.

L = Projected length of upwind zone from average ground level of crest in wind direction.

S = A factor obtained from Fig. 8.4.

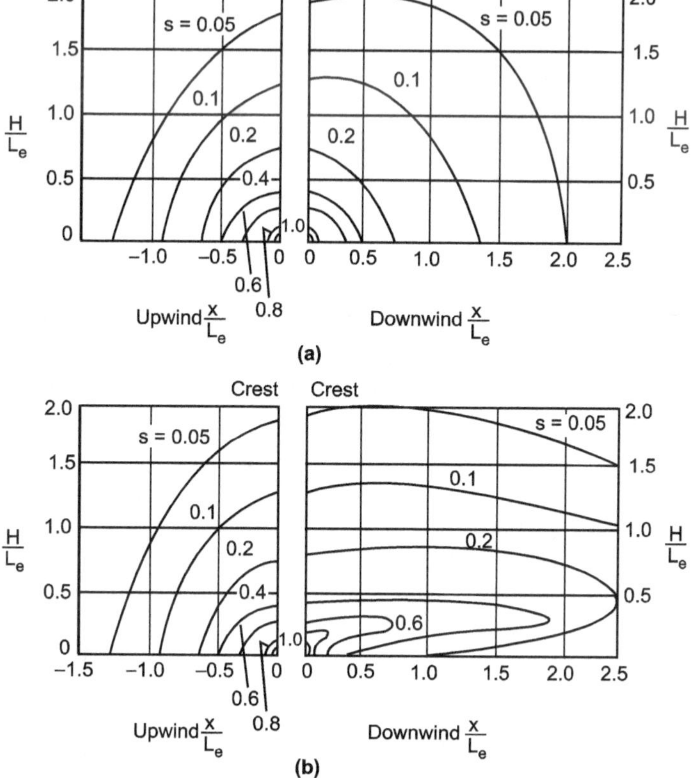

Fig. 8.4: Factors for Hill and Ridge

8.9 PROCEDURE FOR CALCULATION OF WIND LOAD

1. After knowing the factors k_1, k_2 and k_3, V_z is calculated as:

$$V_z = V_b \times k_1 \times k_2 \times k_3$$

2. Design wind pressure 'p_d' is calculated by the relation

$$p_d = 0.6 \, V_z^2 .$$

3. Wind load F on a building by static wind method is given by

$$F = C_f \times A_e \times p_d - \text{for the building as a whole.} \qquad \text{...(a)}$$

$$F = (C_{pe} - C_{pi}) \times A \times p_d - \text{for individual structural elements as roofs walls, glazings}$$

and their fixings.   ... (b)

where, C_f = Force co-efficient for the building (Table 8.7 to 8.9).

C_{pe}, C_{pi} = Force co-efficients for exterior and interior of the building.

(Table 8.10 to 8.12)

A_e, A = Effective area of the structure, member, A_e is taken less than total area due to openings and shielding.

Table 8.7: Wind Force Co-efficient C_f for Single Frames

Solidarity sections ratio, ϕ	C_f for flat sided members
0.1	1.9
0.2	1.8
0.3	1.7
0.4	1.7
0.5	1.6
0.75	1.6
1.00	2.0

Table 8.8: Wind Shielding Factor, for Multiple Frames

Solidarity ratio	Frame spacing ratio				
	< 0.5	1.0	2.0	4.0	> 8.0
0	1.0	1.0	1.0	1.0	1.0
0.1	0.9	1.0	1.0	1.0	1.0
0.2	0.8	0.9	1.0	1.0	1.0
0.3	0.7	0.8	1.0	1.0	1.0
0.4	0.6	0.7	1.0	1.0	1.0
0.5	0.5	0.6	0.9	1.0	1.0
0.7	0.3	0.6	0.8	0.9	1.0
1.0	0.3	0.6	0.6	0.8	1.0

$$\text{Frame Spacing Ratio} = \frac{\text{c/c distance of frames or girders}}{\text{Least overall dimension of frame}}$$

Table 8.9: Force Coefficient for Towers of Flat-Sided Members

Solidarity ratio, ϕ	Force coefficient for square tower	Equilateral triangular tower
0.1	3.8	3.1
0.2	3.3	2.7
0.3	2.8	2.3
0.4	2.3	1.9
0.5	2.1	1.5

Table 8.10: External Pressure Coefficients, C_{pe} for Pitched Roof of Rectangular Clad Buildings

Building height ratio	Roof angle α	Wind angle 0°		Wind angle 90°	
		EF	GH	EG	FH
$h/w \leq 1/2$	0	− 0.8	− 0.4	− 0.8	− 0.4
	5	− 0.9	− 0.4	− 0.8	− 0.4
	10	− 1.2	− 0.4	− 0.8	− 0.6
	20	− 0.4	− 0.4	− 0.7	− 0.6
	30	0	− 0.4	− 0.7	− 0.6
	45	+ 0.3	− 0.5	− 0.7	− 0.6
	60	+ 0.7	− 0.6	− 0.7	− 0.6
$1/2 < h/w \leq 3/2$	0	− 0.8	− 0.6	− 1.0	− 0.6
	5	− 0.9	− 0.6	− 0.9	− 0.6
	10	− 1.1	− 0.6	− 0.8	− 0.6
	20	− 0.7	− 0.5	− 0.8	− 0.6
	30	− 0.2	− 0.5	− 0.8	− 0.8
	45	+ 0.2	− 0.5	− 0.8	− 0.8
	60	+ 0.6	− 0.5	− 0.8	− 0.8
$3/2 < h/w < 6$	0	− 0.7	− 0.6	− 0.9	− 0.7
	5	− 0.7	− 0.6	− 0.8	− 0.8
	10	− 0.7	− 0.6	− 0.8	− 0.8
	20	− 0.8	− 0.6	− 0.8	− 0.8
	30	− 1.0	− 0.5	− 0.8	− 0.7
	40	− 0.2	− 0.5	− 0.8	− 0.7
	50	+ 0.2	− 0.5	− 0.8	− 0.7
	60	+ 0.5	− 0.5	− 0.8	− 0.7

- **Note:** h is the height to eaves or parapet and w is the lesser dimension of the building.

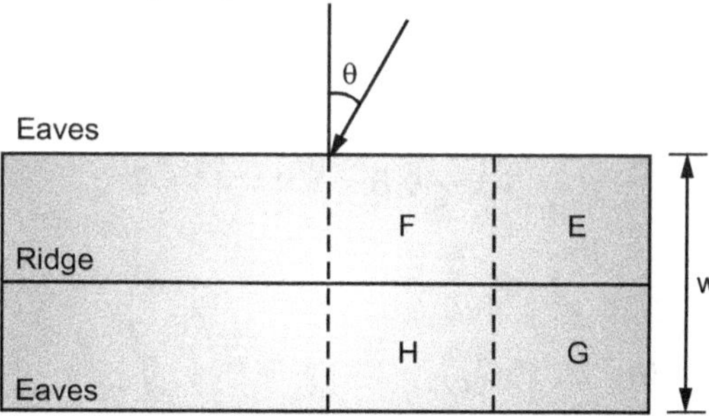

Fig. 8.5: Sloping Roof Plan

Table 8.11: External Pressure Coefficients, C_{pe} for Walls of a Rectangular Clad Building

Building height ratio	Building plan ratio	Wind angle $\theta°$	C_{pe} for surface			
			A	B	C	D
$\frac{h}{w} \leq \frac{1}{2}$	$1 < \frac{l}{w} \leq \frac{3}{2}$	0	+ 0.7	− 0.2	− 0.5	− 0.5
		90	− 0.5	− 0.5	+ 0.7	− 0.2
	$\frac{3}{2} < \frac{l}{w} < 4$	0	+ 0.7	− 0.25	− 0.6	− 0.6
		90	− 0.5	− 0.5	+ 0.7	− 0.1
$\frac{1}{2} < \frac{h}{w} \leq \frac{3}{2}$	$1 \leq \frac{l}{w} \leq \frac{3}{2}$	0	+ 0.7	− 0.25	− 0.6	− 0.6
		90	− 0.6	− 0.6	+ 0.7	− 0.25
	$\frac{3}{2} \leq \frac{l}{w} < 4$	0	+ 0.7	− 0.3	− 0.7	− 0.7
		90	− 0.5	− 0.5	+ 0.7	− 0.1
$\frac{3}{2} < \frac{h}{w} < 6$	$1 < \frac{l}{w} \leq \frac{3}{2}$	0	+ 0.8	− 0.25	− 0.8	− 0.8
		90	− 0.8	− 0.8	+ 0.8	− 0.25
	$\frac{3}{2} \leq \frac{l}{w} \leq 4$	0	+ 0.7	− 0.4	− 0.7	− 0.7
		90	− 0.5	− 0.5	+ 0.8	− 0.1

- **Note:** h is the height to eaves or parapet, l is the greater horizontal dimension of a building and w is the lesser horizontal dimension of a building.

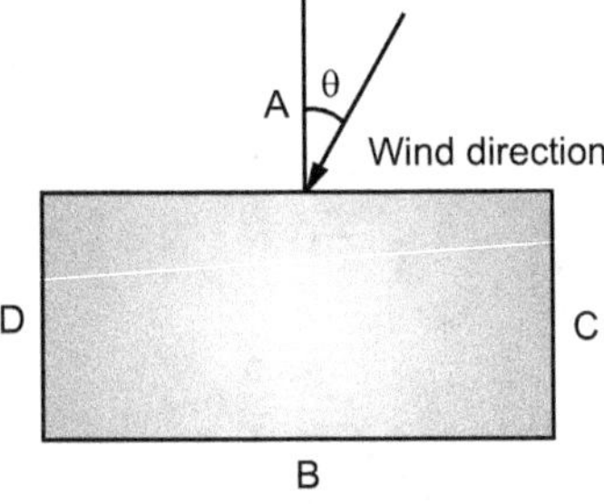

Fig. 8.6: Building Plan

Table 8.12: Internal Pressure Coefficients, C_{pi}

Permeability	Wall opening	C_{pi}
Low	upto 5% of wall area	± 0.2
Medium or Normal	5 – 20% of wall area	± 0.5
High	> 20% of wall area	± 0.7

SOLVED EXAMPLES

Ex. 8.1: *Find out the wind pressure to be accounted for designing a sloping roof of a shed having span 15 m and its pitch $\frac{1}{4}$. The height of the shed may be considered as 6 m with permeability. The building is situated in Mumbai.*

Assume, probability factor k_1 = 1.0, terrain/height/structure size factor k_2 = 0.8, topography factor k_3 = 1.0.

Sol.: Basic wind pressure at Mumbai, V_b = 44 m/s

∴ Design wind speed,

$$V_z = k_1 \cdot k_2 \cdot k_3 \cdot V_b = 1.0 \times 0.8 \times 1.0 \times 44$$

$$= 35.2 \text{ m/s}$$

Design wind pressure,

$$p_d = 0.6 \, V_z^2 = 0.6 \times (35.2)^2$$

$$= 743.42 \text{ N/m}^2 \simeq 750 \text{ N/m}^2.$$

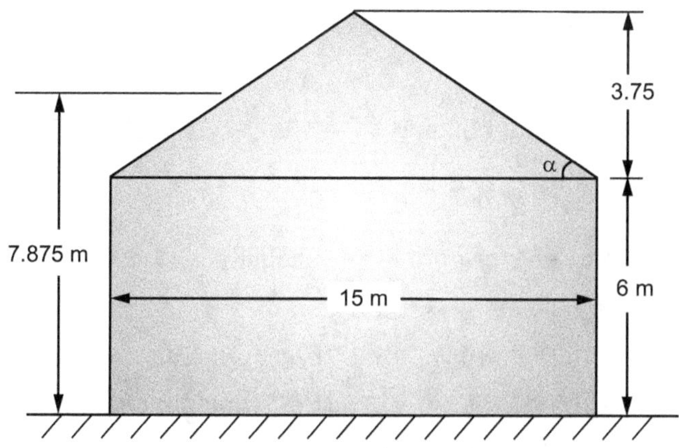

Fig. 8.7

Rise $= \dfrac{1}{4} \times$ span $= \dfrac{1}{4} \times 15 = 3.75$ m

Mean height of roof above G.L. $= 6 + \dfrac{3.75}{2} = 7.875$ m

Basic wind pressure, $p = 750$ N/m^2

Slope of roof, $\alpha = \tan^{-1}\dfrac{1}{2} = 26.56°$

Case I: Wind normal to ridge

From Table 8.10 and 8.12, assuming $\dfrac{h}{w} < \dfrac{1}{2}$. $\left(\because \dfrac{h}{w} = \dfrac{6}{15} = 0.4 \right)$

External wind pressure coefficient 'C_{pe}' will be as given below.

Slope	For wind angle 0°		Wind angle 90°	
	Windward	**Leeward**	**Near Gable end** $\left(\dfrac{1}{4} \text{ length of bldg.}\right)$	**Internal bays** $\left(\text{mid } \dfrac{1}{2} \text{ length}\right)$
20°	− 0.4	− 0.4	− 0.7	− 0.6
30°	0	− 0.4	− 0.7	− 0.6

On windward slope, $C_{pe} = -\left[0.4 - 0.4 \times \dfrac{6.6}{10}\right] = -0.1376$

On Leeward slope, $C_{pe} = -0.4$

Internal air pressure co-efficient for low permeability, $C_{pi} = \pm 0.2$ [Refer Table 8.12]

Combined wind pressure = (External wind pressure $\pm$ Internal wind pressure)

Windward slope $(- 0.1376 - 0.2) \times 750 = -253.2$ N/m^2 (uplift)

$(- 0.1376 + 0.2) \times 750 = +46.8$ N/m^2 (downward)

Consider pulling pressure as –ve, and pushing pressure as +ve.

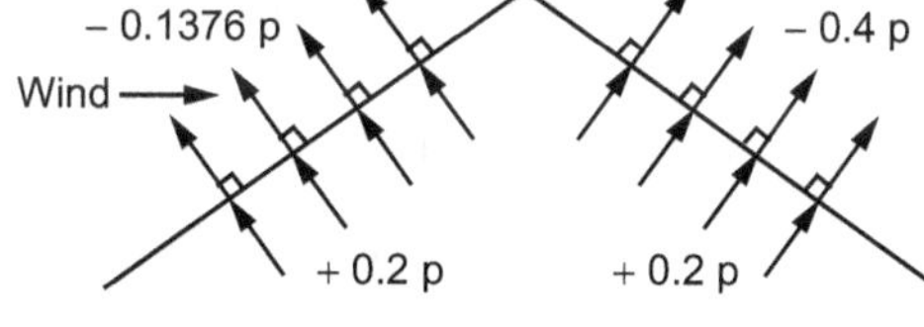

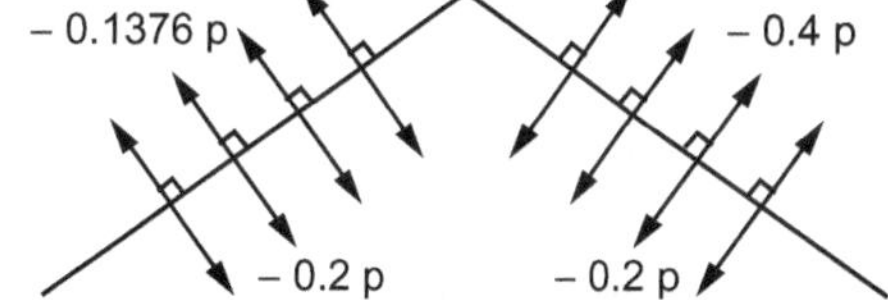

    (a) With +ve Internal Pressure   **(b) With –ve Internal Pressure**

Fig. 8.8

Leeward slope $(- 0.4 - 0.2) \times 750 = -450$ N/m^2 (uplift)

$(- 0.4 + 0.2) \times 750 = -150$ N/m^2 (uplift)

Case II: Wind Parallel to ridge

External wind pressure co-efficient C_{pe}.

On both slopes for $\frac{1}{4}^{th}$ length of building $= - 0.7$.

On both slopes for mid $\frac{1}{2}$ length of building $= - 0.6$.

Internal air pressure co-efficient for normal permeability, $C_{pi} = \pm 0.2$.

Combined external + internal wind pressure

Near Gable end $(- 0.7 - 0.2) \times 750 = - 675$ N/m^2 (uplift)

$$(- 0.7 + 0.2) \times 750 = - 375 \text{ N/m}^2 \text{ (uplift)}$$

Internal bays $(- 0.6 - 0.2) \times 750 = - 600$ N/m^2 (uplift)

$$(- 0.6 + 0.2) \times 750 = - 300 \text{ N/m}^2 \text{ (uplift)}$$

- The roof should be designed for the worst effect of wind. On studying the calculations, it is evident that the roof will be subjected to maximum uplift, when the wind blows parallel to ridge and the internal pressure is +ve.

- Hence roof will be designed for a pressure of 675 N/m^2 on both roof slopes.

8.10 PROCEDURE OF DESIGNING A STEEL ROOF TRUSS

1. From the span, roofing material, lighting etc. available decide the type of truss.

2. Decide spacing of truss between $\frac{1}{5}$ to $\frac{1}{3}$ of the span.

3. Calculate dead load (DL), live load (LL), snow load (SL), and wind load (WL) if any on the truss. Sometimes weight of purlin is given in N/m^2 on plan area and sometimes it is given per m plan length. Thus, appropriate plan area or length shall be multiplied.

4. Find the forces developed in each member for each type of load either graphically or analytically.

5. Find maximum force in each member for combination of loads like:
   (a) 1.5 (DL + LL), (b) 1.2 (DL + LL + WL), (c) 0.9 DL + 1.5 WL

6. Group the member having equal forces for design.

7. Design the member with following specifications:

(A) Compression Members:

(i) Use double angle section for top chord member and design it as continuous member and for struts use single angle and design it as discontinuous structure.

(ii) For single angle section, use an equal angle and for double angle section use unequal angles with longer legs vertical.

(iii) Minimum size of angle for top chord member is 50 × 50 × 6 mm and for struts is also 50 × 50 × 6 mm., so as to avoid web buckling.

(iv) Minimum thickness of gusset plate 6 mm and minimum nominal diameter of bolts is 16 mm.

(v) Use minimum two bolts at each end.

(vi) Slenderness ratio of the member should not exceed 180.

(vii) Effective length = 0.7 L to L

(viii) As the member of the truss may be subjected to reversal of stresses, the designed compression member shall be checked for tension and vice-versa.

(B) Tension Members

(i)   Use double angle sections for the main tie and single angle iron for other ties.

(ii)   Minimum size of angles for the main tie is $50 \times 50 \times 6$ mm and that for other ties is also $50 \times 50 \times 6$ mm. Because single angle ties have twisting tendency and produce eccentric forces at joints, double angle sections are preferred.

(iii)   Where a tie is subjected to reversal of loads, its slenderness ratio should not exceed 350.

(C) Design of End Bearings

(i)   When trusses are supported on steel column, end connections are designed as hinged joints and should be strong enough to resist the uplift of the truss.

(ii)   When trusses are supported on R.C.C. columns, masonry walls, bearing plate should be provided to distribute the load evenly.

(iii)   One end of bearing may be fixed and other sliding bearing by passing down to it through slotted holes in the bearing plates.

(iv)   The bolts should be sufficient to resist the uplift.

Ex. 8.2: *A pratt type roof truss is required for workshop building with following details, Span = 10 m, Slope 30°, spacing of truss 4 m, Roofing AC sheet 160 N/m², Weight of purlin 100 N/m, Weight of truss 100 N/m². The height of eaves is 5.5 m and is situated near Delhi. Take risk factor k_1 = 1.0, height and size factor k_2 = 0.88, topography factor k_3 = 1.0, For Delhi basic wind speed, V_b = 47 m/s. Find the panel point loads for DL, LL and WL.*

Sol.:   Sloping area $= \dfrac{5}{\cos 30} \times 4 \times 2 = 46.12$ m²

                 Plan area $= 10 \times 4 = 40$ m²

Loads

(i) Live load for purlin $= 750 - (30 - 10) \times 20 = 350$ N/m²

          Minimum live load $= 400$ N/m² $\therefore$ O.K.

            L.L. for trusses $= \dfrac{2}{3} \times 400 = 266.67$ N/m²

(ii) Wind load:

       Design wind speed $V_z = V_b \times k_1 \times k_2 \times k_3$

                      $= 7 \times 1.0 \times 0.88 \times 1.0$

                      $= 41.36$ m/s

     Design wind pressure $p_z = 0.6\, V_z^2 = 0.6 \times (41.36)^2$

                      $= 1026.4$ N/m²

From Table 8.10 and 8.12, wind load is evaluated assuming low permeability as follows:

$$\frac{h}{w} = \frac{6.50}{10.00} = 0.65$$

i.e. $\dfrac{1}{2} < \dfrac{h}{w} < \dfrac{3}{2}$

Slope	For wind angle 0°		For wind angle 90° on both slopes	
	Windward	Leeward	Near Gable end (For outer $\frac{1}{4}^{th}$ length of building)	Internal bays (For mid $\frac{1}{2}$ length of building)
30°	$- 0.2$	$- 0.5$	$- 0.8$	$- 0.8$

         Total pressure $= (C_{pe} - C_{pi})\ p_z$

	Windward slope	**Leeward slope**
Wind normal to ridge		
(a) With internal pressure + 0.2 p	$(- 0.2 - 0.2^*) \times 1026.4$ $= - 410.56$ N/m^2	$(- 0.5 - 0.2) \times 1026.4$ $= - 718.48$ N/m^2
(b) With internal pressure − 0.2 p	$(- 0.2 + 0.2) \times 1026.4$ $= 0$	$(0.5 + 0.2) \times 1026.4$ $= - 307.92$ N/m^2
Wind parallel to ridge	outer $\frac{1}{4}^{th}$ length of building	mid $\frac{1}{2}^{th}$ length of building
(a) With internal pressure + 0.2 p	$(- 0.8 - 0.2) \times 1026.4$ $= - 1025.4$ N/m^2	$(- 0.8 - 0.2) \times 1026.4$ $= - 1025.4$ N/m^2
(b) With internal pressure − 0.2 p	$(- 0.8 + 0.2) \times 1026.4$ $= - 615.84$ N/m^2	$(- 0.8 + 0.2) \times 1026.4$ $= - 615.84$ N/m^2

***Negative because effect of positive internal pressure +0.2 p will uplift the roof.**

Thus maximum wind load = − 1026.4 N/m^2 (uplift on both slopes)

Panel Point Loads

(i)   Dead load

$$\begin{aligned}
\text{For A.C. sheets} &= 160 \times 46.12^* &&= 7380 \text{ N} \\
\text{Purlins} &= 100 \times 8 \times 4 &&= 3200 \text{ N} \\
\text{Trusses} &= 100 \times 40 &&= 4000 \text{ N} \\
\text{Total D.L.} & &&= 14580 \text{ N}
\end{aligned}$$

$$\text{Dead load on each top panel} = \frac{14580}{6} = 2430 \text{ N}$$

$$\therefore \quad \text{Dead load at end panel points} = \frac{2430}{2} = 1215 \text{ N}$$

(ii)   Imposed load for trusses on top panel points = 266.67 × 40 = 10667

$$\text{Imposed load per panel} = \frac{10667}{6} = 1778 \text{ N}$$

$$\therefore \quad \text{Imposed load at end panel points} = \frac{1778}{2} = 889 \text{ N}.$$

(iii)   Wind load on each of the top panel = − 1026.4 × 1.93 × 4 = − 7924 N (Uplift)

$$\therefore \quad \text{Wind load at end panel points} = \frac{-7924}{2} = 3962 \text{ N}$$

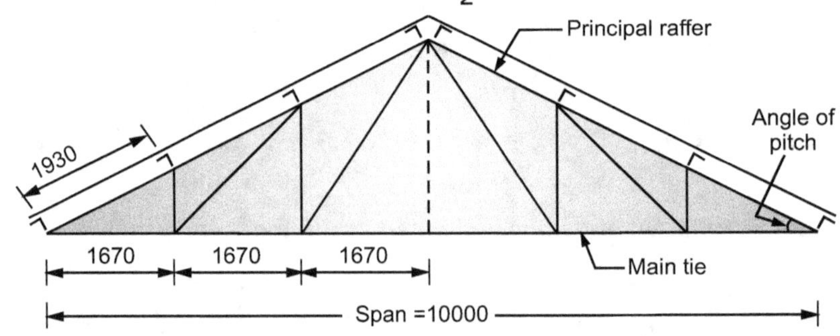

Fig. 8.9

Ex. 8.3: *A truss of 20 m span has 4.5 m rise. There are 5 panels on each side. Trusses are spaced at 3.5 m/c. The design wind pressure is 1200 N/m^2. There are 20% openings. Calculate the panel point load for wind load.*

Wind load co-efficient

$$\text{For } \left. \begin{array}{l} 20^\circ = - 0.4 \text{ p} \\ 30^\circ = - 0 \text{ p} \end{array} \right\} \begin{array}{l} \text{Windward} \\ \text{side} \end{array} \quad \& \quad \left. \begin{array}{l} - 0.4 \text{ p} \\ - 0.4 \text{ p} \end{array} \right\} \begin{array}{l} \text{Leeward} \\ \text{side} \end{array}$$

**Consider inclined area in case of roofing material.*

Slope	For wind angle 0°		For wind angle 90° on both slopes	
	Windward slope	Leeward slope	For outer $\frac{1}{4}^{th}$ length of building	For mid $\frac{1}{2}^{th}$ length of building
20°	– 0.4 p_z	– 0.4 p_z	– 0.7 p_z	– 0.6 p_z
30°	0	– 0.4 p_z	– 0.7 p_z	– 0.6 p_z

Sol.:

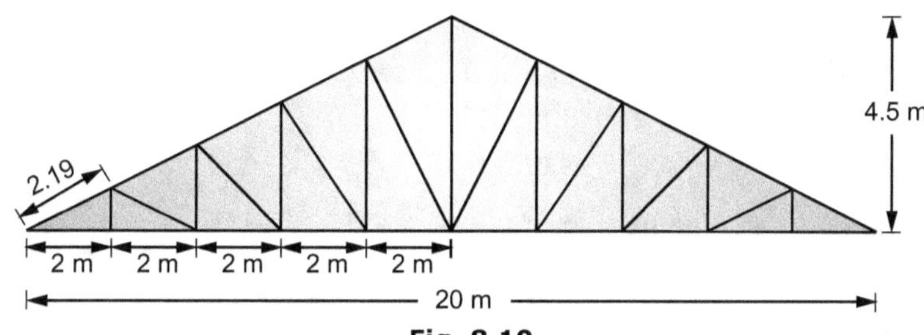

Fig. 8.10

$$\theta = \tan^{-1}\frac{4.5}{10} = 24.23°$$

External wind pressure on windward side

$$= -\left[0.4 - \frac{0.4 - 0}{10} \times 4.23\right] p_z = -0.23\ p_z$$

External wind pressure on Leeward side $= -0.4\ p_z$

Internal wind pressure for 20% wall opening $C_{pi} = \pm 0.5\ p_z$, $p_z = 1200\ N/m^2$

Total wind pressure $= \left[C_{pe} - C_{pi}\right]\ p_z$

Solved Examples

Ex. 8.4: *A hall of size 15 m × 30 m is provided with Fink type steel roof trusses at 3.75 m c/c. Calculate panel point load for dead load and live load cases from following data.*

(i) Unit weight of roof covering = 175 N/m²

(ii) Self weight of purlin =100 N/m²

(iii) Weight of bracing = 60 N/m²

(iv) Rise to span ratio = $\frac{1}{5}$

(Assume additional data required if any)

Ans. DL and LL on Panel point of Fink Truss

Given: Hall of size = 15 m × 30 m

∵   Span of Truss = L = 15 m

Spacing of Truss = S = 3.75 m c/c

Pitch of Truss $= \frac{1}{5} = \frac{R}{L}$

∵   Slope of Truss $\theta = \tan^{-1}\left(2 \times \frac{1}{5}\right) = 21.80°$

Assume above truss with 8 panels

Panel Point Plan Area $= \frac{15}{8} \times 3.75 = 7.03\ m^2$

DL calculations: DL on plan area of Roof Truss is as below:

1.   Wt. of roof covering material = 175 N/m²

2.   Self Wt. of Purlin $= 100$ N/m^2

3.   Wt. of bracing $= 60$ N/m^2

4.   Self Wt. of roof truss $= \left(\dfrac{\text{Span}}{3} + 5\right) 10$

$$= \left(\dfrac{15}{3} + 5\right) 10$$

$$= 100 \text{ N/m}^2$$

$$\text{Total DL} = 435 \text{ N/m}^2$$

$$\text{DL on each interior panel point} = 7.03 \times 4.35$$

$$= 3058.05 \text{ N}$$

$$(A) = 3.06 \text{ kN}$$

$$\text{DL on each end panel point} = \dfrac{3.06}{2}$$

$$= 1.53 \text{ kN}$$

LL calculations: (min. 400 N/m^2)

$$\text{LL of purlins} = 750 - (21.80 - 10)\,20$$

$$= 514 \text{ N/m}^2$$

$$\text{LL of Truss supporting purlin} = \dfrac{2}{3}\,(514)$$

$$\text{LL of Truss supporting purlin} = 342.62 \text{ N/m}^2$$

$$\text{LL on each interior panels} = (A) = 7.03 \times 342.62$$

$$\text{LL on each interior panels} = (A) = 2408.9 \text{ N} = 2.41 \text{ kN}$$

$$\text{LL on each end panels} = \dfrac{2.41}{2}$$

$$= 1.205 \text{ kN}$$

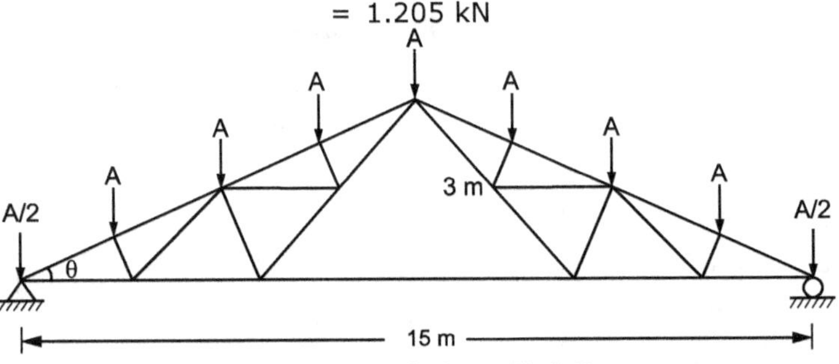

Fig. 8.11: DL and LL on Fink Truss
for DL (A) $= 3.06$ kN
for LL (A) $= 2.41$ kN

Note: Any other assumption for no. of panels may change answer.

8.11 DESIGN OF ANGLE PURLINS

- Purlins are the members spanning on the roof frames running generally through top chord joints. They support roof covering and span of purlin is the spacing between adjacent trusses.
- When angles are used, the connections of the purlin to the rafter are made by using cleat angles as shown in Fig. 8.12 (a) for channel section purlin. (See Fig. 8.12 (b)).
- The gravity loads act vertically through the c.g. of the purlin whereas wind load act normal to the roof slopes.
- The purlin is subjected to bi-axial bending because it is not laterally supported. If the section is unsymmetrical it has to resist lateral torsional buckling.

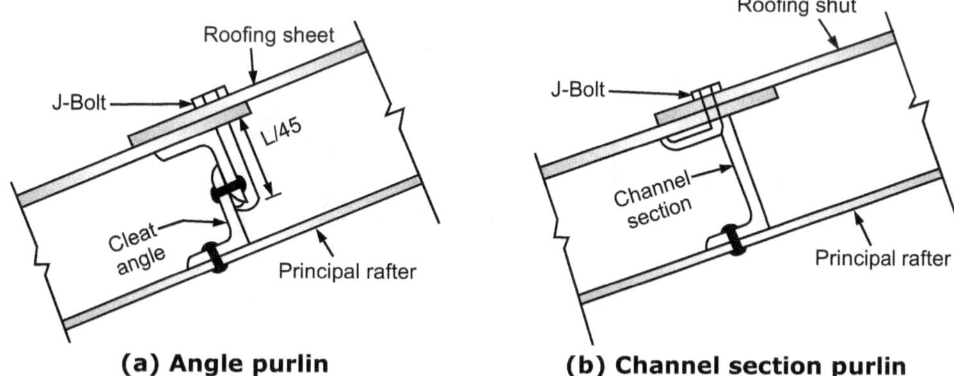

(a) Angle purlin   **(b) Channel section purlin**

Fig. 8.12

- As per IS: 800-2007 angle purlins may be designed as per following step-by-step procedure:
- Consider the span of purlin as c/c distance between the trusses. Also find the angle of pitch.
   1. Calculate the gravity load due to sheeting, self-weight of purlin and live load. This is calculated on plan area. Also calculate wind load which is normal to roof. Generally wind load is considered negative because it uplifts the sheeting and purlin due to combination of internal and external wind pressure.
   2. Calculate the components of these loads parallel to roof (along xx axis) and normal to roof (along yy axis).
   3. Compute the factored loads due to following combinations:
      (a)   Load combination 1 = 1.5 (DL + LL)
      (b)   Load combination 2 = 1.5 (DL + WL)
      (c)   Load combination 3 = 1.2 (DL + LL + WL)
      Find the critical load combination and the thus calculate wy and wx.
   4. Calculate bi-axial moments $M_x = \dfrac{w_y L^2}{10}$ and $M_y = \dfrac{w_x L^2}{10}$.
   5. Select the angle section satisfying following requirements:

      (a) The width of angle pulrin (parallel to roof covering) should not be less than $\dfrac{L}{60}$.

      (b) The depth of angle pulrin (normal to roof covering) should not be less than $\dfrac{L}{45}$.

      Classify the section on the basis of limiting breadth to thickness and depth to thickness ratios.
   6. Calculate the design moment: For bending @ xx axis

      $M_{dx} = \beta_b \times Z_p \dfrac{f_y}{\gamma_{mo}}$ or $1.2 \, Z_e \times \dfrac{f_y}{\gamma_{mo}}$ whichever is less.

      $\beta_b = 1$ for plastic and compact sections.

      For semi-compact section $M_{dx} = Z_e \dfrac{f_y}{\gamma_{mo}}$.

      Similarly for bending @ yy axis, calculate M_{dy}.
   7. Check for bi-axial bending for outstanding toe of purlin using *interaction formula.*

      $\dfrac{M_x}{M_{dx}} + \dfrac{M_y}{M_{dy}} \leq 1$

Ex. 8.5: *Design an equal angle purlin for a roof truss having following details:*

(i)   Spacing of trusses = 3.5 m

(ii) Span of truss = 15 m, rise = 3m.

(iii) Spacing of purlins = 1.35 m.

(iv) Weight of G.I. sheets = 130 N/m².

(v) Wind load on purlin = 1170 N/m².

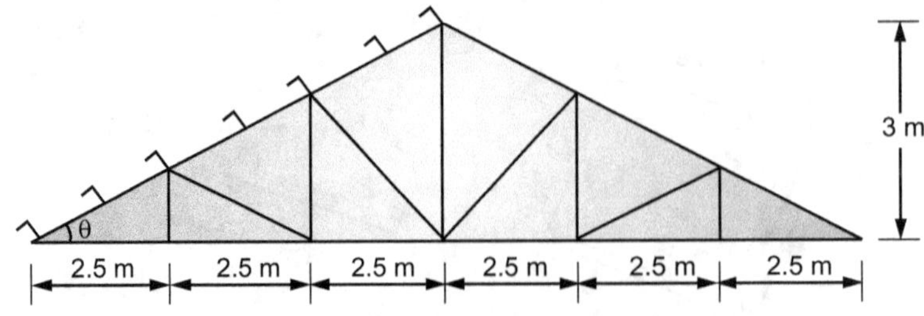

Fig. 8.13

Sol.: $\theta = \tan^{-1}\left(\dfrac{3}{7.5}\right) = 21.80°$ $\therefore$ $\cos\theta = 0.928$ and $\sin\theta = 0.371$

1. Load calculation: (For 1.35 m spacing of purlins)

    (a)   D.L. of G.I. Sheet $= 130 \times 1.35$ $= 176$ N/m

            Self Wt. of purlin   (assumed) $= 100$ N/m

                                      Total $= 276$ N/m

    (b)   LL $= 750 - 20\,(\theta - 10) = 750 - 20\,(21.80 - 10)$

                                    $= 514$ N/m^2

  $\therefore$ LL on purlin per metre $= 514 \times 1.35 \cos\theta$

                                     $= 643.94$ say 644 N/m

    (c)   Wind load $= -1170 \times 1.35 \approx -1580$ N/m (Normal to roof).

2. Components of load along xx axis (parallel to roof) and along yy axis (normal to roof).

    (a) DL, $w_{dy} = 276 \times \cos 21.80° = 256$ N/m

                     $w_{dx} = 276 \times \sin 21.80° = 102.5$ N/m

    (b) LL, $w_{ly} = 644 \times \cos 21.80° = 598$ N/m

                     $w_{lx} = 644 \times \sin 21.80° = 239$ N/m

    (c) WL, $w_{wy} = -1580$ N/m,

                     $w_{wx} = 0$

3. Factored loads due to combinations

    (a) Load combination 1 $= 1.5\,(DL + LL)$

                     $w_{y1} = 1.5\,(256 + 598) = 1281$ N/m

                     $w_{x1} = 1.5\,(102.5 + 239) = +512$ N/m

    (b) Load combination 2 $= 1.5\,(DL + WL)$

                     $w_{y2} = 1.5\,(256 - 1580) = -1986$ N/m

                     $w_{x2} = 1.5\,(102.5 - 0) = +154$ N/m

    (c) Load combination 3 $= 1.2\,(DL + LL + WL)$

                     $w_{y3} = 1.2\,(256 + 598 - 1580) = -871$ N/m

                     $w_{x3} = 1.2\,(102.5 - 239 - 0) = +410$ N/m

Thus among above three combinations $1.5\,(DL + WL)$ is critical.

  $\therefore$ $w_y = 1986$ N/m (absolute value), $w_x = 154$ N/m

4. Bi-axial BMs

$$M_x = \frac{w_y \cdot L^2}{10} = \frac{1986 \times 3.5^2}{10} = 2432.85 \text{ Nm}$$

$$M_y = \frac{w_x \cdot L^2}{10} = \frac{154 \times 3.5^2}{10} = 188.65 \text{ Nm}$$

5. Selection of angle section:

$$\text{Width of angle parallel to roof} = \frac{L}{60} = \frac{3500}{60} = 58.33 \text{ mm}$$

$$\text{Depth of angle normal to roof} = \frac{L}{45} = \frac{3500}{45} = 77.78 \text{ mm}$$

Try ISA $90 \times 90 \times 6$ @ 8.2 kg/m $\simeq$ 82 N/m < 100. $\therefore$ O.K.

Also, $\dfrac{b}{t} = \dfrac{90}{6} = 15 < 15.7 \, \varepsilon \, (\because \, \varepsilon = 1)$

$\dfrac{d}{t} = \dfrac{90}{6} = 15 < 15.7 \, \varepsilon$

Thus this section is semi-compact

$\therefore$ $\beta_b = \dfrac{Z_e}{Z_p}$

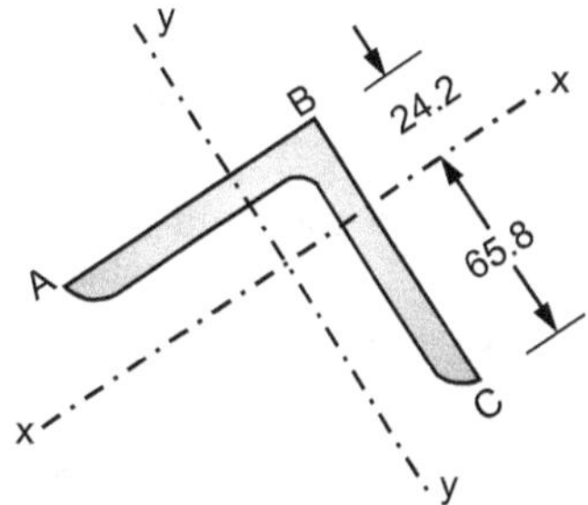

Fig. 8.14: Purlin in Bi-axial bending

$$I_{xx} = I_{yy} = 80.10 \times 10^4 \text{ mm}^4$$

$$Z_{xx \, (Top)} = \frac{I_{xx}}{y_{top}}$$

$$= \frac{80.10 \times 10^4}{24.2}$$

$$= 33.1 \times 10^3 \text{ mm}^3 \text{ for points A and B}$$

$$Z_{xx \, (Bottom)} = \frac{I_{xx}}{y_{bottom}} = \frac{80.10 \times 10^4}{65.80}$$

$$= 12.2 \times 10^3 \text{ mm}^3 \text{ for point C}$$

6. Calculation of design moment

Consider the purlin as a laterally supported beam.

$$M_d = \beta_b \times Z_p \times \frac{f_y}{\gamma_{mo}} \text{ or } 1.2 \times Z_e \times \frac{f_y}{\gamma_{mo}} \text{ whichever is less}$$

$$\beta_b = \frac{Z_e}{Z_p} \text{ for semi-compact sections.}$$

For bending @ xx axis

$\therefore$ $M_{dx} = \dfrac{Z_e}{Z_p} \times Z_p \times \dfrac{f_y}{\gamma_{mo}} \quad \left(Z_{xx} = \dfrac{I_{xx}}{y_{max}} = \dfrac{80.10 \times 10^4}{65.8} = 12173 \text{ mm}^3 \right)$

$$= 12173 \times \frac{250}{1.10} = 2766648 \text{ Nmm}$$

Bending M_x produces compressive stress at C (due to uplift wind force)

For bending @ yy axis

$$M_d = Z_e \times \frac{f_y}{\gamma_{mo}} \quad \left(Z_{yy} = \frac{I_{yy}}{x_{max}} = \frac{80.1 \times 10^4}{24.2} = 33099 \text{ mm}^3 \right)$$

$$M_{dy} = 33099 \times \frac{250}{1.10}$$

$$= 7522539 \text{ Nmm} > 188.65 \times 10^3 \text{ Nmm}$$

Bending moment M_{dy} produces compressive stress at B and C.

7. Check for bi-axial bending for point C of purlin:

$$\frac{M_x}{M_{dx}} + \frac{M_y}{M_{dy}} \leq 1$$

$$\frac{2432.85 \times 10^3}{2766648} + \frac{188.65 \times 10^3}{7522539} = 0.904 < 1 \quad \therefore \text{ O.K.}$$

As all the requirements are satisfied ISA $90 \times 90 \times 6$ mm section is O.K.

Ex. 8.6: *Check suitability of angle purlin*

ISA 100 × 75 × 8 mm with following details:

(i) Dead load = 1.2 kN/m

(ii) Live load = 0.75 kN/m

(iii) Wind load = 2.0 kN/m

(iv) Spacing of trusses = 4 m c/c

Sol.: 1. Load calculations: DL and LL are gravity loads, whereas wind load acts normal to roof. Moreover angle of pitch is also not given. Hence bi-axial bending moments cannot be determined. Hence we will consider all the loads acting normal to roof and find the worst load combination.

2. $\therefore$ $w_{dy} = 1.2$ kN/m, $w_{ly} = 0.75$ kN/m, $w_{wy} = -2.00$ kN/m

3. Factored loads due to combinations

(a) Load combination   1   = 1.5 (DL + LL)

$$w_{y1} = 1.5\,(1.2 + 0.75) = 2.925 \text{ kN/m}$$

(b) Load combination   2   = 1.5 (DL + WL)

$$w_{y2} = 1.5\,(1.20 - 2.0) = -1.20 \text{ kN/m}$$

(c) Load combination   3   = 1.2 (DL + LL + WL)

$$w_{y3} = 1.2\,(1.2 + 0.75 - 2) = -0.06 \text{ kN/m}$$

Thus among above three combinations w_{y1} is critical.

4. Bending moments

$$M_x = \frac{w_y \cdot L^2}{10} = \frac{2.925 \times 4^2}{10} = 4.68 \text{ kNm}$$

5. Suitability of angle selection:

$$\frac{b}{t} = \frac{75}{8} = 9.375 < 9.4\,\varepsilon \qquad\qquad (\because \varepsilon = 1)$$

$$\frac{d}{t} = \frac{100}{8} = 12.5 < 15.7\,\varepsilon$$

Considering least favourable condition i.e. $\frac{d}{t} < 15.7\,\varepsilon$ the section can be classified as semi-compact. For given angle section ISA 100 × 75 × 8 mm.

$$I_{xx} = 131.6 \times 10^4 \text{ mm}^4$$
$$I_{yy} = 63.3 \times 10^4 \text{ mm}^4$$
$$C_{xx} = 31.0 \text{ mm}$$

6. Calculation of design moment:

For bending @ xx axis

$$Z_e = Z_{xx} = \frac{I_{xx}}{y_{max}} = \frac{131.6 \times 10^4}{(100 - 31)} = 19072 \text{ mm}^3$$

$\therefore$
$$M_{dx} = Z_e \cdot \frac{f_y}{\gamma_{mo}} = 19072 \times \frac{250}{1.10} = 4334545 \text{ Nmm} = 4.33 \text{ kNm}$$

As M_{dy} cannot be calculated, neglect the ratio $\dfrac{M_y}{M_{dy}}$ in inter-action formula.

7. Check for interaction:

$$\frac{M_x}{M_{dx}} + \frac{M_y}{M_{dy}} \leq 1$$

$$\frac{4.68}{4.33} + 0 = 1.08 > 1$$

Available ISA 100 × 75 × 8 mm is not suitable.

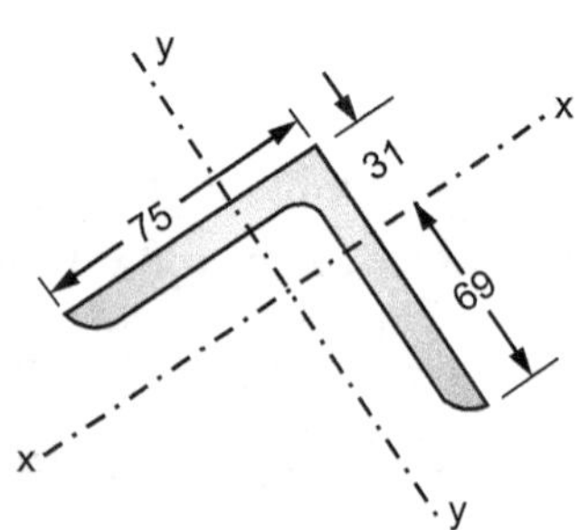

Fig. 8.15

Ex. 8.7: *Design an unequal angle purlin for truss with 6 panel, 12 m, rise 3 m, spacing 4 m c/c. DL = 150 N/m^2, LL = 350 N/m^2, WL = 750 N/m^2 (including self wt.)*

Apply check as per IS 800 - 2007 requirement. Draw neat sketch showing typical connection between purlin and principal rafter.

	Wt. (N/mm)	I_{xx} (mm^4)	C_x (mm)
ISA 100 $\times$ 65 $\times$ 6 mm	74	96.7 $\times$ 10^4	31.9
ISA 100 $\times$ 75 $\times$ 6 mm	78	100.9 $\times$ 10^4	30.1
ISA 90 $\times$ 60 $\times$ 6 mm	67	70.6 $\times$ 10^4	28.7

Sol.: Given: Span of truss = 12 m, Rise = 3 m, c/c distance between truss = 4 m c/c, DL = 150 N/m^2, LL = 350 N/m^2, WL = 750 N/m^2.

$\theta = \tan^{-1}\left(\dfrac{3}{6}\right) = 26.56°$, spacing of purlin $\dfrac{2}{\cos 26.56°} = 2.236$ m.

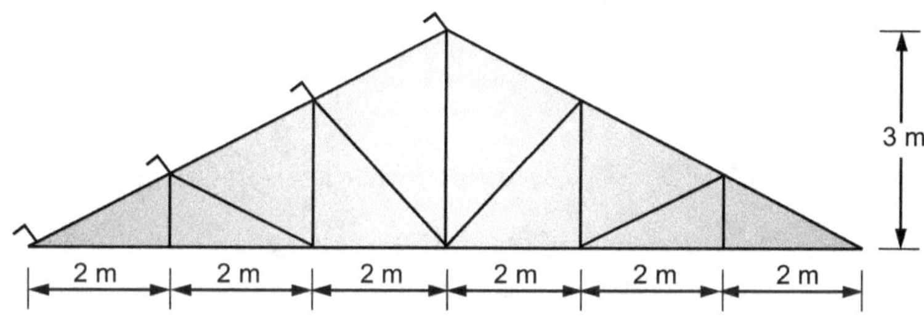

Fig. 8.16

1. **Load calculation** (Assume given DL and LL are on plan area)

 (a) DL = 150 $\times$ 2 = 300 N/m (on plan area)

 (b) LL = 350 $\times$ 2 = 700 N/m (on plan area)

 (c) WL = $-$ 750 $\times$ 2.236 = $-$ 1677 N/m (Normal to roof)

2. **Components of load** along xx (parallel to roof) and along yy (normal to roof) axes.

 (a) DL, w_{dy} = 300 $\times$ cos 26.56 = 268 N/m

       w_{dx} = 300 $\times$ sin 26.56 = 134 N/m

 (b) LL, w_{ly} = 700 $\times$ cos 26.56 = 626 N/m

       w_{lx} = 700 $\times$ sin 26.56 = 313 N/m

 (c) WL, w_{wy} = $-$ 1677 N/m

       w_{wx} = 0

3. **Factored loads due to combinations**

 (a) Load combination 1 = 1.5 (DL + LL)

       w_{y1} = 1.5 (268 + 626) = 1341 N/m

       w_{x1} = 1.5 (134 + 313) = 520.5 N/m

 (b) Load combination 2 = 1.5 (DL + WL)

       w_{y2} = 1.5 (268 $-$ 1677) = $-$ 2113.5 N/m

       w_{x2} = 1.5 (134 $-$ 0) = $+$ 201 N/m

 (c) Load combination 3 = 1.2 (DL + LL + WL)

       w_{y3} = 1.2 (268 + 626 $-$ 1677) = $-$ 939.6 N/m

       w_{x3} = 1.2 (134 + 313 $-$ 0) = $+$ 536.4 N/m

Thus among above three combination 1.5 (DL + WL) is critical.

$$\therefore \qquad w_y = 2113.5 \text{ N/m}$$
$$w_x = 201 \text{ N/m}$$

4. Bi-axial BM

$$M_x = \frac{w_y \cdot L^2}{10} = \frac{2113.5 \times 4^2}{10} = 3381.6 \text{ Nm} = 3.38 \text{ kNm}$$

$$M_y = \frac{w_x \cdot L^2}{10} = \frac{201 \times 4^2}{10} = 321.6 \text{ Nm} = 0.32 \text{ kNm}$$

5. Selection of angle section

Width of angle parallel to roof $= \dfrac{L}{60} = \dfrac{4000}{60} = 66.67$ mm

Depth of angle normal to roof $= \dfrac{L}{45} = \dfrac{4000}{45} = 88.89$ mm

Try ISA $100 \times 75 \times 6$ mm @ 78 N/m

$$\dfrac{b}{t} = \dfrac{100}{6} = 16.66 > 15.7\,\varepsilon \;\Big\}\;\; (\because \varepsilon = 1)$$

$$\dfrac{d}{t} = \dfrac{75}{6} = 12.5 < 15.7\,\varepsilon \;\Big\}\;\; \therefore \text{Classification of section is slender}$$

When the section is classified as slender section, it is susceptible to buckle locally even before reaching yield stress. The design strength for such sections is calculated as per provisions of IS 801 or by deducting width of compression plate element in excess of semi-compact section. Or the section below semi-compact section shall be selected.

Hence the sections available in the list are not suitable, because it has to satisfy width and depth of angle requirements also.

8.12 DESIGN OF BOLTED ROOF TRUSS

Ex. 8.8: *Design a Howe type roof truss which is required for a work shop building at Mumbai with following details:*

Span – 16 m, Rise – 4 m, Spacing of trusses – 4 m, Roofing G.I. sheet, Weight of purlin – 100 N/m, Weight of truss – 100 N/m², Clear height of building – 6 m. Thickness of support = 0.35 m.

Sol.: General Design

Effective span of trusses = 16 m

Spacing of trusses = 4 m

$$\text{Slope of truss, } \theta = \tan^{-1}\left(\frac{\text{Rise}}{L/2}\right) = \tan^{-1}\left(\frac{4}{8}\right)$$

$$\theta = 26.56°$$

Length of principal rafter $= \sqrt{8^2 + 4^2} = 8.94$ m

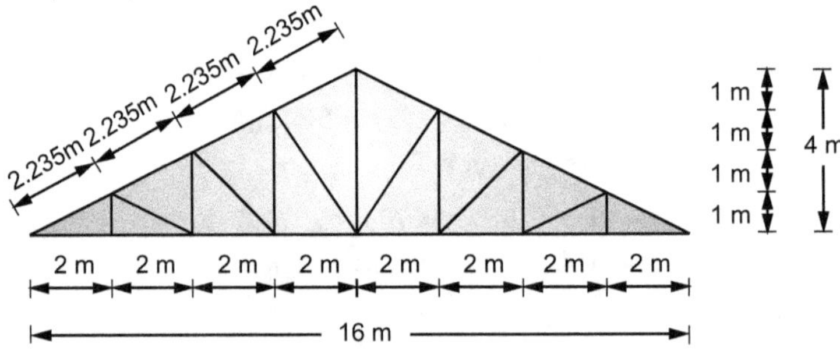

Fig. 8.17

$$\left.\begin{array}{l}\text{Sloping area of roof}\\\text{supported by a truss}\end{array}\right\} = 2 \times 8.94 \times 4 = 71.52 \ m^2$$

Plan area of roof supported by a truss = $16 \times 4 = 64 \ m^2$

Loads:

1. Imposed Load

Imposed load on a truss = $750 - (\theta - 10) \times 20 = 750 - (26.56 - 10) \times 20$
$$= 419 \ N/m^{2*}$$

2. Wind Load Analysis:
Basic wind speed for Mumbai, V_b = 44 m/s, (Table 7.4 and Fig. 8.3).

Taking risk factor k_1 = 1.0 (Table 7.5)

Terrain, height and size factor k_2 = 0.88 (Table 7.6) and topography factor k_3 = 1.00

Design wind speed $V_z = V_b \times k_1 \times k_2 \times k_3 = 44 \times 1 \times 0.88 \times 1 = 38.72 \ m/s$

Design wind pressure $p_z = 0.6 \ V_z^2 = 0.6 \times 38.72^2 = 899.54$ say 900 N/m^2

Mean height of eaves above ground level (assuming the plinth height to be 1.0 m)
$$= \ 6 + 1 = 7 \ m$$

From Tables 7.10 and 7.12, wind load is evaluated assuming normal permeability as follows:

$$\frac{h}{w} \ = \ \frac{6}{16} = 0.375 < \frac{1}{2}$$

$\therefore$ for $\qquad \dfrac{h}{w} < \dfrac{1}{2}$ and $\alpha = 26.56°$

(i) Wind normal to ridge:

$$\text{Windward } C_{pe} \ = \ - \ 0.4 - \frac{(- \ 0.4 - 0)}{10} \ \times (26.56 - 20) = - \ 0.1376$$

$$\text{Leeward } \ C_{pe} \ = \ - \ 0.4$$

Internal air pressure coefficient for normal permeability,
$$C_{pi} \ = \ \pm \ 0.2$$

Combined external + internal wind pressure $= (C_{pe} - C_{pi}) \cdot p_z$

Windward slope $\qquad (-0.1376 - 0.2) \ 899.54 = - \ 303.84 \ N/m^2$ (uplift)

$\qquad\qquad\qquad\qquad (-0.1376 + 0.2) \ 900 = 55.16 \ N/m^2$ (downward)

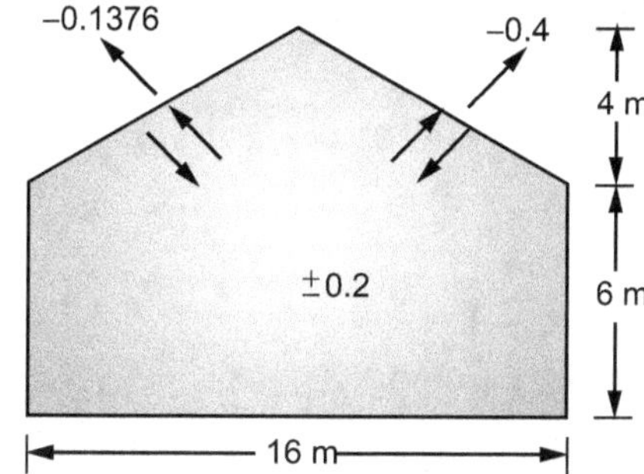

Fig. 8.18

Leeward slope $\qquad\qquad (- \ 0.4 - 0.2) \ 900 = - \ 540 \ N/m^2$ (uplift)

$\qquad\qquad\qquad\qquad (- \ 0.4 + 0.2) \ 900 = - \ 180.0 \ N/m^2$ (uplift)

(ii) Wind parallel to ridge:
External pressure co-efficient C_{pe},

On both slopes for $\frac{1}{4}^{th}$ length of building = $- \ 0.7$

On both slopes for mid $\frac{1}{2}$ length of building = $- \ 0.6$

Internal air pressure co-efficient for normal permeability, $C_{pi} = \pm 0.2$

Combined external + Internal wind pressure on both slopes for $\frac{1}{4}^{th}$ length of building

$$(- 0.7 - 0.2) \times 900 = - 810 \text{ N/m}^2 \text{ (uplift)}*$$
$$(- 0.7 + 0.2) \times 900 = - 450 \text{ N/m}^2 \text{ (uplift)}$$

On both slopes for mid $\frac{1}{2}$ length of building

$$(- 0.6 - 0.2) \times 900 = - 720 \text{ N/m}^2 \text{ (uplift)}$$
$$(- 0.6 + 0.2) \times 900 = - 360 \text{ N/m}^2 \text{ (uplift)}$$

Thus, maximum wind pressure for design is 55.16 N/m^2 (downward) and 810 N/m^2 (uplift).

Design of Purlins

1. Load calculations (for 2.235 m spacing of purlins)

(a)   D.L. of G.I. Sheet $= 142 \times 2.235$ $= 317$ N/m

     Self Weight of purlin (assumed) $= 100$ N/m

                                      Total $= 417$ N/m (on plan area)

(b) $LL* = 419 \times 2 \times 1 = 838$ N/m (on plan area)

(c) $WL** = - 810 \times 2.235 = - 1810$ N/m (normal to roof)

2. Components of load along xx axis (parallel to roof) and along yy axis (normal to roof)

(a) DL, $w_{dy} = 417 \cos 26.56° = 373$ N/m

            $w_{dx} = 417 \sin 26.56° = 186$ N/m

(b) LL, $w_{ly} = 838 \cos 26.56° = 750$ N/m

            $w_{lx} = 838 \sin 26.56° = 375$ N/m

(c) WL, $w_{wy} = - 1810$ N/m

            $w_{wx} = 0$

3. Factored loads due to combinations

(a) Load combination $1 = 1.5$ (DL + LL)

         $w_{y1} = 1.5 (373 + 750) = 1684.5$ N/m

         $w_{x1} = 1.5 (186 + 375) = 841.5$ N/m

(b) Load combination $2 = 1.5$ (DL + WL)

         $w_{y2} = 1.5 (373 - 1810) = - 2155.5$ N/m

         $w_{x2} = 1.5 (186 - 0) = + 279$ N/m

(c) Load combination

         $3 = 1.2$ (DL + LL + WL)

         $w_{y3} = 1.2 (373 + 750 - 1810) = - 824.4$ N/m

         $w_{x3} = 1.2 (186 + 375 - 0) = + 673.2$ N/m

Hence among above three combination 1.5 (DL + WL) is critical.

$\therefore$ $w_y = 2156$ N/m (absolute value)

         $w_x = 279$ N/m

4. Bi-axial BM

$$M_x = \frac{w_y \cdot L^2}{10} = \frac{2156 \times 4^2}{10} = 3449.6 \text{ Nm} = 3.45 \text{ kNm}$$

$$M_y = \frac{w_x \cdot L^2}{10} = \frac{279 \times 4^2}{10} = 446.4 \text{ Nm} = 0.45 \text{ kNm}$$

5. Selection of angle section

Width of angle parallel to roof $= \dfrac{L}{60} = \dfrac{4000}{60} = 66.67$ mm

Depth of angle normal to roof $= \dfrac{L}{45} = \dfrac{4000}{45} = 88.89$ mm

Try ISA $100 \times 75 \times 8$ mm @ 103 N/m satisfying above requirements.

$$\left.\begin{array}{l} \dfrac{b}{t} = \dfrac{75}{8} = 9.375 < 9.4\,\varepsilon \quad (\because \varepsilon = 1) \\[2mm] \dfrac{d}{t} = \dfrac{100}{8} = 12.50 < 15.7\,\varepsilon \end{array}\right\} \therefore \text{Classification of section is semi-compact.}$$

For ISA $100 \times 75 \times 8$ mm, $I_{xx} = 131.6 \times 10^6$ mm^4

$I_{YY} = 63.3 \times 10^4$ mm^4, $C_{xx} = 31.0$ mm, $C_{yy} = 18.7$ mm

$$Z_{xx\,(Top)} = \frac{I_{xx}}{y_{top}} = \frac{131.6 \times 10^4}{(31)} = 42452 \text{ mm}^3$$

$$Z_{xx\,(Bottom)} = \frac{I_{xx}}{y_{bottom}} = \frac{131.6 \times 10^4}{(100 - 31)} = 19072 \text{ mm}^3$$

$$Z_{yy} = \frac{I_{yy}}{x_{max}} = \frac{63.3 \times 10^4}{18.7} = 33850 \text{ mm}^3$$

6. Calculation of design moment for bending @ xx-axis

Consider the purlin as a laterally supported beam.

$$M_{dx} = Z_e \times \frac{f_y}{\gamma_{mo}}$$

$$= 19072 \times \frac{250}{1.10} \times (10^{-6}) = 4.33 \text{ kNm}$$

For bending @ yy axis

$$\therefore \qquad M_{dy} = Z_{yy}\,\frac{f_y}{\gamma_{mo}}$$

$$= 33850 \times \frac{250}{1.10}\,(10^{-6}) = 7.69 \text{ kNm}$$

7. Check for bi-axial bending for outstanding toe of purlin

$$\frac{M_x}{M_{dx}} + \frac{M_y}{M_{dy}} \leq 1$$

$$\frac{3.45}{4.33} + \frac{0.45}{7.69} = 0.85 < 1 \quad \therefore \text{ O.K.}$$

Design of Roof Truss

1. Dead load (assumed to be acting on top panel points)

Due to G.I. sheet $= 142 \times$ inclined area

$$= 142 \times 71.52 \quad = 10156 \text{ N}$$

Due to purlins $=$ No. of purlins $\times$ spacing of trusses $\times$ weight/metre

$$= 10 \times 4 \times 103 \quad = 4120 \text{ N}$$

Due to truss $\left[\left(\dfrac{16}{3} + 5\right)\right] \times 10 \approx 103.5$ N/m^2 on plan area)

$$= 103.5 \times 64 \quad = \underline{6624 \text{ N}}$$

$$\text{Total dead load} = 20900 \text{ N}$$

Dead load on each top panel $= \dfrac{20900}{8} = 2612.5$ N say 2.62 kN

Dead load on end panel point $= \dfrac{2612.5}{2} = 1306.75$ N say 1.31 kN

2. Live load (on plan area)

$$\text{Live load on the truss} = \frac{2}{3} \times 419 \times 64$$

$$= 17877 \text{ N}$$

$$\therefore \quad \text{Live load on each top panel point} = \frac{17877}{8}$$

$$= 2235 \text{ N say } 2.23 \text{ kN}$$

$$\text{Live load on end panel point} = \frac{2235}{2}$$

$$= 1117.5 \text{ N say } 1.115 \text{ kN}$$

3. Wind Load Calculations:

$\therefore$ Wind load on each top panel point = $- 810 \times 2.235 \times 4 = - 7242$ N (uplift) = $- 7.24$ kN

Wind load on end panel point $= -\dfrac{7242}{2} = - 3621$ N $\approx - 3.62$ kN

The force diagrams for truss under dead load and negative wind load are shown in Figs. 8.19 and 8.20 respectively. Forces due to live load are calculated by multiplying forces due to dead load by $\dfrac{2.23}{2.62} = 0.8511$.

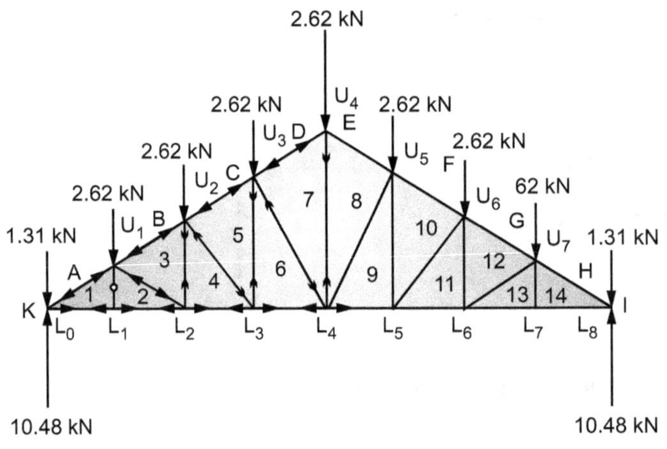

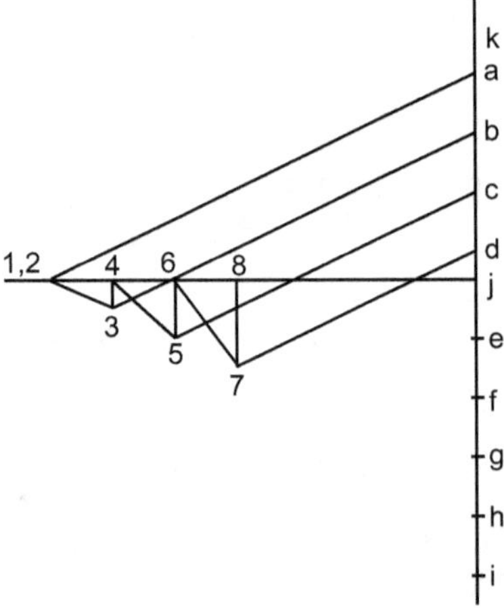

Fig. 8.19: Force Diagram for Dead Loads

(Scale 1 cm = 1.25 kN)

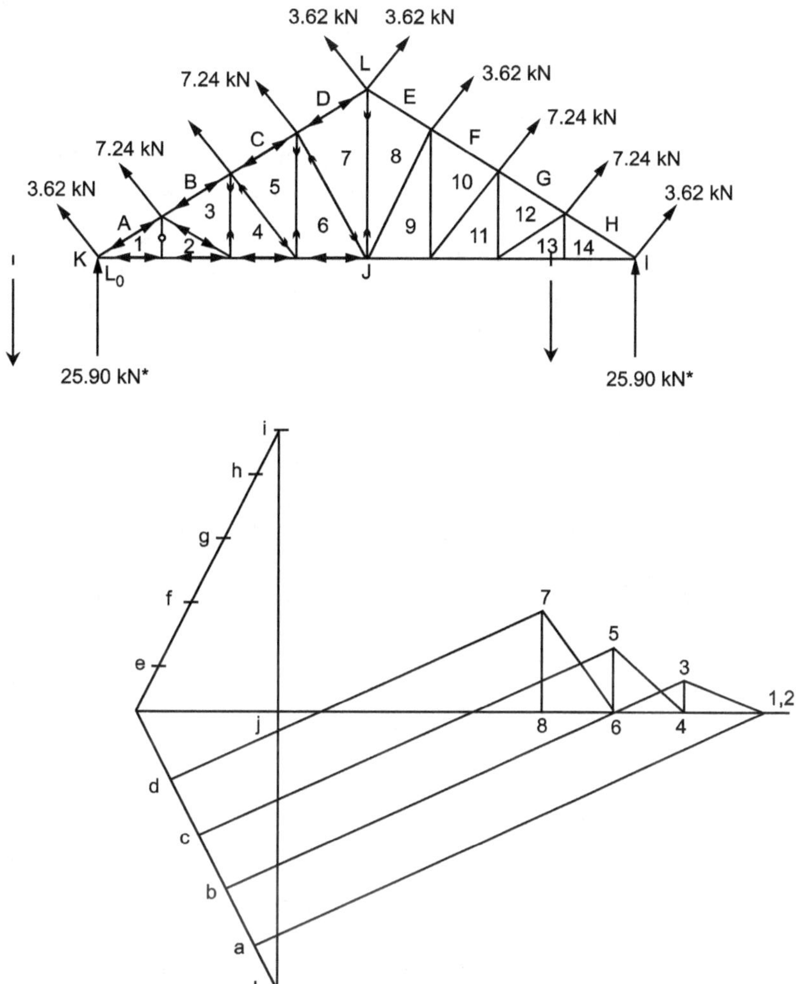

Fig. 8.20: Force Diagram for Wind Loads

4. Combination of loads:

Due to symmetric loading only half portion of the force diagram need be drawn. The forces in each of the members in trusses, are shown in Table 8.13. Following combinations were considered to obtain design loads.

(i)   1.5 (DL + LL)

(ii) 1.2 (DL + LL + WL)

(iii) 0.9 DL + 1.5 WL

----

* Reaction due to wind load = $4 \times 7.24 \times \cos 26.56° = 25.90$ kN

Table 8.13: Force in Members of Truss in kN
(Tension +ve, Compression –ve)

Members	Dead load DL	Live load LL	Wind load WL	Load combination 1.5 (DL + LL)	Load combination 1.2 (DL + LL + WL)	Load combination 0.9 (DL + 1.5 WL)	Design load (kN)	Length of member (m)
Principal Rafter								
L_0U_1, L_8U_7	– 20.40	– 17.40	+ 51.84	– 56.70	+ 16.85	+ 59.40		2.235*
U_1U_2, U_6U_7	– 17.50	– 14.90	+ 45.24	– 48.60	+ 15.41	+ 52.11	– 56.70	"
U_2U_3, U_5U_6	– 14.60	– 12.40	+ 40.70	– 40.50	+ 16.44	+ 47.91	+ 59.60	"
U_3U_4, U_4U_5	– 11.70	– 10.00	+ 35.10	– 32.55	+ 16.08	+ 42.12		"
Main Tie								
L_0L_1, L_7L_8	+ 18.10	+ 15.40	– 44.42	+ 50.25	– 13.10	– 50.34		2.00
L_1L_2, L_6L_7	+ 18.10	+ 15.40	– 44.42	+ 50.25	– 13.10	– 50.34	+ 50.25	"
L_2L_3, L_5L_6	+ 15.60	+ 13.30	– 28.35	+ 43.35	+ 0.66	–28.48	– 50.34	"
L_3L_4, L_4L_5	+ 13.00	+ 11.10	– 19.91	+ 36.15	+ 5.03	–18.16		"
Main Vertical Tie								
U_4L_4	+ 3.90	+ 3.30	– 12.15	+ 10.80	– 5.94	– 14.72	+ 10.80 – 14.72	4.00
Main Inclined Strut								
U_3L_4, U_5L_4	– 4.70	– 4.00	+ 14.72	– 13.05	+ 7.22	+ 17.85	– 13.05 + 17.85	3.605*
Minor Vertical Ties								
L_3U_3, L_5U_5	+ 2.60	+ 2.20	– 8.27	+ 7.20	– 4.16	– 10.07	+ 7.20	3.00
L_2U_2, L_6U_6	+ 1.30	+ 1.15	– 4.05	+ 3.68	– 1.92	– 4.91	– 10.07	2.00
Minor Inclined Struts								
L_3U_2, L_5U_6	– 3.60	– 3.10	+ 11.64	– 10.05	+ 5.93	+ 14.22	– 10.05	2.828*
L_2U_1, L_6U_7	– 2.90	– 2.50	+ 8.95	– 8.10	+ 4.26	+ 10.82	+ 10.82	2.235*

Design of Members

1. Principal Rafter (Members L_0U_1, U_1U_2, U_2U_3, U_3U_4, U_4U_5, U_6U_7 and U_7L_8)

Design as a compression member and check for tension

$$\text{Maximum compression} = -56.70 \text{ kN}$$
$$\text{Maximum tension} = +59.40 \text{ kN}$$
$$\text{Length of member} = 2.235 \text{ m}$$
$$\text{Effective length, KL} = 0.85 \, L = 0.85 \times 2.235 = 1.90 \text{ m}$$

---

* These lengths are calculated by using pythagorus theorem.

Using a minimum double angle section 2 – ISA 50 × 50 × 6 mm connected on both sides of 8 mm gusset plate by 16 mm diameter rivets as shown in Fig. 8.21.

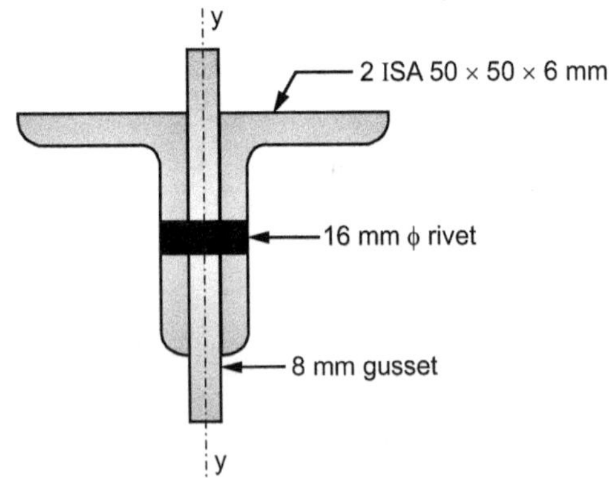

Fig. 8.21

Provide tacking bolts along the length of member.

$$r_{min} = r_{xx} = 15.1 \text{ mm}$$

$$\lambda = \frac{KL}{r} = \frac{1900}{15.1} = 125.83 < 180 \quad \therefore \text{ O.K}$$

From Table 4.4 by interpolation (Consider buckling class as C)

kL/r	f_{cd}
120	83.7
130	74.3

$$f_{cd} = 83.7 - \frac{83.7 - 74.3}{10} \times 5.83 = 78.22 \text{ MPa}$$

$$A_g = 2 \times 568 = 1136 \text{ mm}^2$$

$$\text{Design compressive strength} = f_{cd} \times A_g = 78.22 \times 1136$$

$$= 88858 \text{ N} = 88.86 \text{ kN} > 56.70 \text{ kN}$$

Check for tension: $\quad A_n = A_g - \text{deduction for holes}$

$$A_n = 2 [568 - 18 \times 6] = 920 \text{ mm}^2$$

Design tensile strength by net section rupture,

$$T_{dn} = \frac{\alpha \cdot A_n \cdot f_u}{\gamma_{m1}} = \frac{0.8 \times 920 \times 410}{1.125} = 241400 \text{ N} = 241.40 \text{ kN}$$

Design tensile strength by gross-section yielding,

$$T_{dg} = \frac{A_g \cdot f_y}{\gamma_{m0}} = \frac{1136 \times 250}{1.10} = 258180 \text{ N} = 258.18 \text{ kN}$$

$\therefore$ Design tensile strength = Least of T_{dn} and T_{dg} = 241.40 kN > 59.40 kN

$\therefore$ O.K.

2. Main tie (Members L_0L_1, L_1L_2, L_2L_3, L_3L_4, L_4L_5, L_5L_6, L_6L_7, L_7L_8)

Design as a tension member and check for compression

$$\text{Maximum tension} = + 50.25 \text{ kN}$$

$$\text{Maximum compression} = - 50.34 \text{ kN}$$

$$\text{Length of member, } L = 2.00 \text{ m}$$

Try 2 – ISA 50 $\times$ 50 $\times$ 6 mm with tacking bolts along the length

$$A_n = 2 [568 - 18 \times 6] = 920 \text{ mm}^2$$

Design tensile strength by gross-section yielding

$$T_{dg} = \frac{A_g \cdot f_y}{\gamma_{m0}} = \frac{1136 \times 250}{1.10} = 258180 \text{ N} = 258.18 \text{ kN}$$

Design tensile strength by net section rupture,

$$T_{dn} = \frac{\alpha A_n f_u}{\gamma_{m1}} = \frac{0.8 \times 920 \times 410}{1.25} = 241400 \text{ N} = 241.40 \text{ kN}$$

$\therefore$ Design tensile strength = 241.40 kN > 50.25 kN $\quad \therefore$ O.K.

Check for compression

$$r_{min} = r_{xx} = 15.1 \text{ mm}$$

$$\text{Effective length, } l = 0.85 L = 0.85 \times 2.00 = 1.70 \text{ m}$$

$$\lambda = \frac{l}{r_{min}} = \frac{1700}{15.1} = 112.58 < 350 \quad \therefore \text{ O.K.}$$

From Table 4.4, by interpolation, (Consider buckling class c)

kL/r	f_{cd}
110	94.6
120	83.7

$$f_{cd} = 94.6 - \frac{94.6 - 83.7}{10} \times 2.58$$

$$= 91.79 \text{ MPa}$$

Design compressive load $= f_{cd} \times A_g = 91.79 \times 2 \times 568$

$$= 10427 \text{ N} = 104.27 \text{ kN} > 50.34 \text{ kN}$$

3. Main vertical tie member ($U_4 L_4$)

Normally carries tension, but due to wind the member will be subjected to reversal of stresses i.e. compression.

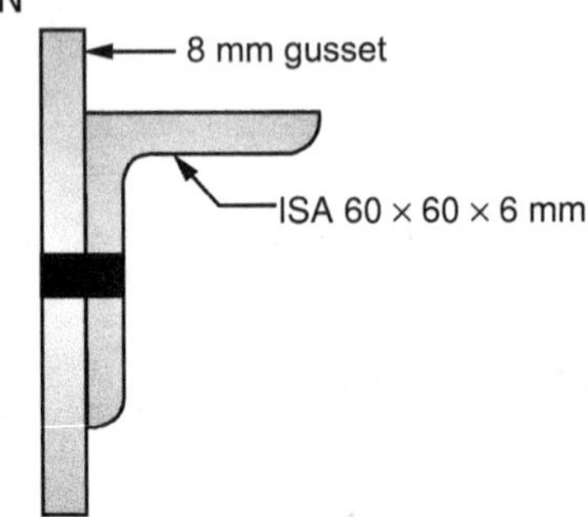

Maximum tension $= + 10.80$ kN

Maximum compression $= - 14.72$ kN

Length of member, L $= 4.00$ m

Try single angle 60 × 60 × 6 mm connected by two bolts of 16 mm diameter.

Fig. 8.22

Check for design strength T_d of trial section:

(i) Design tensile strength by gross-section yield consideration

$$T_{dg} = \frac{A_g \cdot f_y}{\gamma_{mo}} = \frac{684 \times 250}{1.10} = 155454 \text{ N} = 155.45 \text{ kN}$$

(ii) Design tensile strength by net-section rupture consideration

Net area of connected leg,

$$A_{nc} = \left[L - d_h - \frac{t}{2}\right] t = \left[60 - 18 - \frac{6}{2}\right] 6 = 234 \text{ mm}^2$$

Gross area of outstanding leg,

$$A_{go} = [b - t/2] t = \left[60 - \frac{6}{2}\right] \times 6 = 342 \text{ mm}^2$$

Outstanding leg width,

$$w = b - \frac{t}{2} = 60 - \frac{6}{2} = 57 \text{ mm}$$

$$b_s = w + g - t$$
$$= 57 + 35^* - 6 = 85 \text{ mm (see table B2)}$$

$$L_c = 1 \times 50 = 50 \text{ mm}$$

$$\beta = 1.4 - 0.076 \frac{w}{t} \times \frac{f_y}{f_u} \times \frac{b_s}{L_c}$$

$$= 1.4 - 0.076 \times \frac{57}{6} \times \frac{250}{410} \times \frac{86}{50} = 0.64$$

$$0.9 \frac{f_u}{f_y} \times \frac{\gamma_{m0}}{\gamma_{m1}} = 0.9 \times \frac{410}{250} \times \frac{1.10}{1.25} = 1.29$$

$$\beta = 0.64 < 1.29 \text{ but } \ngtr 0.7 \qquad \therefore \text{ Take } \beta = 0.7$$

$$T_{dn} = \frac{0.9 A_{nc} f_u}{\gamma_{m1}} + \frac{\beta \cdot A_{go} f_y}{\gamma_{m0}}$$

$$= \frac{0.9 \times 234 \times 410}{1.25} + \frac{0.7 \times 342 \times 250}{1.10}$$

$$= 123486 \text{ N}$$

$$= 123.48 \text{ kN}$$

(iii) Design tensile strength by block shear

Assume p = 50 mm and e = 40 mm

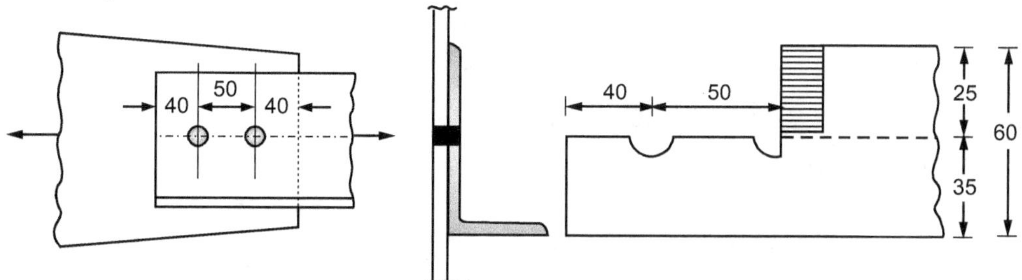

Fig. 8.23

A_{vg} = Minimum gross area in shear along bolt line = $L_v \cdot t$

$\quad = [1 \times 50 + 40] \times 6 = 540 \text{ mm}^2$

A_{vn} = Minimum net area in shear along bolt line

$\quad = \left[L_v - \left(1 + \dfrac{1}{2} \right) d_h \right] t$

$\quad = [1 \times 50 + 40 - 0.5 \times 18] \times 6 = 486 \text{ mm}^2$

A_{tg} = Minimum gross area in tension from bolt hole to toe of angle

$\quad = L_t \cdot t = 25 \times 6 = 150 \text{ mm}^2$

A_{tn} = Minimum net area in tension from bolt hole to toe of angle

$\quad = [L_t - 0.5 \, d_h] \, t$

$\quad = [25 - 0.5 \times 18] \times 6 = 96 \text{ mm}^2$

$T_{db_1} = \dfrac{A_{vg} f_y}{\sqrt{3} \, \gamma_{m0}} + \dfrac{0.9 \, A_{tn} \cdot f_u}{\gamma_{m1}}$

$\quad = \dfrac{540 \times 250}{\sqrt{3} \times 1.10} + \dfrac{0.9 \times 96 \times 410}{1.25}$

$\quad = 99196 \text{ N} = 99.19 \text{ kN}$

$T_{db_2} = \dfrac{A_{tg} f_y}{\gamma_{m0}} + \dfrac{0.9 \, A_{vn} \cdot f_u}{\sqrt{3} \, \gamma_{m1}}$

$\quad = \dfrac{150 \times 250}{1.10} + \dfrac{0.9 \times 486 \times 410}{\sqrt{3} \times 1.25}$

$\quad = 116922 \text{ N} = 116.92 \text{ kN}$

$\therefore \quad T_{db} = 99.19 \text{ kN}$

Hence the design tensile strength of the angle

$\quad = $ Least of $T_{dg}, T_{dn}, T_{db} = 99.19 \text{ kN} >> (+ 10.80 \text{ kN} \quad \therefore \quad \text{O.K.})$

Check for compression member:

For the selected discontinuous angle ISA 60 × 60 × 6 mm with minimum two bolts at each end, $r_{min} = r_{vv} = 11.5$ mm and kL = 1 × 4.00 = 4 m = 4000 mm.

Also $k_1 = 0.2$, $k_2 = 0.35$ and $k_3 = 20$

Imperfection factor, $\alpha = 0.49$ (class c buckling)

$$\lambda_{vv} = \frac{\left(\dfrac{l}{r_{vv}} \right)}{\varepsilon \sqrt{\dfrac{\pi^2 E}{250}}} = \frac{\left(\dfrac{4000}{11.5} \right)}{(1) \sqrt{\dfrac{\pi^2 \times 2 \times 10^5}{250}}} = 3.91$$

$$\lambda_\phi = \frac{(b_1 + b_2)}{\varepsilon \sqrt{\dfrac{\pi^2 E}{25}} \times 2t} = \frac{(60 + 60)}{(1) \sqrt{\dfrac{\pi^2 \times 2 \times 10^5}{250}} \times 2 \times 6} = 0.1125$$

$$\lambda_e = \sqrt{k_1 + k_2 \, \lambda_{vv}^2 + k_3 \, \lambda_\phi^2}$$

$$= \sqrt{0.2 + 0.35 \times 3.91^2 + 20 \times 0.1125^2} = 2.409$$

$$\phi = 0.5\,[1 + \alpha\,(\lambda_e - 0.2) + \lambda_e^2]$$

$$= 0.5\,[1 + 0.49\,(2.409 - 0.2) + 2.409^2] = 3.943$$

∴ Design compression stress,

$$f_{cd} = \frac{\dfrac{f_y}{\gamma_{m0}}}{\phi + \sqrt{\phi^2 - \lambda_e^2}} = \frac{\dfrac{250}{1.1}}{3.943 + \sqrt{3.943^2 - 2.409^2}}$$

$$= 32.17 \text{ N/mm}^2$$

∴ Design compressive load

$$= f_{cd} \times A_g = 32.17 \times 684 = 22004 \text{ N} = 22.00 \text{ kN} > 14.72 \text{ kN}$$

4. Main inclined strut member (U_3L_4, U_5L_4)

Maximum compression = – 13.05 kN

Maximum tension = + 17.85 kN

Length of member, L = 3.605 m

Assume $\quad f_{cd}$ = 43.6 N/mm^2 (stress corresponding to λ = 180)

$$A_{approx} = \frac{P_u}{f_{cd}} = \frac{13.05 \times 10^3}{43.6} = 299.31 \text{ mm}^2$$

Effective length, $\quad kL = 0.85 \times 3605 = 3064$ mm

Select minimum size of angle having

$$r_{min} = \frac{3064}{180} = 17.02 \text{ mm}$$

Referring steel table try ISA 90 × 90 × 6 mm, having r_v = 17.5 mm and A = 1047mm^2.

Also k_1 = 0.2, k_2 = 0.35 and k_3 = 20 (Assuming bolts ≥ 2 with fixed end fixity) Refer table 4.6.

Imperfection factor, α = 0.49 (class c buckling)

$$\lambda_{vv} = \frac{\left(\dfrac{l}{r_{vv}}\right)}{\varepsilon \sqrt{\dfrac{\pi^2 E}{250}}} = \frac{\left(\dfrac{3064}{17.5}\right)}{(1)\sqrt{\dfrac{\pi^2 \times 2 \times 10^5}{250}}} = 1.97$$

$$\lambda_\phi = \frac{(b_1 + b_2)}{\varepsilon \sqrt{\dfrac{\pi^2 E}{25}} \times 2t} = \frac{(90 + 90)}{(1)\sqrt{\dfrac{\pi^2 \times 2 \times 10^5}{250}} \times 2 \times 6} = 0.1688$$

$$\lambda_e = \sqrt{k_1 + k_2 \, \lambda_{vv}^2 + k_3 \, \lambda_\phi^2}$$

$$= \sqrt{0.2 + 0.35 \times 1.97^2 + 20 \times 0.1688^2} = 1.4588$$

$$\phi = 0.5\,[1 + \alpha\,(\lambda_e - 0.2) + \lambda_e^2]$$

$$= 0.5\,[1 + 0.49\,(1.4588 - 0.2) + 1.4588^2] = 1.8724$$

∴ Design compression stress,

$$f_{cd} = \frac{\dfrac{f_y}{\gamma_{m0}}}{\phi + \sqrt{\phi^2 - \lambda_e^2}} = \frac{\dfrac{250}{1.1}}{1.8724 + \sqrt{1.8724^2 - 1.4588^2}}$$

$$= 74.60 \text{ N/mm}^2$$

∴   Design compressive load

$$= f_{cd} \times A_g = 74.60 \times 1047 = 78106 \text{ N}$$

$$= 78.10 \text{ kN} >> 13.05 \text{ kN}$$

(**Note:** Eventhough the load carrying capacity is much more the section should satisfy the criteria for maximum slenderness ratio).

Check for tension:

(i) Design tensile strength by gross-section yielding consideration

$$T_{dg} = \frac{A_g \cdot f_y}{\gamma_{m0}} = \frac{1047 \times 250}{1.10} = 237954 \text{ N} = 237.95 \text{ kN}$$

(ii) Design tensile strength by net-section rupture consideration

Net area of connected leg,

$$A_{nc} = \left[L - d_h - \frac{t}{2} \right] t$$

$$= \left[90 - 18 - \frac{6}{2} \right] \times 6 = 414 \text{ mm}^2$$

Gross area of outstanding leg,

$$A_{go} = [b - d_h]\, t = [90 - 18] \times 6 = 432 \text{ mm}^2$$

Outstanding leg width, $w = b - \dfrac{t}{2} = 90 - \dfrac{6}{2} = 87$ mm

$$p_s = w + g - t = 87 + 55^* - 6 = 136 \text{ mm (see table B2)}$$

$$L_c = 2 \times 50 = 100 \text{ mm}$$

(distance b_s is more hence increased no. of bolts i.e. 3 otherwise β will be negative)

$$\beta = 1.4 - 0.076 \frac{w}{t} \times \frac{f_y}{f_u} \times \frac{b_s}{L_c}$$

$$= 1.4 - 0.076 \times \frac{87}{6} \times \frac{250}{410} \times \frac{136}{100} = 0.48$$

$$0.9 \frac{f_u}{f_y} \times \frac{\gamma_{m0}}{\gamma_{m1}} = 0.9 \times \frac{410}{250} \times \frac{1.10}{1.25} = 1.29$$

$$\beta = 0.48 < 1.29 \text{ but} \ngtr 0.7 \qquad \therefore \text{ Take } \beta = 0.7$$

$$T_{dn} = \frac{0.9\, A_{nc}\, f_u}{\gamma_{m1}} + \frac{\beta \cdot A_{go}\, f_y}{\gamma_{m0}}$$

$$= \frac{0.9 \times 414 \times 410}{1.25} + \frac{0.7 \times 432 \times 250}{1.10}$$

$$= 190940 \text{ N} = 190.94 \text{ kN}$$

(iii) Design tensile strength by block shear consideration.

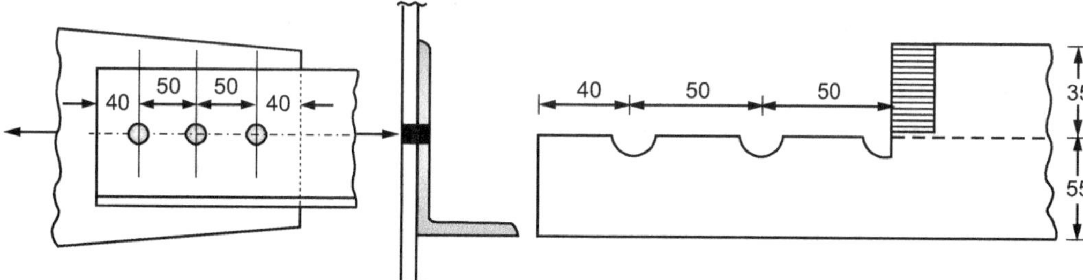

Fig. 8.24

$$A_{vg} = \text{Minimum gross area in shear along bolt line} = L_v \cdot t$$
$$= [2 \times 50 + 40] \times 6 = 840 \text{ mm}^2$$

$$A_{vn} = \text{Minimum net area in shear along bolt line}$$
$$= \left[L_v - \left(2 + \frac{1}{2}\right) d_h \right] t$$
$$= [2 \times 50 + 40 - 0.5 \times 18] \times 6 = 786 \text{ mm}^2$$

$$A_{tg} = \text{Minimum gross area in tension from bolt hole to toe of angle}$$
$$= L_t \cdot t = 35 \times 6 = 210 \text{ mm}^2$$

$$A_{tn} = \text{Minimum net area in tension from bolt hole to toe of angle}$$
$$= [L_t - 0.5\, d_h]\, t$$
$$= [35 - 0.5 \times 18] \times 6 = 156 \text{ mm}^2$$

$$T_{db_1} = \frac{A_{vg}\, f_y}{\sqrt{3}\, \gamma_{m0}} + \frac{0.9\, A_{tn} \cdot f_u}{\gamma_{m1}}$$
$$= \frac{840 \times 250}{\sqrt{3} \times 1.10} + \frac{0.9 \times 156 \times 410}{1.25} = 156272 \text{ N} = 156.27 \text{ kN}$$

$$T_{db_2} = \frac{A_{tg}\, f_y}{\gamma_{m0}} + \frac{0.9\, A_{vn} \cdot f_u}{\sqrt{3}\, \gamma_{m1}}$$
$$= \frac{210 \times 250}{1.10} + \frac{0.9 \times 786 \times 410}{\sqrt{3} \times 1.25} = 181688 \text{ N} = 181.68 \text{ kN}$$

$\therefore \qquad T_{db} = \text{Least of } T_{db_1} \text{ and } T_{db_2} = 156.27 \text{ kN}$

$\therefore$ Design strength of angle = Least of T_{dg}, T_{dn}, T_{db} = 156.27 kN

$>> 17.85 \text{ kN} \quad \therefore \text{ Angle section is O.K.}$

5. Other Minor Members:

Since other members are not severely loaded and length of member is small, a single angle ISA 50 × 50 × 6 mm will be sufficient.

Design of connections:

Use 4.6 grade 16 mm diameter bolts and 8 mm thick gusset plate.

- Design strength of a bolt in single shear

$$= \frac{1}{\gamma_{mb}} \left[\frac{f_{ub}\, A_n}{\sqrt{3}} \right] = \frac{1}{1.25} \left[\frac{400}{\sqrt{3}} \times 0.78 \times \left(\frac{\pi}{4} \times 16^2 \right) \right]$$
$$= 28795 \text{ N} = 28.97 \text{ kN}$$

- Design strength of a bolt in double shear

$$= 2 \times 28975 = 57950 \text{ N} = 57.95 \text{ kN}$$

- Design bearing strength of a bolt

$$= \frac{1}{\gamma_{mb}} [2.5\, k_b \cdot d \cdot t \cdot f_{ub}]$$

k_b is smaller of $\left[\dfrac{e}{3 d_o};\ \dfrac{p}{3 d_o} - 0.25;\ \dfrac{f_{ub}}{f_u};\ 1.0 \right]$

i.e. $\left[\dfrac{40}{3 \times 18} = 0.74;\ \dfrac{50}{3 \times 18} - 0.25 = 0.67;\ \dfrac{400}{410} = 0.975;\ 1.0 \right]$

Hence $k_b = 0.67$

$\therefore$ Design bearing strength of 6 mm plate

$$= \frac{1}{1.25} [2.5 \times 0.67 \times 16 \times 6 \times 400] = 51456 \text{ N} = 51.45 \text{ kN}$$

Design bearing strength of 8 mm plate

$$= \frac{1}{1.25} [2.5 \times 0.67 \times 16 \times 8 \times 400] = 68608 \text{ N} = 68.60 \text{ kN}$$

The following table shows the sections adopted and number of bolts

Table 8.14

Member	Design load (P_u) (kN)	Section provided	Bolt value (B_v) (kN)	Number of bolts $= \dfrac{P_u}{B_v}$
Principal rafter	− 56.70 + 59.40	2 ISA × 50 × 50 × 6	57.95	02
Main tie	+ 50.25 − 50.34	2 ISA × 50 × 50 × 6	57.95	02
Main vertical tie	+10.80 − 14.72	ISA 60 × 60 × 6	28.97	02
Main inclined strut	− 13.05 + 17.85	ISA 90 × 90 × 6	28.97	03 (as required for β requirement)
All other members	−	ISA 50 × 50 × 6	28.97	02

Note: *Minimum number of bolts required to connect any member to gusset plate = 2, even though the requirement as per formula $\dfrac{P_u}{B_v}$ may be < 1.*

Design of bearing plate:

$$\text{End reaction due to DL + LL} = \frac{20900 + 17877}{2} = 19388.5 \text{ N}$$

$$= 19.39 \text{ kN (downward)}$$

$$\text{End reaction due to DL + WL} = \frac{20900}{2} - 25900 = -15450 \text{ N}$$

$$= 15.45 \text{ kN (uplift)}$$

Number of 16 mm diameter bolts required to transmit Load to shoe angles

$$= \frac{\text{Factored downward load}}{\text{Design strength of bolt}}$$

$$= \frac{1.5 \times 19.39}{57.95} = 0.5 \text{ say 2 (minimum)}$$

Assume safe bearing pressure on masonry = 0.80 N/m^2

$$\therefore \quad \text{Bearing area required} = \frac{19.390 \times 10^3}{0.8} = 24237.5 \text{ mm}^2$$

$$\text{Width of bearing plate} = 75 + 75 + 8 = 158 \text{ mm}$$

(Assuming shoe angles of 75 × 75 × 8 mm)

$$\text{Length of bearing plate} = \frac{24237.5}{158} = 153.4 \text{ mm say 200 mm}$$

∴ Provide bearing plate of size 158 × 200 mm

$$\text{Actual bearing pressure intensity} = \frac{19390}{158 \times 200} = 0.614 \text{ N/mm}^2$$

Maximum B.M. for a 1 mm wide strip of bearing plate considering the critical section for bending at the exterior edge of the vertical leg of shoe angle.

$$M_x = 0.614 \times \frac{(75 - 6)^2}{2} = 1461.63 \text{ Nmm}$$

$$\text{Factored moment,} \quad M_d = 1.5 \times 1461.63 = 2192.44 \text{ Nmm}$$

$$\text{Design bending strength} = \frac{f_y \cdot Z_p}{1.1} = \frac{250}{1.1}\left[1.5 \times \frac{t^2}{6}\right]$$

$$2192.44 = 56.82 \, t^2$$

$$\therefore \quad t = 6.21 \text{ mm say 8 mm}$$

∴ Provide a bearing plate of thickness 8 mm with 2 – 20 mm diameter bolts sufficient to resist the uplift force.

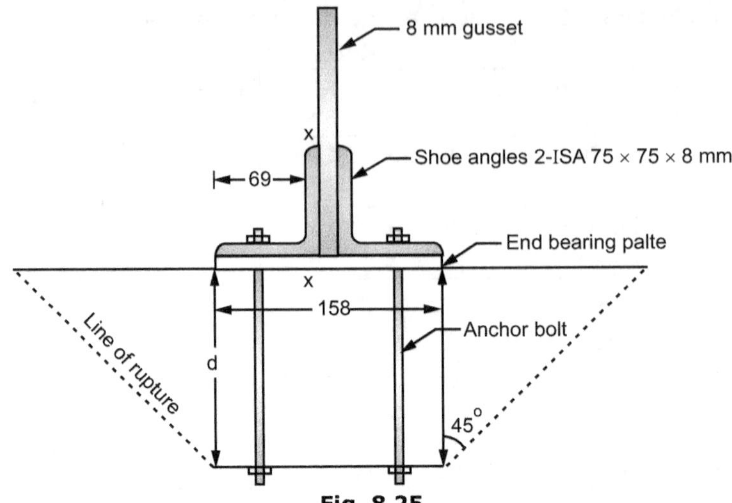

Fig. 8.25

The length of bolt should be such that the weight of masonry is more than the uplift force = 15.45 kN.

Provide 2 bolts of 20 mm diameter.

The bolts should be embedded in the masonry to a depth 'd' such that the wall may hold the truss down against the net uplift force. Assuming line of rupture at 45° to the horizontal and specific weight of masonry = 20 kN/m^3

Weight of masonry between rupture lines (See Fig. 8.31).

$$15.45 = \frac{1}{2}\,[0.158 + (0.158 + 2d)] \cdot d \times 0.35 \times 20$$

$$15.45 = [0.316 + 2d] \times 3.5\,d$$
$$4.414 = 0.316\,d + 2d^2$$

or $d^2 + 0.158\,d - 2.207 = 0$

∴ $\quad\quad\quad\quad\quad\quad\quad d = 1.408$ m say 1.45 m

The base angles are provided with 22 mm wide × 28 mm long slots to permit expansion of truss due to temperature stresses.

The details of trusses are shown in Fig. 8.26.

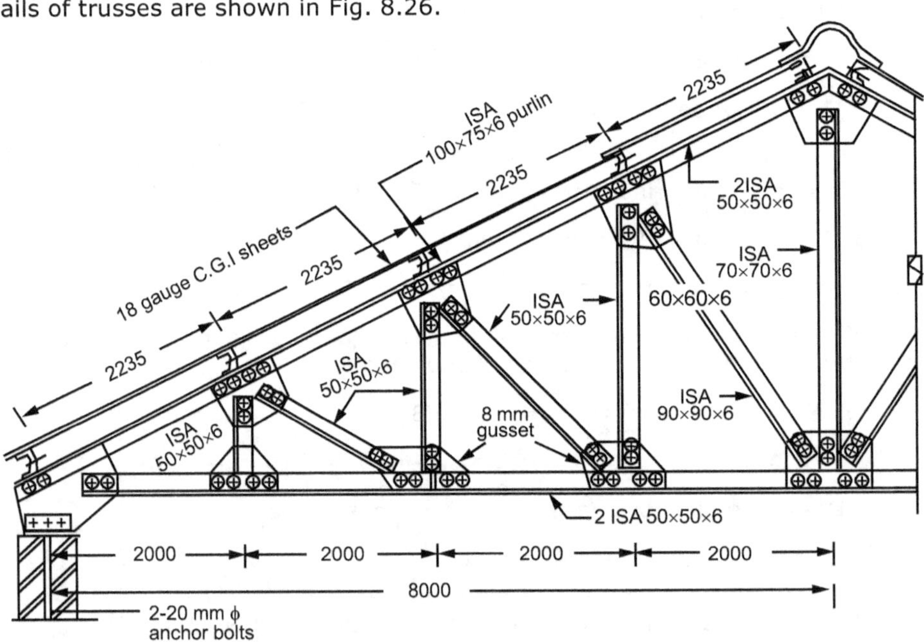

Fig. 8.26

8.13 DESIGN OF WELDED ROOF TRUSS

Ex. 8.9: *A fink type roof truss is required for a workshop building at Alibaug with the following details:*

1. *Span of truss – 18 m*

2. *Pitch of truss – $\frac{1}{5}$ of span*

3. *Spacing of truss – 4 metres c/c*
4. *Weight of A.C. sheet roofing – 162 N/m² (16.2 kg/m²)*
5. *Weight of purlin – 200 N/m*
6. *Weight of truss – 120 N/m² (Plan area)*
7. *Design wind pressure = – 1500 N/m².*
8. *Clear height = 6 m, height of plinth = 1 m*
9. *Width of support = 400 mm.*

For openings with medium permeability, the wind pressure for different conditions should be considered from the following table only:

External pressure coefficients	Wind normal to ridge		Wind parallel to ridge	
Slope	Windward slope	Leeward slope	Outer quarter length	Mid half length
20°	– 0.4 p	– 0.4 p	– 0.7 p	– 0.6 p
30°	0	– 0.4 p	– 0.7 p	– 0.6 p
Internal pressure coefficients	± 0.5 p			

(a) Determine panel point loads consisting of dead load, live load and wind load.
(b) Explain the steps of design of angle section purlin as per IS 800.
(c) Draw to scale the elevation of roof truss showing arrangement of purlins and rafters.
(d) Design principal rafter, main tie, main vertical tie, main inclined struts with welded end connection.
(e) Design end connections with suitable welds.

Sol.: General Design:

$$\text{Effective span, } L = 18 \text{ m}$$
$$\text{Spacing of trusses} = 4 \text{ m}$$
$$\text{Rise of truss} = \frac{1}{5} \text{ span} = \frac{1}{5} \times 18 = 3.6 \text{ m}$$
$$\text{Slope of truss, } \theta = \tan^{-1}\left(\frac{\text{Rise}}{L/2}\right) = \tan^{-1}\left(\frac{3.6}{9}\right)$$
$$\theta = 21.80°$$
$$\text{Length of principal rafter} = \sqrt{9^2 + 3.6^2} = 9.70 \text{ m}$$

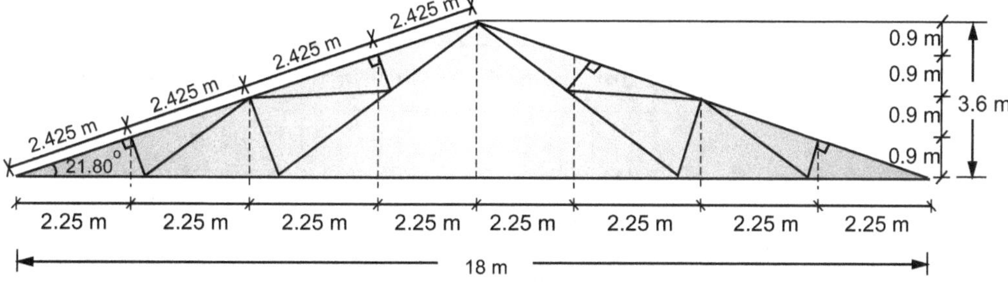

Fig. 8.27

$$\text{Sloping area of roof supported by a truss} = 2 \times 9.70 \times 4 = 77.60 \text{ m}^2$$
$$\text{Plan area of roof supported by a truss} = 18 \times 4 = 72 \text{ m}^2$$

Loads:

1. Imposed load:

$$\text{Imposed load on truss} = 750 - (\theta - 10) \times 20$$
$$= 750 - (21.80 - 10) \times 20$$
$$\therefore \quad LL = 514 \ N/m^{2*}$$

2. Wind load:

Basic wind speed for Alibaug (Near Mumbai) is $V_b = 47$ m/s. (See wind map of India in Fig. 8.3).

Taking risk factor $k_1 = 1.00$ (Table 7.5).

Terrain, height and size factor $k_2 = 0.99$ (Table 7.6) (considering category 1 and class C building) topography factor, $k_3 = 1.00$.

$$\text{Design wind speed } V_z = V_b \times k_1 \times k_2 \times k_3$$
$$\therefore \quad V_z = 47 \times 1 \times 0.99 \times 1$$
$$V_z = 45.53 \ m/s$$

Design wind pressure,
$$p_z = 0.6 \ V_z^2$$
$$\therefore \quad p_z = 0.6 \times (45.53)^2$$
$$p_z = 1299.02 \text{ say } 1300 \ N/m^2$$

Height of eaves above G.L.
$$h = \text{Clear height} + \text{Plinth height}$$
$$h = 6 + 1 = 7 \ m$$

Width or span of truss,
$$w = 18 \ m$$
$$\therefore \quad \frac{h}{w} = \frac{7}{18} = 0.39$$

For
$$\frac{h}{w} < \frac{1}{2} \quad \text{and} \quad \theta = 21.80°$$

Use of given table in this example statement can be made. Otherwise refer Tables 8.10 and 8.12.

External wind pressure co-efficient on windward side

$$C_{pe_w} = -\left[0.4 - \frac{0.4 - 0}{10} \times 1.80\right] = -0.328$$

External wind pressure co-efficient on Leeward side
$$C_{pe_l} = -0.4$$

Internal wind pressure coefficient $C_{pi} = \pm 0.5$

$$\text{Total wind pressure} = [C_{pe} - C_{pi}] \ p_z$$

Wind normal to ridge	Windward slope	Leeward slope
(a) Internal pressure + 0.5 p_z	$= (-0.328 - 0.5) \ p_z$ $= -0.828 \times 1300$ $= -1076 \ N/m^2$	$= (-0.4 - 0.5) \ p_z$ $= -0.9 \times 1300$ $= -1170 \ N/m^2$
(b) Internal pressure − 0.5 p_z	$= (-0.328 + 0.5) \ p_z$ $= +0.172 \times 1300$ $= +224 \ N/m^2$	$= (-0.4 + 0.5) \ p_z$ $= +0.1 \times 1300$ $= +130 \ N/m^2$
Wind parallel to ridge	**Outer $\frac{1}{4}^{th}$ length of building**	**Mid $\frac{1}{2}$ length of building**
(a) Internal pressure + 0.5 p_z	$= (-0.7 - 0.5) \ p_z$ $= -1.2 \times 1300$ $= -1560 \ N/m^2$	$= (-0.6 - 0.5) \ p_z$ $= -1.1 \times 1300$ $= -1430 \ N/m^2$
(b) Internal pressure − 0.5 p_z	$= (-0.7 + 0.5) \ p_z$ $= -0.2 \times 1300$ $= -260 \ N/m^2$	$= (-0.6 + 0.5) \ p_z$ $= -0.1 \times 1300$ $= -130 \ N/m^2$

The maximum wind pressure for design is $+130 \ N/m^2$ (downward) and $-1560 \ N/m^2$ (uplift)**.

Design of Angle Purlins

$$\text{Spacing of purlins} = 2.425 \text{ m}$$
$$\text{Weight of A.C. sheet} = 162 \text{ N/m}^2$$

Load on purlin per metre length

(i) $\qquad\qquad$ Weight of A.C. sheet $= 162 \times 2.425 = 392.85$ N/m

(ii) $\qquad\qquad$ Self weight of purlin $= 200$ N/m

$$\text{Total dead load} = 592.85 \text{ N/m say } 593^* \text{ N/m}$$
$$\text{Live load} = 514^* \times 2.25 \times 1 = 1155.5 \text{ N/m}$$
$$\text{Wind load on inclined area} = -1560^{**} \times 2.425 \times 1 = -3783 \text{ N/m}$$

---

* (on plan area)

** (normal to roof)

2. Components of load along xx axis (parallel to roof) and along yy axis (normal to roof).

(a) DL, $\qquad\qquad w_{dy} = 593 \cos 21.80° = 551$ N/m

$\qquad\qquad\qquad\qquad w_{dx} = 593 \sin 21.80° = 220$ N/m

(b) LL, $\qquad\qquad w_{ly} = 1156 \cos 21.80° = 1073$ N/m

$\qquad\qquad\qquad\qquad w_{lx} = 1156 \sin 21.80° = 429$ N/m

(c) WL, $\qquad\qquad w_{wy} = -3783$ N/m,

$\qquad\qquad\qquad\qquad w_{wx} = 0$

3. Factored loads due to combinations:

(a) Load combination 1 $= 1.5$ (DL + LL)

$\qquad\qquad\qquad w_{y1} = 1.5 (551 + 1073) = 2436$ N/m

$\qquad\qquad\qquad w_{x1} = 1.5 (220 + 429) = 973.5$ N/m

(b) Load combination 2 $= 1.5$ (DL + WL)

$\qquad\qquad\qquad w_{y2} = 1.5 (551 - 3783) = -4848$ N/m

$\qquad\qquad\qquad w_{x2} = 1.5 (220 - 0) = +330$ N/m

(c) Load combination 3 $= 1.2$ (DL + LL + WL)

$\qquad\qquad\qquad w_{y3} = 1.2 (551 + 1073 - 3783) = -2590.8$ N/m

$\qquad\qquad\qquad w_{x3} = 1.2 (220 + 429 - 0) = +778.8$ N/m

Hence the critical combination is combination 2 i.e. 1.5 (DL + WL)

$\therefore \qquad\qquad w_y = 4848$ N/m (absolute value), $w_x = 330$ N/m

4. Bi-axial BM

$$M_x = \frac{w_y \cdot L^2}{10} = \frac{4848 \times 4^2}{10} = 7756.8 \text{ Nm} = 7.76 \text{ kNm}$$

$$M_y = \frac{w_x \cdot L^2}{10} = \frac{330 \times 34^2}{10} = 363 \text{ Nm} = 0.36 \text{ kNm}$$

5. Selection of angle selection:

$$\text{Width of angle parallel to roof} = \frac{L}{60} = \frac{4000}{60} = 66.67 \text{ mm}$$

$$\text{Depth of angle normal to roof} = \frac{L}{45} = \frac{4000}{45} = 88.89 \text{ mm}$$

Try ISA 125 $\times$ 75 $\times$ 10 mm @ 14.9 kg/m (146 N/m) satisfying above requirements.

Also, $\quad \dfrac{b}{t} = \dfrac{75}{10} = 7.5 < 9.4 \, \varepsilon \quad$ ($\therefore \varepsilon = 1$)

$\qquad\quad \dfrac{d}{t} = \dfrac{125}{10} = 12.5 < 15.7 \, \varepsilon \quad \therefore$ Classification of section is semi-compact.

For ISA 125 × 75 × 10 mm, $I_{xx} = 300.3 \times 10^4$ mm^4

$I_{yy} = 81.6 \times 10^4$ mm^3, $C_{xx} = 42.4$ mm, $C_{yy} = 17.6$ mm

$$Z_{xx\,(Top)} = \frac{I_{xx}}{y_{top}} = \frac{300.3 \times 10^4}{42.4} = 70825 \text{ mm}^3$$

$$Z_{xx\,(Bottom)} = \frac{I_{xx}}{y_{bottom}} = \frac{300.3 \times 10^4}{(125 - 42.4)} = 36356 \text{ mm}^3$$

$$Z_{yy} = \frac{I_{yy}}{x_{max}} = \frac{81.6 \times 10^4}{17.6} = 46364 \text{ mm}^3$$

6. Calculation of design moment

Consider the purlin as a laterally supported beam.

For bending @ xx axis

$$M_{dx} = Z_e \cdot \frac{f_y}{\gamma_{mo}} = 36356 \times \frac{250}{1.10}(10^{-6}) = 8.263 \text{ kNm}$$

For bending @ yy axis

$$M_{dy} = Z_{yy} \cdot \frac{f_y}{\gamma_{mo}} = 46364 \times \frac{250}{1.10}(10^{-6}) = 10.537 \text{ kNm}$$

7. Check for bi-axial bending for outstanding toe of purlin.

$$\frac{M_x}{M_{dx}} + \frac{M_y}{M_{dy}} \leq 1$$

$$\frac{7.76}{8.263} + \frac{0.36}{10.537} = 0.973 < 1 \qquad \therefore \text{ O.K.}$$

Design of Roof Truss

1. Dead load (assumed to act on top panel points)

$$\text{Due to AC sheet } = 162 \times \text{inclined area}$$

$$= 162 \times 77.60 = 12572 \text{ N}$$

$$\text{Due to purlins } = \text{No. of purlins} \times \text{spacing of trusses} \times \text{weight/metre}$$

$$= 10 \times 4 \times 101 = 4040 \text{ N}$$

$$\text{Due to truss}\left[\left(\frac{L}{3} + 5\right) \times 10 = \left(\frac{18}{3} + 5\right) 10 = 110 \text{ N/m}^2, \text{ but given value is } 120 \text{ N/m}^2 \text{ on plan area}\right]$$

$$= 120 \times 72 = 8640$$

$$\text{Total dead load } = 25252 \text{ N}$$

$$\text{Dead load on each top panel } = \frac{25252}{8} = 3155.5 \text{ N} \simeq 3.16 \text{ kN}$$

$$\text{Dead load on end panel point } = \frac{3156.5}{2} = 1578.25 \text{ N} \simeq 1.58 \text{ kN}$$

2. Live load (on plan area):

$$\text{Live load on truss } = \frac{2}{3} \times 514 \times 72 = 24672 \text{ N}$$

$$\text{Live load on each top panel point } = \frac{24672}{8} = 3084 \text{ N} \simeq 3.08 \text{ kN}$$

$$\text{Live load on end panel point } = \frac{3084}{2} = 1542 \text{ N} \simeq 1.54 \text{ kN}$$

3. Wind load:

$$\text{Wind load on each top panel point } = -1560 \times 2.425 \times 4 = -15132 \text{ N}$$

$$= -15.14 \text{ kN (uplift)}$$

$$\text{Wind load on each end point } = -\frac{15132}{2} = -7566 \text{ N} = -7.57 \text{ kN (uplift)}$$

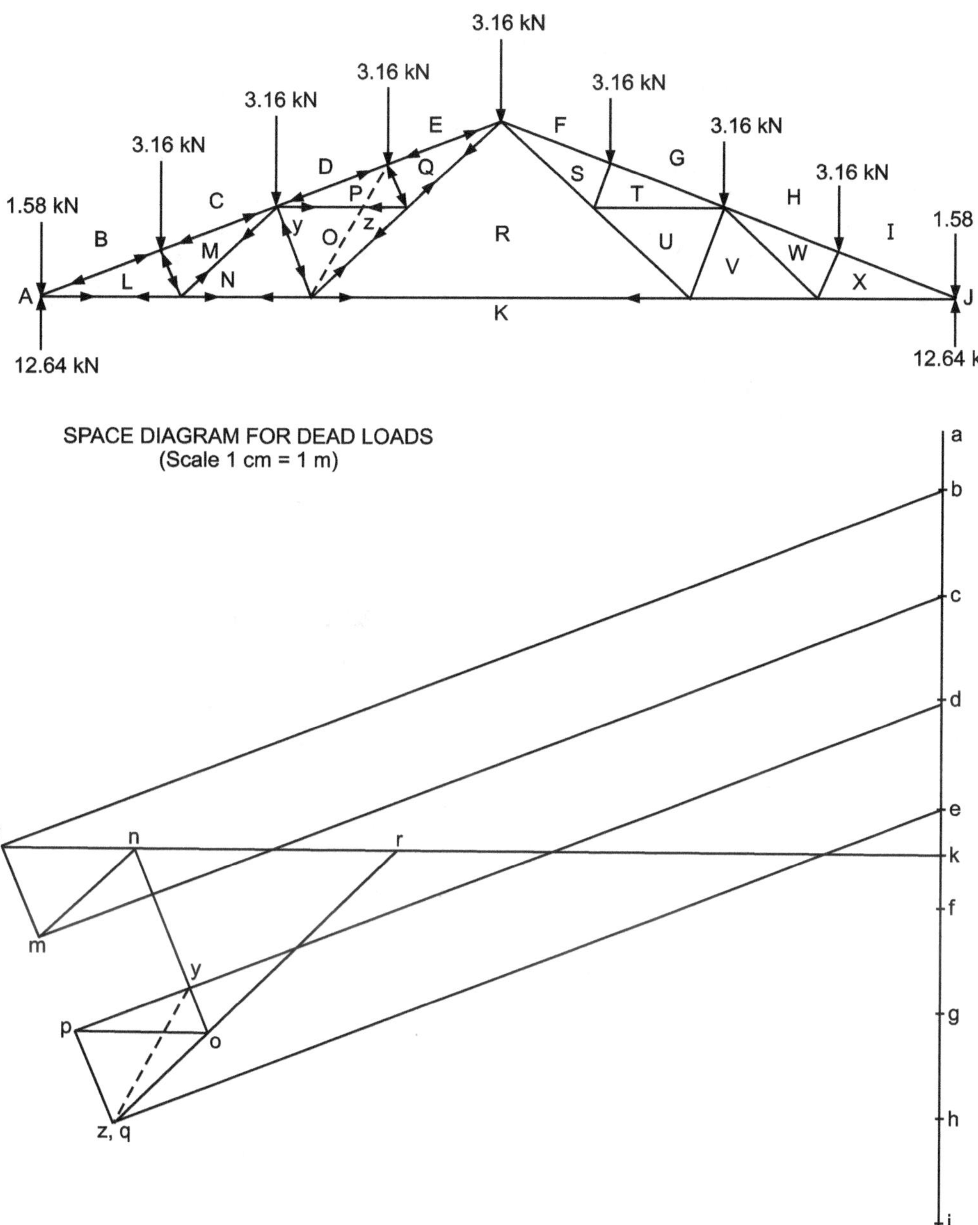

SPACE DIAGRAM FOR DEAD LOADS
(Scale 1 cm = 1 m)

FORCE DIAGRAM FOR DEAD LOADS
(Scale 1 cm = 2 kN)

Fig. 8.28

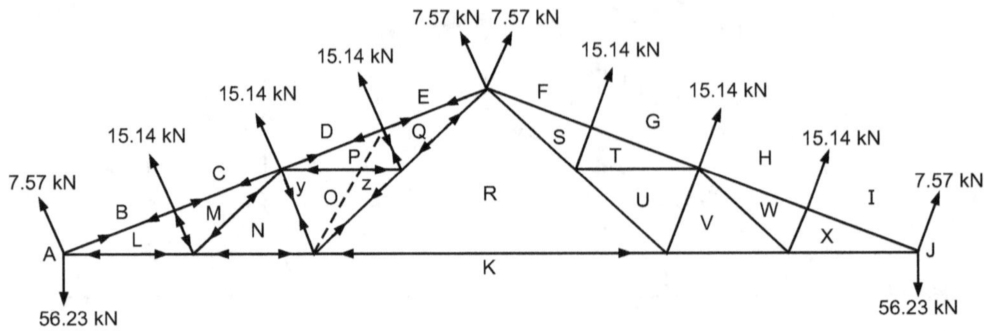

SPACE DIAGRAM FOR WIND LOADS
(Scale 1 cm = 1 m)

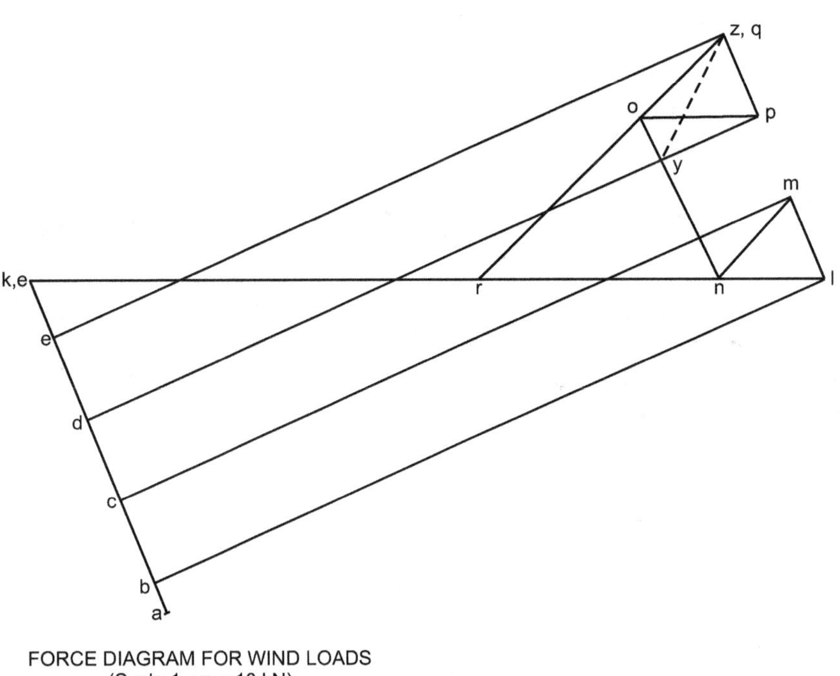

FORCE DIAGRAM FOR WIND LOADS
(Scale 1 cm = 10 kN)

Fig. 8.29

The force diagrams for truss under dead load and negative wind load are shown in Fig. 8.28 and 8.29 respectively.

Forces due to live load are calculated by multiplying forces due to dead load by $\dfrac{3.08}{3.16} = 0.9747$.

4. Combination of loads:

Due to symmetric loading only half portion of force diagram need be drawn. The forces in each of the members are shown in Table 8.15.

Table 8.15: Forces in Members of Truss
(Tension +ve, Compression – ve)

Members	Dead load DL	Live load LL	Wind load WL	Load combination 1.5 (DL + LL)	Load combination 1.2 (DL + LL + 1.5 WL)	Load combination 0.9 DL + 1.5 WL	Design Load (kN)	Length of member (m)
Principal Rafter								
BL , IX	− 31.00	− 30.20	+ 127.00	− 91.80	+ 78.96	+ 162.6		2.425
CM , HW	− 29.00	− 28.30	+ 127.00	− 85.95	+ 83.64	+ 164.4	− 91.80	2.425
DP , GT	− 27.80	− 27.10	+ 127.00	− 82.35	+ 86.52	+ 165.5	+ 166.40	2.425
EQ , FS	− 26.80	− 26.10	+ 127.00	− 79.35	+ 88.92	+ 166.4		2.425
Main tie								
LK , XK	+ 28.30	+ 27.60	− 136.50	+ 83.85	− 96.72	− 179.3		2.612
NK , VK	+ 24.40	+ 23.80	− 115.00	+ 72.30	− 80.16	− 150.5	+ 83.85	2.612
RK	+ 15.50	+ 15.10	− 74.00	+ 45.90	− 52.08	− 97.05	− 179.3	2.612
Major inclined ties								
QR , SR	+ 11.80	+ 11.50	− 60.00	+ 34.95	− 44.04	− 79.38	+ 34.95	2.612
OR , UR	+ 7.90	+ 7.70	− 40.00	+ 23.40	− 29.28	− 52.89	− 79.38	2.612
Major Strut								
NO , UV	− 5.00	− 5.80	+ 31.00	− 16.20	+ 24.24	+ 42.00	− 16.20 + 42.00	1.94
Minor ties								
MN , WV	+ 3.80	+ 3.70	− 19.00	+ 11.25	− 13.80	− 25.08	+ 11.25	2.612
PO , TU							− 25.08	2.612
Minor struts								
LM , WX	− 3.00	− 2.90	+ 15.00	− 8.85	+ 10.92	+ 19.80	− 8.85	0.97
PQ , ST							+ 19.80	0.97

DESIGN OF MEMBERS

1. Principal Rafter (Members BL, CM, DP, EQ, IX, HW, GT and FS)

Design as a compression member and check for tension.

$$\text{Maximum compression} = -91.80 \text{ kN}$$

$$\text{Maximum tension} = +166.40 \text{ kN}$$

$$\text{Length of member} = 2.425 \text{ m}$$

Effective length l_e = 0.85 L = 0.85 × 2.425 = 2.06 m = 2060 mm

Using 2 ISA − 55 × 55 × 6 mm connected on both sides of 8 mm thick gusset plate by 3 mm fillet welds as shown in Fig. 8.30.

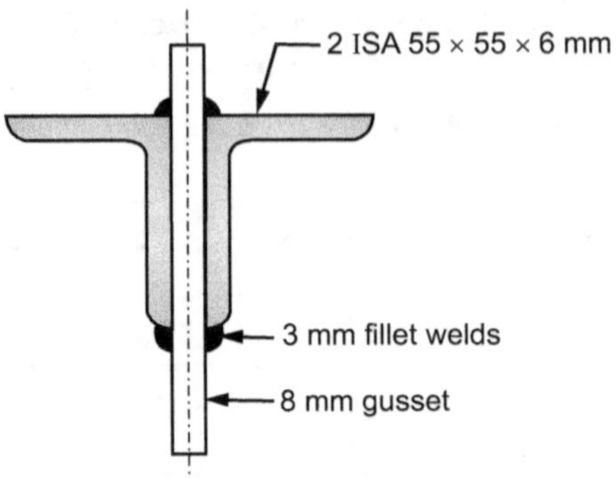

Fig. 8.30

Provide intermittent stitch welds along the length of the member.

$$r_{min} = r_{xx} = 16.6 \text{ mm}$$

$$\lambda = \frac{l_e}{r_{min}} = \frac{2060}{16.6} = 124.09 < 180 \quad \therefore \text{ O.K.}$$

From Table 4.4, by interpolation,

kL/r	f_{cd}
120	83.7
130	74.3

$$f_{cd} = 83.7 - \frac{83.7 - 74.3}{10} \times 4.09 = 79.85 \text{ MPa}$$

$$A_g = 2 \times 626 = 1252 \text{ mm}^2$$

Design compressive strength $= f_{cd} \times A_g$

$$= \frac{79.85 \times 1252}{1000} = 99972 \text{ N}$$

$$= 99.97 \text{ kN} > 91.80 \text{ kN} \quad \therefore \text{ O.K.}$$

- **Check for Tension:** $A_{net} = 2\,[A_g] = 2[626]$

or $\qquad\qquad A_n = 1252 \text{ mm}^2$ (No deduction since ends are welded)

Design tensile strength by net section rupture,

$$T_{dn} = \frac{\alpha\, A_n\, f_u}{\gamma_{m1}} = \frac{0.8 \times 1252 \times 410}{1.25} = 328520 \text{ N} = 328.52 \text{ kN}$$

Design tensile strength by gross section yielding,

$$T_{dg} = \frac{A_g f_y}{\gamma_{m0}} = \frac{1252 \times 250}{1.1} = 284540 \text{ N} = 284.54 \text{ kN}$$

$\therefore$ Design tensile strength $=$ Least of T_{dn} and T_{dg}

$$= 284.54 \text{ kN} > 166.40 \text{ kN} \quad \therefore \text{ OK.}$$

2. Main tie (Members LK, NK, RK, XK, VK):

Design as a tension member and check for compression.

$$\text{Maximum tension} = + 83.85 \text{ kN}$$

$$\text{Maximum compression} = - 179.3 \text{ kN}$$

$$\text{Length of member, } L = 2.612 \text{ m}$$

Try 2 ISA 75 × 75 × 6 mm on both sides of gusset with intermittent stitch fillet welds along the length,

$$A_{net} = 2\,[A_g]$$

or $\qquad\qquad A_n = 2 \times 866 = 1732 \text{ mm}^2 \qquad (\because \text{ Ends are welded})$

Design tensile strength by net section rupture,

$$T_{dn} = \frac{\alpha\, A_n\, f_u}{\gamma_{m1}} = \frac{0.8 \times 1732 \times 410}{1.25} = 454470 \text{ N} = 454.47 \text{ kN}$$

Design tensile strength by gross-section yielding,

$$T_{dg} = \frac{A_g f_y}{\gamma_{m0}} = \frac{1732 \times 250}{1.10} = 393630 \text{ N} = 393.63 \text{ kN}$$

$\therefore$ Design tensile strength = Least of T_{dn} and T_{dg}

$$= 393.63 \text{ kN} > 83.85 \text{ kN} \quad \therefore \text{ OK.}$$

Strength is much higher than required. This is to satisfy the strength requirement in compression also.

- **Check for compression:**

$$r_{min} = r_{xx} = 23.0 \text{ mm}$$

Effective length, $\quad l_e = 0.85\, L = 0.85 \times 2.612$

$$l_e = 2.22 \text{ m} = 2220 \text{ mm}$$

$$\lambda = \frac{l_e}{r_{min}} = \frac{2220}{23.0} = 96.52$$

From Table 4.4, by interpolation,

kL/r	f_{cd}
90	121
100	107

$$f_{cd} = 121 - \frac{121 - 107}{10} \times 6.52 = 111.87 \text{ MPa}$$

Design Compressive strength $= f_{cd} \times A_g = \dfrac{117.87 \times 2 \times 866}{1000}$

$$= 204.15 \text{ kN} > 179.30 \text{ kN} \quad \therefore \text{ O.K.}$$

3. Major Inclined Ties (QR, OR, SR, UR):

This member normally carries tension but due to wind the member will be subjected to reversal of stresses i.e. compression.

$$\text{Maximum tension} = +\,34.95 \text{ kN}$$
$$\text{Maximum compression} = -\,79.38 \text{ kN}$$
$$\text{Length of member, } L = 2.612 \text{ m}$$

Try single angle 65 × 65 × 6 mm connected by 3 mm welds.

Check for design strength T_d of trial section:

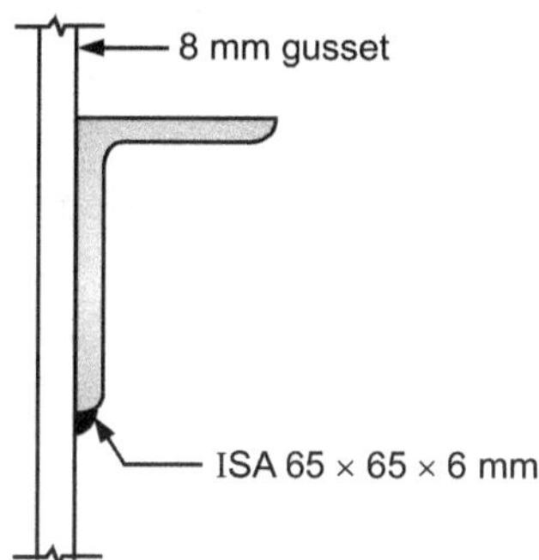

Fig. 8.31

(i) Design tensile strength by gross-section yield consideration

$$T_{dg} = \frac{A_g \cdot f_y}{\gamma_{mo}}$$

$$= \frac{744 \times 250}{1.10} = 169090 \text{ N} = 169.09 \text{ kN}$$

(ii) Design tensile strength by net-section rupture consideration

Net area of connected leg,

$$A_{nc} = \left[L - \frac{t}{2}\right] t = \left[65 - \frac{6}{2}\right] \times 6 = 372 \text{ mm}^2$$

Gross area of outstanding leg,

$$A_{go} = \left[b - \frac{t}{2}\right] \cdot t = \left[65 - \frac{6}{2}\right] \times 6 = 372 \text{ mm}^2$$

$$T_{dn} = \frac{0.9 \, A_{nc} \cdot f_u}{\gamma_{m1}} + \frac{\beta \, A_{go} \, f_y}{\gamma_{m0}}$$

$$w = \text{Outstand leg width} = 65 \text{ mm}$$

$$b_s = w = 65 \text{ mm}$$

$$L_c = \text{Average length of bottom and top welds}$$

$$\text{Length of weld required} = \frac{\text{Maximum axial force}}{\text{Strength of weld/mm length}}$$

$$= \frac{79.38 \times 10^3}{f_{wd} \times t_t} = \frac{79.38 \times 10^3}{157.8 \times (0.7 \times 3)}$$

$$\left(f_{wd} = \frac{f_u}{\sqrt{3} \cdot \gamma_{mw}} = \frac{410}{\sqrt{3} \times 1.5} = 157.8 \text{ N/mm}^2\right) = 239.54 \text{ say } 240$$

$$\therefore \quad L_c = \frac{240}{2} = 120 \text{ mm}$$

$$\beta = 1.4 - 0.076 \, \frac{w}{t} \times \frac{f_y}{f_u} \times \frac{b_s}{L_c}$$

$$= 1.4 - 0.076 \times \frac{65}{6} \times \frac{250}{410} \times \frac{65}{120} = 1.128$$

$$0.9 \, \frac{f_u}{f_y} \times \frac{\gamma_{m0}}{\gamma_{m1}} = 0.9 \times \frac{410}{250} \times \frac{1.10}{1.25} = 1.30$$

$$\therefore \quad \beta = 1.128 \ (< 1.30 \ > 0.70) \quad \therefore \quad \text{acceptable}$$

$$\therefore \quad T_{dn} = \frac{0.9 \times 372 \times 410}{1.25} + \frac{1.128 \times 372 \times 250}{1.10}$$

$$= 205181 \text{ N} = 205.18 \text{ kN}$$

(iii) Design tensile strength by block shear consideration:

$$A_{vg} = \text{Minimum gross area in shear along weld line} = L_v \cdot t$$

$$= \text{Total length of weld} \times t = 240 \times 6 = 1440 \text{ mm}^2$$

$$A_{vn} = \text{Minimum net area in shear along weld line} = L_v \cdot t$$

$$= 1440 \text{ mm}^2$$

$$A_{tg} = \text{Minimum gross area in tension for connected leg} = L_t \cdot t$$

$$= 65 \times 6 = 390 \text{ mm}^2$$

$$A_{tn} = \text{Minimum net area in tension for connected leg } L_t \cdot t = 390 \text{ mm}^2$$

$$T_{db_1} = \frac{A_{vg} \, f_y}{\sqrt{3} \, \gamma_{m0}} + \frac{0.9 \, A_{tn} \cdot f_u}{\gamma_{m1}}$$

$$= \frac{1440 \times 250}{\sqrt{3} \times 1.10} + \frac{0.9 \times 390 \times 410}{1.25} = 255420 \text{ N}$$

$$T_{db_2} = \frac{A_{tg} \, f_y}{\gamma_{m0}} + \frac{0.9 \, A_{vn} \cdot f_u}{\sqrt{3} \, \gamma_{m1}}$$

$$= \frac{390 \times 250}{1.10} + \frac{0.9 \times 1440 \times 410}{\sqrt{3} \times 1.25} = 334061 \text{ N}$$

$$\therefore \quad T_{db} = \text{Least of } T_{db_1} \text{ and } T_{db_2} = 255420 \text{ N} = 255.42 \text{ kN}$$

$\therefore$ Design tensile strength of the angle = Minimum of T_{dg}, T_{dn}, T_{db}

$$= 169.09 \text{ kN} > > 34.95 \text{ kN}$$

(Even though greater design tensile strength is obtained, the selected angle should satisfy design compressive strength requirement also).

Check for compression:

For the selected discontinuous angle $65 \times 65 \times 6$ mm connected with welds, $r_{min} = r_{vv} = 19.8$ and $kL = 1 \times 2.612 = 2.612$ m $= 2612$ mm.

Also $k_1 = 0.2$, $k_2 = 0.35$ and $k_3 = 20$

Imperfection factor, $\alpha = 0.49$ (class c buckling)

$$\lambda_{vv} = \frac{\left(\frac{l}{r_{vv}}\right)}{\varepsilon\sqrt{\frac{\pi^2 E}{250}}} = \frac{\left(\frac{2612}{19.8}\right)}{(1)\sqrt{\frac{\pi^2 \times 2 \times 10^5}{250}}} = 1.4846$$

$$\lambda_\phi = \frac{(b_1 + b_2)}{\varepsilon\sqrt{\frac{\pi^2 E}{25}} \times 2t} = \frac{(65 + 65)}{(1)\sqrt{\frac{\pi^2 \times 2 \times 10^5}{250}} \times 2 \times 6} = 0.1219$$

$$\lambda_e = \sqrt{k_1 + k_2\,\lambda_{vv}^2 + k_3\,\lambda_\phi^2}$$

$$= \sqrt{0.2 + 0.35 \times 1.4846^2 + 20 \times 0.1219^2} = 1.1263$$

$$\phi = 0.5\,[1 + \alpha\,(\lambda_e - 0.2) + \lambda_e^2]$$

$$= 0.5\,[1 + 0.49\,(1.1263 - 0.2) + 1.1263^2] = 1.3612$$

$\therefore$ Design compression stress,

$$f_{cd} = \frac{\dfrac{f_y}{\gamma_{m0}}}{\phi + \sqrt{\phi^2 - \lambda_e^2}} = \frac{\dfrac{250}{1.1}}{1.3612 + \sqrt{1.3612^2 - 1.1263^2}} = 106.92 \text{ MPa}$$

$\therefore$ Design compressive load

$$= f_{cd} \times A_g = \frac{106.92 \times 744}{1000} = 79550 \text{ N}$$

$$= 79.55 \text{ kN} > 79.38 \text{ kN} \quad \therefore \text{ safe.}$$

4. Major struts (NO, UV)

Generally, this member carries compression due to DL + LL, but when there are wind storms, wind pressure increases and then it carries tension.

Maximum compression $= -16.20$ kN

Maximum tension $= +42.00$ kN

Length of member, $L = 1.94$ m

Try a single ISA $50 \times 50 \times 6$ mm with intermittent fillet stitch welds.

$$A_g = 568 \text{ mm}^2$$

$$r_{min} = r_v = 9.6 \text{ mm}$$

Effective length, $l_e = 0.85\,L = 0.85 \times 1.94$

$$l_e = 1.649 \text{ m} = 1649 \text{ mm}$$

$$\lambda = \frac{l_e}{r_{min}} = \frac{1649}{9.6} = 171.77 \not> 180 \quad \therefore \text{ O.K.}$$

From Table 4.4, by interpolation,

kL/r	f_{cd}
170	48.1
180	43.6

$$= 48.1 - \frac{48.1 - 43.6}{10} \times 1.77 = 47.30 \text{ MPa}$$

Design compressive load,

$$P_{cd} = f_{cd} \times A_g$$

$$P_c = \frac{47.30 \times 568}{1000} = 26868 \text{ N}$$

$$P_c = 26.86 \text{ kN} > 16.20 \text{ kN}$$

(**Note:** Load carrying capacity is more, but minimum size of section is $50 \times 50 \times 6$ mm. Moreover, the section should satisfy the criteria for maximum slenderness ratio $\not> 180$).

Check for tension:

(i) Design tensile strength by gross-section yielding consideration

$$T_{dg} = \frac{A_g \cdot f_y}{\gamma_{m0}}$$

$$= \frac{568 \times 250}{1.10} = 129090 \text{ N} = 129.09 \text{ kN}$$

(ii) Design tensile strength by net-section rupture consideration

Net area of connected leg,

$$A_{nc} = \left[L - \frac{t}{2}\right] t = \left[50 - \frac{6}{2}\right] \times 6 = 282 \text{ mm}^2$$

Gross area of outstanding leg,

$$A_{go} = \left[b - \frac{t}{2}\right] t = \left[50 - \frac{6}{2}\right] \times 6 = 282 \text{ mm}^2$$

$$T_{dn} = \frac{0.9 \, A_{nc} \cdot f_u}{\gamma_{m1}} + \frac{\beta \, A_{go} \, f_y}{\gamma_{m0}}$$

$$w = \text{Outstand leg width} = 50 \text{ mm}$$

$$b_s = w = 50 \text{ mm}$$

$$\text{Length of weld required} = \frac{\text{Maximum axial force}}{\text{Strength of weld/mm length}} = \frac{42 \times 10^3}{157.8 \times (0.7 \times 3)}$$

$$= 126.74 \text{ mm} \quad \text{say } 130 \text{ mm}$$

$$L_c = \text{Average length of bottom and top welds}$$

$$= \frac{130}{2} = 65 \text{ mm}$$

$$\beta = 1.4 - 0.076 \frac{w}{t} \times \frac{f_y}{f_u} \times \frac{b_s}{L_c}$$

$$= 1.4 - 0.076 \times \frac{50}{6} \times \frac{250}{410} \times \frac{50}{65} = 1.103$$

$$0.9 \frac{f_u}{f_y} \times \frac{\gamma_{m0}}{\gamma_{m1}} = 0.9 \times \frac{410}{250} \times \frac{1.10}{1.25} = 1.298$$

$\therefore \qquad \beta = 1.103 \ (< 1.30 \quad > 0.7)$ is acceptable

$$\therefore \qquad T_{dn} = \frac{0.9 \times 282 \times 410}{1.25} + \frac{1.103 \times 282 \times 250}{1.10}$$

$$= 153348 \text{ N} = 153.35 \text{ kN}$$

(iii) Design tensile strength by block shear consideration.

$$A_{vg} = \text{Minimum gross area in shear along weld line}$$

$$= \text{Total length of weld} \times t = 130 \times 6 = 780 \text{ mm}^2$$

$$A_{vn} = \text{Minimum net area in shear along weld line} = 780 \text{ mm}^2$$

$$A_{tg} = \text{Minimum gross area in tension of connected leg}$$

$$= 50 \times 6 = 300 \text{ mm}^2$$

$$A_{tn} = \text{Minimum net area in tension of connected leg} = 300 \text{ mm}^2$$

$$T_{db_1} = \frac{A_{vg}\, f_y}{\sqrt{3}\, \gamma_{m0}} + \frac{0.9\, A_{tn} \cdot f_u}{\gamma_{m1}}$$

$$= \frac{780 \times 250}{\sqrt{3} \times 1.10} + \frac{0.9 \times 300 \times 410}{1.25} = 190908 \text{ N}$$

$$T_{db_2} = \frac{A_{tg}\, f_y}{\gamma_{m0}} + \frac{0.9\, A_{vn} \cdot f_u}{\sqrt{3}\, \gamma_{m1}}$$

$$= \frac{300 \times 250}{1.10} + \frac{0.9 \times 780 \times 410}{\sqrt{3} \times 1.25} = 201120 \text{ kN}$$

$\therefore \quad T_{db} = $ Least of T_{db_1} and $T_{db_2} = 190908$ N $= 190.90$ kN

$\therefore$ Design strength of angle = Minimum of T_{dg}, T_{dn}, $T_{db} = 129.09$ kN > 42 kN

5. Other minor members:

Since other minor ties, MN, PO, WV, TU and minor struts LM, PQ, WX, ST are not severely loaded and length of members is also small, a single angle of minimum size ISA 50 × 50 × 6 mm is sufficient.

- **Design of connections:**

Min. size $= 3$ mm.

Max. size $= \dfrac{3}{4} \times t = \dfrac{3}{4} \times 6 = 4.5$ mm

$\therefore$ Use 3 mm size fillet welds.

Table 8.16: Size of Fillet Weld = 3 mm For All Members

Member	Design load (kN)	Section provided (mm)	Load on each angle (kN)	Effective length of weld L (mm)	C_x (mm)	Length of weld along toe x_1 (mm)	Length of weld along heel x_2 (mm)
Principal Rafter	− 91.80 + 166.40	2 ISA 55 × 55 × 6	83.20	255	15.7	75	180
Main tie	+ 83.85 − 179.30	2 ISA 75 × 75 × 6	89.65	275	20.6	75	200
Major inclined ties	+ 34.95 − 79.38	ISA 65 × 65 × 6	79.38	245	18.1	70	175
Major strut and other minor members	− 16.20 + 42.00	ISA 50 × 50 × 6	42.00	140	14.5	50	90

[**Note:** By carrying actual calculations, effective length of weld may come less by 10 to 20 mm than value shown in above table. L is slightly increased to satisfy the requirement, that length of side weld should not be less than the depth of the angle].

Let us consider principal rafter for example.

Force in each angle $= \dfrac{1}{2} \times 166.40 = 83.20$ kN

Design stress in weld,

$$f_{wd} = \frac{f_u}{\sqrt{3} \cdot \gamma_{mw}} = \frac{410}{\sqrt{3} \times 1.5} = 157.8 \text{ N/mm}^2$$

Effective length of weld required

$$L = \frac{P_d}{f_{wd} \times t_t} = \frac{83.20 \times 10^3}{157.8 \times 0.7 \times 3} = 251.07 \cong 255 \text{ mm}$$

Force in weld/mm length $= f_{wd} \times 1 \times t_b$

$$= 157.8 \times (0.7 \times 3) = 331.38 \text{ N/mm}$$

C_x for ISA 55 × 55 × 6 $= 15.7$ mm

Taking moment @ heel edge of angle

$$331.38 \, x_1 \times 55 \; = \; 832.00 \times 15.7$$

$\therefore$ $\qquad x_1 \; = \; 71.67$ mm say 75 mm

(75 mm is taken because $x_1 \not< $ depth of angle)

$\therefore$ $\qquad x_2 \; = \; 255 - 75$

$$= \; 180 \text{ mm}$$

Table 8.16 shows the sections adopted, length of weld required along toe and heel edge of the angles.

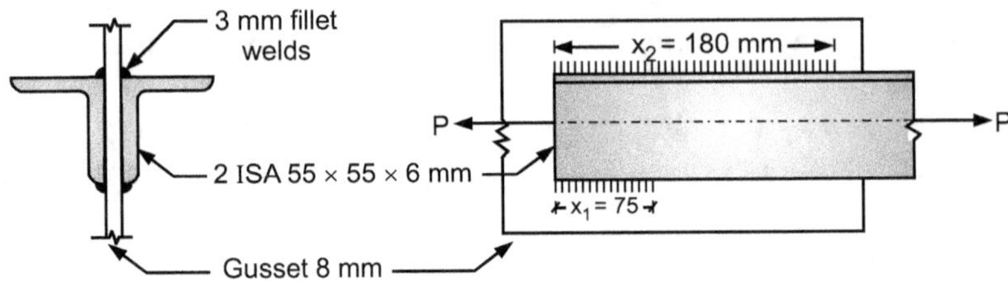

Fig. 8.32

Design of Bearing Plate:

$$\text{End reaction due to D.L. + L.L.} = \frac{25252 + 24672}{2}$$

$$= 24962 \text{ N}$$

$$= 24.96 \text{ kN}$$

$$\text{End reaction due to D.L. + W.L.} = 12.64 - 56.23$$

$$= -43.59 \text{ kN (Uplift)}$$

Assuming permissible bearing pressure on masonry = 0.8 N/mm^2

$$\text{Bearing area required} \;\; = \frac{24.96 \times 10^3}{0.8}$$

$$= 31200 \text{ mm}^2$$

$$\text{Provide shoe angle} \; = \; 75 \times 75 \times 6 \text{ mm}$$

$\therefore$ $\qquad \text{Width required} \; = \; 75 + 75 + 8$

$$= 158 \text{ mm}$$

$$\text{Length required} \;\; = \frac{31200}{158}$$

$$= 197.47 \text{ say } 200 \text{ mm}$$

$\therefore$ Provide 158 × 200 mm plate.

Intensity of pressure under bearing plate

$$= \frac{24.96 \times 10^3}{158 \times 200}$$

$$= 0.79 \text{ N/mm}^2$$

Considering the critical section of plate as the exterior edge of the vertical leg of the shoe angle (See Fig. 8.39). BM in plate, per mm length

$$M \; = \; 0.79 \times \frac{69^2}{2} = 1880.60 \text{ Nmm}$$

Factored moment, $\qquad M_d \; = \; 1.5 \times 1880.60$

$$= 2820.90 \text{ Nmm}$$

Design bending strength,

$$\frac{f_y Z_p}{1.10} = \frac{250}{1.10} \times \left(1.5 \frac{t^2}{6}\right)$$

$$2820.90 = 56.82\, t^2$$

$$\therefore \qquad t = 7.80 \text{ say } 8 \text{ mm}$$

$\therefore$ Use a bearing plate of size $158 \times 200 \times 8$ mm with $2 - 20$ mm diameter anchor bolts sufficient to resist uplift force.

The length of bolt should be such that the weight of the masonry is more than uplift force.

Assuming the line of rupture in the masonry to be at $45°$ and specific weight of masonry 20 kN/m^3,

$$\text{Weight of masonry} = \frac{1}{2}\,[0.158 + (0.158 + 2d)]\, d \times 0.35 \times 20$$

$$= \left(\frac{0.316 + 2d}{2}\right) d \times 0.35 \times 20$$

$$= (0.158 + d)\, d \times 0.35 \times 20 \quad \text{(Wt. of trapezoidal portion)}$$

$$\text{Uplift force} = 43.59 \text{ kN}$$

$$\text{Equating, } (0.158 + d)\, d \times 0.35 \times 20 = 43.590$$

$$(0.158 + d)\, d = 5.23$$

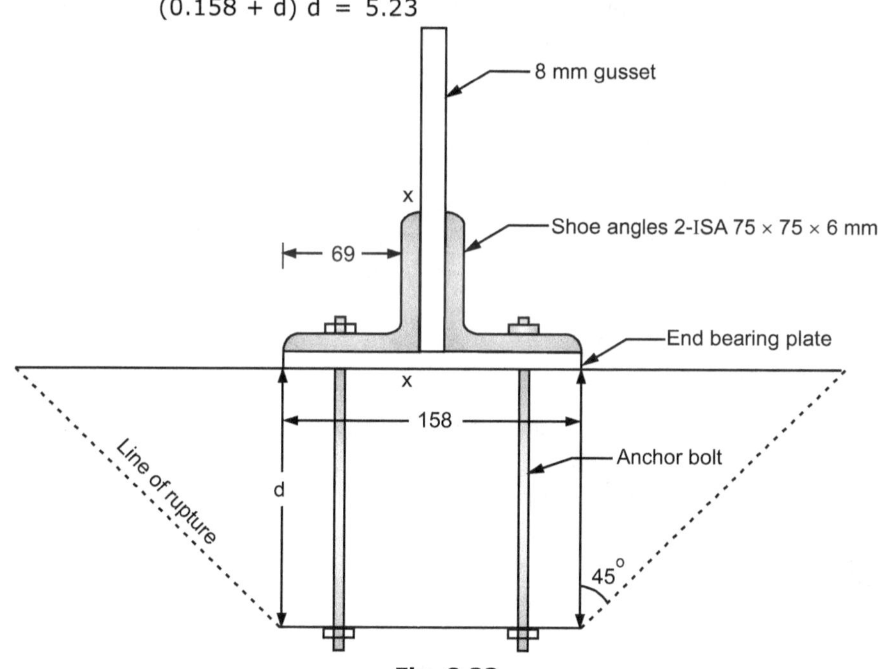

Fig. 8.33

$$d^2 + 0.158\, d - 5.23 = 0$$

$$\therefore \qquad d = \frac{-0.158 \pm \sqrt{(0.158)^2 - 4 \times 1\,(-6.23)}}{2 \times 1}$$

$$d = \frac{-0.158 \pm 4.994}{2}$$

$$d = 2.418 \text{ m}$$

The base angles are provided with 22 mm wide × 28 mm long slots to permit expansion of truss due to temperature stresses.

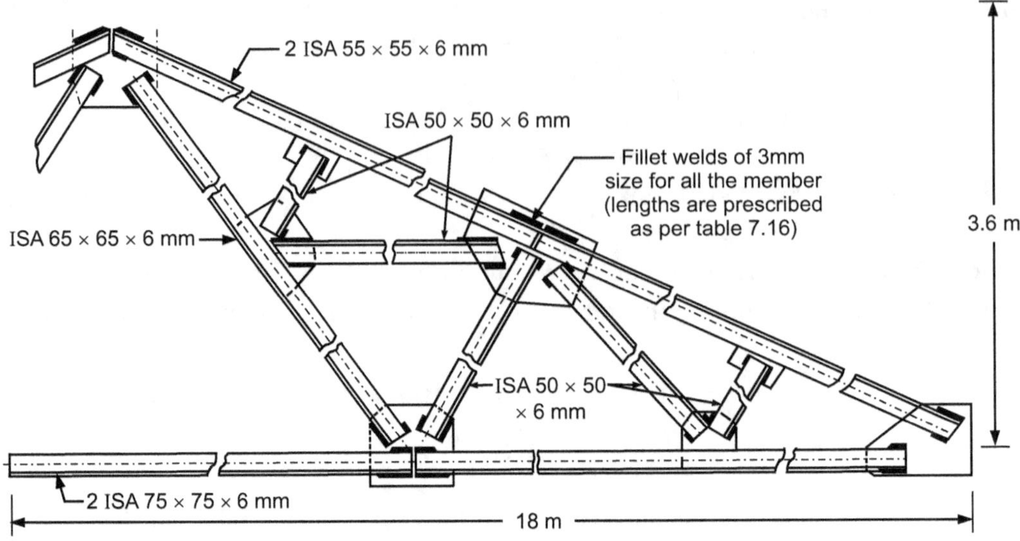

Fig. 8.34: Details of Fink Truss with Welded Joints

Important Points

- *A **truss** is defined as a framed structure composed of members connected to each other at their ends and forming triangles which lie in the same plane.*

- The members of a truss are subjected to direct (axial) tensile forces, as the truss is usually loaded at the point of intersection of its members only. No bending forces are considered.

- The steel roof trusses are useful at the places of heavy rainfall, where roofs have to support additional load due to snowfall and for very large spans.

- Types of trusses are - Howe trusses (6 to 30 m span), Pratt trusses (6 m to 100 m span), Fink or French trusses (simple or compound) (6 to 30 m span), North light trusses, Warren girder type trusses etc.

- The type of roof truss to be provided depends mainly upon the pitch of the truss and the span.

- Generally pitch value ranges from $\frac{1}{3}$ to $\frac{1}{5}$.

- Spacing of trusses are generally $\frac{L}{3}$ to $\frac{L}{5}$.

- ***Principal rafter*** is the top chord member of the truss and is usually subjected to compression.

- **Main tie** is the bottom chord member and is usually subjected to tension.

- Dead load and live load are considered over plan area whereas wind load is considered on inclined or sloping area of the area supported by one truss.

- Generally dead loads are of:

$$\text{G.I. sheets} = 100 \text{ to } 150 \text{ N/m}^2$$
$$\text{A.C. sheets} = 120 \text{ to } 160 \text{ N/m}^2$$
$$\text{Purlins} = 80 \text{ to } 120 \text{ N/m}^2$$

Truss, $\qquad W = \left(\frac{L}{3} + 5\right) \times 10 \text{ N/m}^2$

where, $\qquad L = $ Span in metres

- Live load for trusses have pitch angle $> 10°$ is

$$\boxed{LL = 750 - (\theta - 10) \times 20} \nleq 400 \text{ N/m}^2$$

where, $\qquad \theta = $ Angle of pitch

For trusses $\qquad LL = \frac{2}{3}$ of above LL

whereas for purlins $\qquad LL = $ LL as given above

- Wind loads depends on wind pressure which is calculated from wind map of India which gives a basic maximum wind speed 'V_b' in m/sec.

- Design wind speed

$$\boxed{V_z = V_b \times k_1 \times k_2 \times k_3}$$

where,

V_b = Basic wind speed in m/s

k_1 = Probability factor or risk coefficient

k_2 = Terrain, height, structure size factor

k_3 = Topography factor

- Design wind pressure

$$\boxed{p_d = 0.6\, V_z^2}$$

Wind load = $(C_{pe} - C_{pi}) \times A \times p_d$

where,

C_{pe} = Exterior wind co-efficient

C_{pi} = Interior wind co-efficient

A = Effective area of one panel point

p_d = Design wind pressure

- **Design steps of Angle purlins**

(i) $\qquad M_x = \dfrac{w_y L^2}{10}$ and $M_y = \dfrac{w_x L^2}{10}$

where, $\qquad w_y, w_x$ = Components of critical load in y and x direction

(ii) Try a section satisfying following requirements.

Width of angle purlin $\not< \dfrac{L}{60}$

Depth of angle purlin $\not< \dfrac{L}{45}$

(ii) Calculate design moment

For bending @ xx-axis:

$$M_{dx} = \beta_b \times Z_p \times \frac{f_y}{\gamma_{mo}} \text{ or } 1.2 \times Z_e \times \frac{f_y}{\gamma_{mo}} \text{ whichever is less}$$

β_b = 1 for plastic and compact sections

$$M_{dx} = Z_e \cdot \frac{f_y}{\gamma_{mo}} \text{ for semi-compact section}$$

Similarly find M_{dy} for bending @ yy axis.

(iv) Check for bi-axial bending interaction formula $\dfrac{M_x}{M_{dx}} + \dfrac{M_y}{M_{dy}} \le 1$.

Practice Questions

1. (i) The minimum size of angle used in roof truss .... and minimum number of rivets are ....

(ii) The intensity of live load in a pitched roof varies according to ... of the roof and for $\theta = 23^\circ$, live load intensity is ......

2. Attempt the following:

(i) State four types of trusses.

Ans. Four types of trusses are:

(1) Howe truss   (2) Pratt truss

(3) Simple fink truss   (4) North light truss

(ii) State suitable span length for above trusses.

Ans. (1) Howe truss - 6 to 24 m   (2) Pratt truss - 6 to 30 m

(3) Simple fink truss: 6 to 9 m   (4) North light truss - 8 m to 10 m

(iii) What are the loads for which roof trusses are designed?

Ans. (1) Dead load   (2) Imposed or live load

(3) Wind load   (4) Snow load

(iv) Draw a neat labeled sketch of HOWE truss of 12 m with 8 panels naming all important components.

Ans.

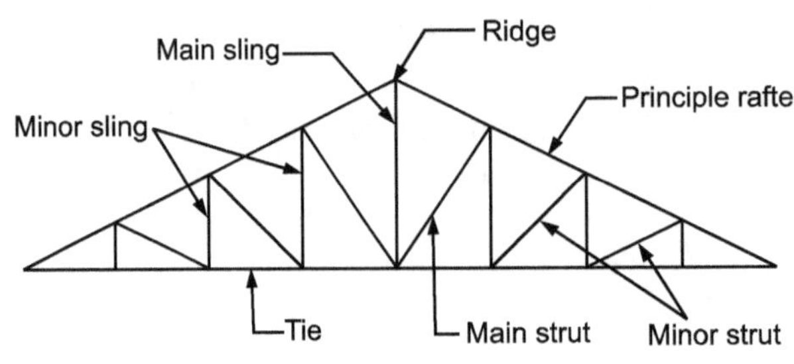

Fig. 8.35

3. State any eight types of trusses.

Ans. Types of trusses:

1. Simple fink truss	2. Simple fan truss
3. Compound fink truss	4. Pratt truss
5. Howe truss	6. King post truss
7. North light truss	8. Warren girder trusses

4. Define truss. Draw neat sketch of any six panel truss showing main tie, principal rafter, Angle of pitch and span. State two uses of steel roof trusses.

Ans. For definition and uses of trusses see Article 7.1. For sketch see Fig. 8.9.

5. Write the steps of design of angle purlin by I.S. code.

6. Draw sketches of Howe type and Pratt type truss point, panel, principal rafters and all members in showing pitch, rise, panel, principal, rafters and all members in showing pitch, rise, panel one of the above types.

7. Draw neat sketches of HOWE and NORTH LIGHT trusses. Mark panel, panel point, rafter and tie in any one truss.

8. Draw a neat sketch of roof truss showing component parts. Also define the component parts.

9. Draw neat sketch of six panel truss showing main tie, principal rafter, pitch and span. Also state any two uses of steel roof truss.

10. Draw line sketches of: (i) Compound fink truss, (ii) North light truss. State their span limits.

11. Draw neat sketch (detailed) of a truss support joint and panel point joint (any one) showing arrangement of members. (Sketch should include gusset plate connected with angles with the help of rivets/bolts).

Ans.

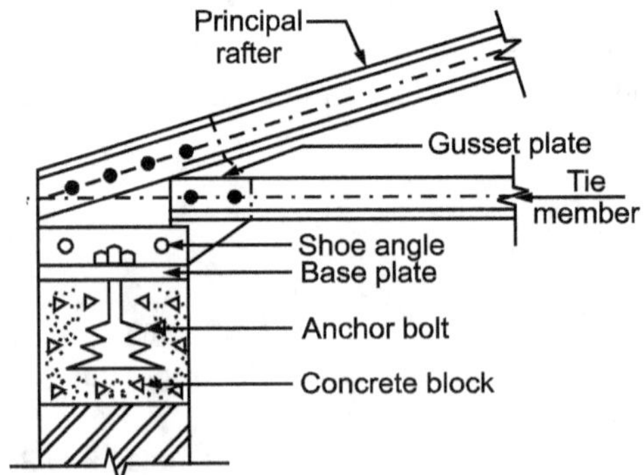

Fig. 8.36

12. Define truss. Draw neat sketch of any six panel truss showing main tie, principal rafter, Angle of pitch and span. State two uses of steel roof trusses.

13. Draw neat sketch of PRATT and FINK type trusses. Mark panel, panel point, rafter and tie in any one truss.

14. (i) Why purlins are provided? State whether purlin is a tie, a strut or a flexure member.

    (ii) What is the permeability of a structure? For the normal permeability, what is the value of internal wind pressure?

    (iii) The wind pressure depends on the slope of the roof for windward and leeward sides. Justify or otherwise the given statement.

    (iv) How live load for a truss is calculated?

15. What do you understand by the term panel point load ? How the panel point load for dead load is calculated?

16. Define steel roof truss and write any three advantages of using steel roof truss.

17. Write any four selection criteria of type of roof truss. Also define the term pitch and slope of roof truss.

18. State the necessity of purlins in Trusses. State the different checks to be taken while designing the purlin (No formulae).

19. What is purlin? State the procedure for design of purlin.

20. Draw neat sketches of connection of an angle purlin with principal rafter at panel point.

21. Define purlin. Also state IS code provisions for angle purlin design.

22. (a) A roof of a hall 16 m × 32 m clear dimensions is to be supported by trusses placed at 4 m c/c. A.C. sheet roofing with 20° slope is to be provided. Supported wall thickness is 400 mm.

    (i) Propose a suitable type of roof trusses.

    (ii) Determine D.L. per panel point.

    (iii) Determine L.L. per panel point.

[Ans. Howe truss may be proposed Assume Wt. of AC sheet = 156 N/m^2, Wt. of purlin = 100 N/m, Wt. of truss = $\left(\dfrac{16}{3} + 5\right) \times 10 \simeq 110$ N/m. Therefore on each top panel point, LL = 2.93 kN, DL = 2.63 kN]

    (b) Design an angle purlin for D.L. + L.L. as per I.S. method for roof in (a) above. Also draw a neat sketch of purlin principal rafter joint. Assume spacing of purlin as 1.8 m c/c. Use I.S. specifications.

[Ans. ISA 100 × 75 × 6 mm]

23. (a) Design an angle purlin for a trussed roof for the following data:

    Spacing of roof trusses = 5m

    Load on purlin = 1.5 kN/m

    Permissible stress in lending = 165 MPa.

    Design for uniaxial bending only.

[Ans. M = 3.75 × 10^6 N/mm, $Z_{reqd.}$ = 22727 mm^3, therefore ISA 125 × 95 × 6 mm purlin satisfies leg requirements also]

    (b) Calculate dead load and live load per panel point for the roof truss having following details:

    (i) Span = 10 m

    (ii) Rise = 1.5 m

(iii)   No. of panels = 8

(iv)   Spacing of trusses = 4 m

(v)   Wt. of purlins = 150 N/m^2

(vi)   Wt. of G.I. sheets = 100 N/m^2

(vii)   Wt. of roof truss = $\left(\dfrac{\text{Span in m}}{3} + 5\right) \times 10$ N/m^2

(viii)   Live load = 750 N/m^2 less 20 N/m^2 for every degree increase in slope over 10° subject to minimum 400 N/m^2.

[Ans. On each top panel point D.L. = 1.67 kN and L.L. = 2.28 kN]

24.   Calculate dead load and live load per panel point for a truss shown in Fig. 8.43. Details of truss.

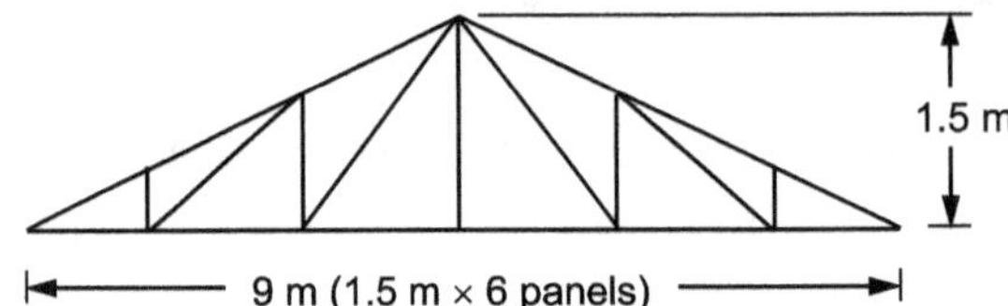

Fig. 8.37

Spacing of truss = 3 m, Wt. of A.C. sheets = 100 N/m^2, Wt. of purlins = 150 N/m^2, Wt. of bracing = 100 N/m^2, Wt. of truss = $\left(\dfrac{L}{5} + 3\right) \times 10$ N/m^2, where L is span in m. Live load = 750 N/m^2 less 20 N/m^2 for every degree increase in slope above 10° slope.

[Ans. DL = 1.716 kN, LL = 1.744 kN per panel point]

25.   Calculate live load and wind load per panel point for a truss shown in Fig. 8.44. Spacing of truss = 3.5 m, External wind pressure = 0.5 p, Internal wind pressure = ± 0.2 p. Design wind pressure = 1500 N/m^2. Live load is 750 N/m^2 less 20 N/m^2 per degree increase in slope above 10°.

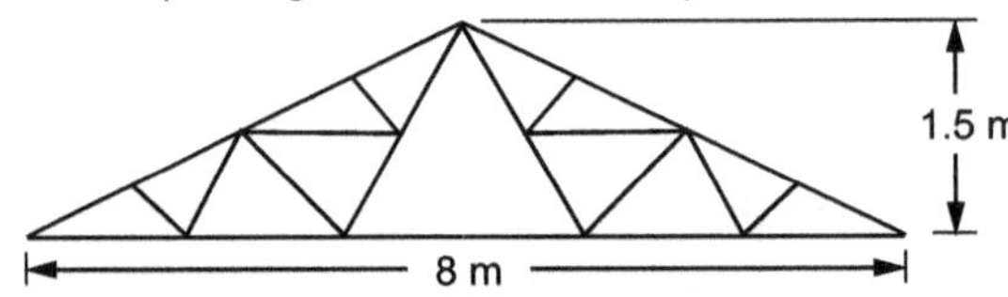

Fig. 8.38

[Ans. LL = 1.258 kN, WL = − 3.923 kN per panel point]

26.   (a) Calculate dead load per panel point of the roof truss having details given below:

(i)   Span = 20 m

(ii)   Rise = 3 m.

(iii)   Spacing = 4 m.

(iv)   No. of panels = 10

(v)   Wt. of roof covering = 120 N/m^2 plan area.

(vi)   Wt. of purlins = 100 N/m^2.

(vii)   Wt. of bracings = 80 N/m^2.

(viii)   Wt. of roof truss = $\left(\dfrac{L}{3} + 5\right) \times 10$ N/m^2.

[Ans. DL per panel point = 3.333 kN]

(b) Design an angle purlin from following data:

    (i)   Spacing of trusses = 3.5 m

    (iii)   u.d.*l.* on purlin = 2 kN/m

Check the section for minimum length of angle. Take σ_{bc} = 165 MPa.

27. Calculate live load and wind load per panel point for steel roof truss having following details:

    (i)   Span = 20 m.

    (ii)   Rise = 4 m.

    (iii)   No. of panels = 10.

    (iv)   Spacing of trusses = 3 m.

    (v)   Live load = 750 N/m² less 20 N/m² for every degree increase in slope above 10° subjected to minimum 400 N/m².

    (vi)   Co-efficient of External wind pressure = – 0.6 p.

    (vii)   Co-efficient of Internal wind pressure = ± 0.3 p.

    (viii) Design wind pressure (p) = 1200 N/m².

[Ans. LL = 2.056 kN, WL = – 5.98 kN per panel point]

28. (a)   A roof of a hall 12 m × 18 m effectively is supported by trusses placed at 3 m c/c. A.C. sheet roofing is to be provided with the following details:

Rise to span ratio is $\frac{1}{4}$

Wt. of roof truss   = 100 N/m².

Wt. of A.C. sheets   = 165 N/m².

Wt. of purlins   = 150 N/m².

Wt. of bracings   = 20 N/m².

Determine the load per panel due to dead load and due to live load.

[Ans. Assume 6 panels of 2 m length, DL = 2.610 kN, LL = 1.676 kN]

(b)   An angle purlin is subjected to total u.d.*l.* of 1200 N/m (inclusive of self wt.) over an effective span of 4 m. Check whether ISA 90 × 90 × 6 mm is adequate or not for this purlin. Also draw neat sketch showing typical connection between purlin and principal rafter. For IS 90906, Z_{xx} = 12.20 cm³.

29. A hall 16 m × 24 m is to be provided with a steel roof truss (fink type). Calculate the panel point load for dead load and live loads. Assume weight of roof covering as 165 N/m², weight of purlin as 200 N/m², weight of truss as 120 N/m². The pitch of truss is $\frac{1}{5}$. For this truss, access is not provided.

30. Total load of 750 N is acting on a purlin of a particular truss. Centre to centre distance between purlins is 1.6 m and span of purlin is 4 m. Check whether ISA 100 × 75 × 6 mm is adequate as a purlin or not.

For ISA 100 × 75 × 6 mm,

    Z_{xx} = 14.4 × 10³ mm³.

    Z_{yy} = 8.5 × 10³ mm³.

31. A hall 14 m × 24 m is to be provided with a roof truss. Calculate the panel point dead load and live load. Suggest a suitable type of truss and assume A.C. sheet roof covering. Rise to span ratio is $\frac{1}{4}$.

32. A purlin of a truss has following details: Centre to centre distance between purlins = 1.5 m, spacing of trusses = 4.5 m, load on purlin = 800 N/m^2.

Check whether ISA 100 × 75 × 6 mm is adequate as purlin or not.

For on ISA 100 × 75 × 6 mm,

$$Z_x = 14400 \text{ mm}^3.$$
$$Z_y = 8500 \text{ mm}^3.$$

Also draw a neat sketch showing typical connection of purlin with rafter in case of steel roof truss.

33. A roof of a hall 12 m × 18 m effectively is supported by trusses placed at 2 m c/c. AC sheet roofing is to be provided with the following details:

Rise to span ratio is $\frac{1}{4}$.

$$\text{Wt. of roof truss} = 100 \text{ N/m}^2.$$
$$\text{Wt. of AC sheets} = 165 \text{ N/m}^2.$$
$$\text{Wt. of purlins} = 150 \text{ N/m}^2.$$
$$\text{Wt. of bracings} = 20 \text{ N/m}^2.$$

Determine the load per panel due to dead load and due to live load.

34. An angle purlin is subjected to total u.d.*l.* of 1200 N/m (inclusive of self weight) over an effective span of 4 m. Check whether ISA 90 × 90 × 6 mm is adequate or not for this purlin. Also draw neat sketch showing typical connection between purlin and principal rafter.

For ISA 90906, $Z_{xx} = 12.20 \text{ cm}^3$.

35. A house roof truss having 12 m span with pitch of $\frac{1}{6}$ is used for corrugated A.C. sheet roof covering which weighs 175 N/m^2. Consider eight panel lengths along the tie member. The weight of the purlin is 55 N/m^2. Assume self weight of truss as 90 N/m^2. Calculate panel point loads for dead load and live load.

36. An unequal angle section is used for a purlin of a roof truss. The roof truss is 12 m in span. The spacing between roof trusses is 2 m. The slope of roof truss is 1 vertical and 2 horizontal. Considering wind load on roof surface normal to roof as 1 kN/m^2 (upward) and vertical downward load from truss as 200 N/m^2, design the purlin using 1.5 code method.

Angle section	Wt/m. (N/m)	Cross-sectional area (mm^2)	Thickness (mm)	Z_{xx} (mm^3)
ISA 60 × 40	58	737	8	6.5×10^3
ISA 65 × 45	64	817	8	7.7×10^3
ISA 100 × 75	105	1336	8	19.1×10^3
ISA 100 × 75	130	1650	10	23.6×10^3
ISA 125 × 75	121	1538	8	29.4×10^3
ISA 150 × 75	137	1748	8	42.0×10^3

37. A roof of a hall 16 m × 32 m clear dimensions is to be supported by trusses placed at 4 m c/c. A.C. sheet roofing with 20° slope is to be provided. Supporting wall thickness is 300 mm.

   (i)   Propose a suitable type of roof truss.

   (ii)   Determine D.L. per panel point.

   (iii)   Determine L.L. per panel point.

38. A purlin of truss has following details:

$$c/c \text{ distance between purlins} = 2 \text{ m.}$$
$$\text{Spacing of trusses} = 4.5 \text{ m.}$$
$$\text{Load on purlins} = 850 \text{ N/m}^2.$$

Check whether ISA 100 × 75 × 6 mm is adequate as purlin or not.

For one ISA 100 × 75 × 6 mm,

$$Z_{xx} = 14400 \text{ mm}^3.$$
$$Z_{yy} = 8500 \text{ mm}^3.$$

Also draw a neat sketch showing typical connection of purlin with rafter in case of steel roof truss.

39. Decide the panel point loads due to dead loads for a single fan truss of span 12 m and rise 3 m. The truss is provided with AC sheets as a roof covering. Spacing of trusses is 4 m c/c.

40. Calculate suitable size of an angle purlin for a trussed roof with spacing of roof trusses 4.0 m c/c and carrying u.d.*l.* of 2.5 kN/m (including self weight). Take permissible tensile stress due to bending as 165 N/mm^2.

∎∎∎

PLASTIC ANALYSIS

Syllabus

- Plastic Analysis: Analysis of Steel Structures - Methods - Elastic, Plastic and Advanced method of Analysis based on IS: 800-2007 - Idealized Stress Vs Strain curve - Problems. For Structural Steel - Requirements and Assumptions of Plastic Method of Analysis - Formation of Plastic Hinges in Flexural Members - Plastic Moment of Resistance and Plastic Modulus of Sections - Shape Factors of Rectangular/Circular/I/ T-Sections - Collapse Load.

About this Chapter

After reading this chapter students can understand:
- Understand methods of analysis of steel structures.
- Draw idealized stress v/s strain curve.
- Write requirements and assumptions of plastic method of analysis.
- Understand formation of plastic hinges in flexural members.
- Calculate plastic moment of resistance and plastic modulus of different shapes of sections.
- Compute shape factors of rectangular, circular, I and T section and collapse load.

9.1 INTRODUCTION (TO ANALYSIS AND DESIGN OF STEEL STRUCTURES)

- Till around 1960, all structures were designed according to elastic theory. If elastic method of design, the stresses are calculated at working loads using elastic properties of steel, although the maximum stress in extreme fibres reaches the yield stress of the material in bending and the rest of cross-section remains under-stressed.

- The allowable stresses are obtained by dividing yield stress by factor of safety. The structures designed by this method may become heavier than designed by plastic method which uses ultimate load or yield stress as design criteria.

- As steel is a ductile material it can absorb large deformations beyond elastic fracture. In other words, steel possess reserved strength beyond yield strength. This method using this reserved strength is called 'plastic method of design'.

9.2 ANALYSIS AND DESIGN METHODS

9.2.1 Elastic Method

- In elastic method of design, it is assumed that the structural members behave elastically. It is a conventional method of design and is also known as working stress method.

- The stresses due to working loads do not exceed the specified allowable stresses. The working loads are the maximum loads, which are likely to occur in normal use throughout the life time of the structure. The values of allowable stresses are determined by applying adequate factor of safety to the guaranteed minimum yield stress. Hence this method is also called as allowable stress method of design.

- The factor of safety accounts for unpredictable overload, defective material and workmanship etc. The elastic method does not take into account the strength of the material beyond the yield stress.

9.2.2 Plastic Method

- Due to plastic deformations and strain hardening of the material particles which were less stressed will be brought into action, so that the structure is able to resist greater loads.

- In modern designs the above principle is followed and the method based on this principle is called *collapse method of design or plastic design.*

- Structural steel has the ability to resist large deformation without fracture. A large part of this deformation occurs during yielding. A relatively large part of deformation occurs during the strain hardening process.

- Fig. 9.1 shows a typical stress-strain diagram for steel specimen under uni-axial tension. Various stages of loading exhibits mainly three parts of graph: elastic, plastic and strain hardening.

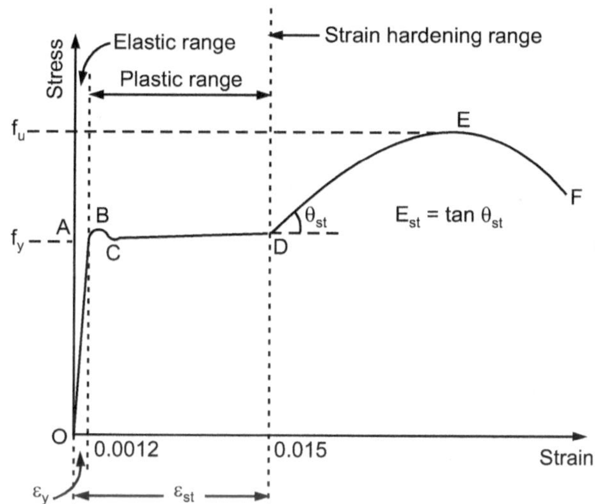

Fig. 9.1: Stress-strain curve for mild steel in uniaxial tension

- Plastic method adopts the ultimate strength as the criterion for design. The material is considered to be stressed beyond the yield stress, in the elastic or plastic range.

- Ultimate loads which causes the collapse are determined by multiplying the working loads by a load factor. In plastic design the importance is given to resistances to bending caused by loads when the members are in plastic range.

- Plastic designs consider the capacity of members to continue offering resistances even after reaching the yield stress.

- Due to ductility property of steel a member is capable of absorbing considerable deformation beyond elastic limit without fracture or collapse.

- The capacity of the member to have this reserve strength after reaching the yield stress is recognized and taken into account in plastic design.

- The sections designed by plastic design are smaller in size than by elastic design and thus the economy can be achieved.

9.2.2 Advanced Method of Analysis

- Structural analysis may be divided into three large main groups. They are static analysis stability and dynamic analysis.

- Static analysis is further sub-divided in static linear analysis and non-linear static analysis.

- Stability analysis deals with structures subjected to compressed time-independent forces. Buckling analysis comes under this.

- Dynamic analysis means that the structures are subjected to time dependent loads, shock and scismic loads as well as moving loads with taking into account the dynamic effects.

- Free vibration analysis determines the natural frequencies (eigen values) and corresponding mode shapes (eigen functions) of vibration. Further stressed-free vibration analysis and Time-history analysis are the other subtypes of dynamic analysis.

- After analyzing the steel structure by any of the above methods as per requirements, the design procedures can be adopted as per IS : 800-2007.

9.3 IDEALISED STRESS V/S STRAIN CURVE

- Fig. 9.1 shows distinctively upper as well as lower yield points. In plastic region, strain rapidly increases and stress-strain relation is almost horizontal in the graph. After strain hardening, increase in stress causes increase in strain, but not linearly. Highest material resistance is shown by ultimate stress and the material no longer resists further load and fails by rupture.

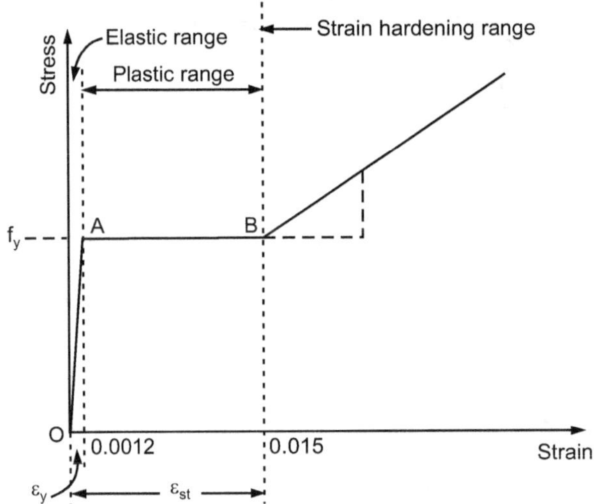

Fig. 9.2: Idealized Stress-Strain Curve for Limit State Design

- In plastic analysis, this graph shown in Fig. 9.1 is much idealized, using the following assumptions:

  (a) Upper and lower yield points are regarded as one point.

  (b) Effect of strain hardening is ignored.

- This makes a very simple stress-strain graph consisting of two lines as shown in Fig. 9.2. After elastic limit, material deforms excessively due to yielding. Yielding of cross-section makes a totally different physically change to cause a failure.

9.4 REQUIREMENTS OF STRUCTURAL STEEL FOR PLASTIC ANALYSIS

- As per IS : 800 - 2007 following requirements shall be satisfied by structural steel.

  1. Yielding stress for the grade of steel used shall not exceed 450 MPa.

  2. Stress-strain characteristics of steel shall be such as to ensure complete plastic moment distribution. These are as follows:

     (i) The stress-strain diagram has a plateau at the yield stress extending for atleast six times the yield strain.

     (ii) The ratio of tensile strength to the yield stress is not less than 1.2.

     (iii) The elongation on the gauge length is not less than 15%.

     (iv) Steel exhibits strain hardening capability.

  3. The member used shall be not rolled or fabricated using hot rolled plates and sections.

9.5 ASSUMPTIONS IN PLASTIC METHOD OF ANALYSIS

- In plastic theory the following assumptions are made:

    (i)   Steel posses the property of ductility and so it can be subjected to deformation in the plastic range without fracture.

    (ii)   The strain distribution across a section is linear i.e., in the case of beams, plane sections remain plane before and after bending.

    (iii)   The stress-strain diagram is modified i.e., steel is assumed to be an ideal elastic-plastic material.

    (iv)   The relation between tensile stress and tensile strain is the same as that between compressive stress and compressive strain.

    (v)   At the stage of loadings, the various loads on the structure bear a constant proportion to each other.

    (vi)   All connections provide satisfactory continuity so as to transmit the plastic moment.

    (vii) Elastic deformation is not considered.

    (viii) The beam is not subjected to any longitudinal force.

9.6 FORMATION OF PLASTIC HINGES IN FLEXURAL MEMBERS

- Consider a simply supported beam with a central point which may be gradually increased from 0 to 'P' as shown in Fig. 9.3 (a).

- As the load increases, the maximum bending moment at mid-span increases proportionately. This causes corresponding increase in moment of resistance and stresses and strains at extreme fibres.

- The moment of resistance is proportional to stresses while the rotation 'θ' is proportional to maximum strain. Thus stress-strain relationship can be proportionately related to moment-rotation relationship as shown in Fig. 9.3 (d).

- As the load continues to increase, a stage is reached when the maximum stress in the outermost fibre reaches the yield stress f_y and the corresponding moment M_y.

- The corresponding rotation is shown in Fig. 9.3 (d) by 'A'. the moment rotation relationship is linear upto this point. With further increase in load (i.e. $M > M_y$ but less than M_p) part of the outer section becomes plastic while the inner portion near the neural axis still remains elastic as shown in Fig. 9.3 (d).

- The rotation continues to increase without increase in moment. The section is therefore, in partially plastic and partially elastic condition.

- As the load further increases a stage is reached when the entire section becomes plastic (Fig. 9.3 (e)) and the section is no more in position to offer any additional moment of resistance and the section simply rotates under constant moment.

- The maximum moment of resistance the section can offer is called plastic moment M_p. The portion of the member near the critical section, simply rotates at constant moment M_p is called plastic hinge.

- Thus plastic hinge may be defined as the yielded zone of the member at which infinite rotation can take place at a constant plastic moment acting at the section.

- In other words, plastic hinge (Fig. 9.3 (c)) is the yield section of the beam, which acts as if it were hinged, except a constant restraining plastic moment.

- The yield length up (Fig. 9.3 (a)) is the length of the beam over which the moment is greater than or equal to the yield moment. The yield length depends on the type of loading and geometry of the cross-section of the structural member.

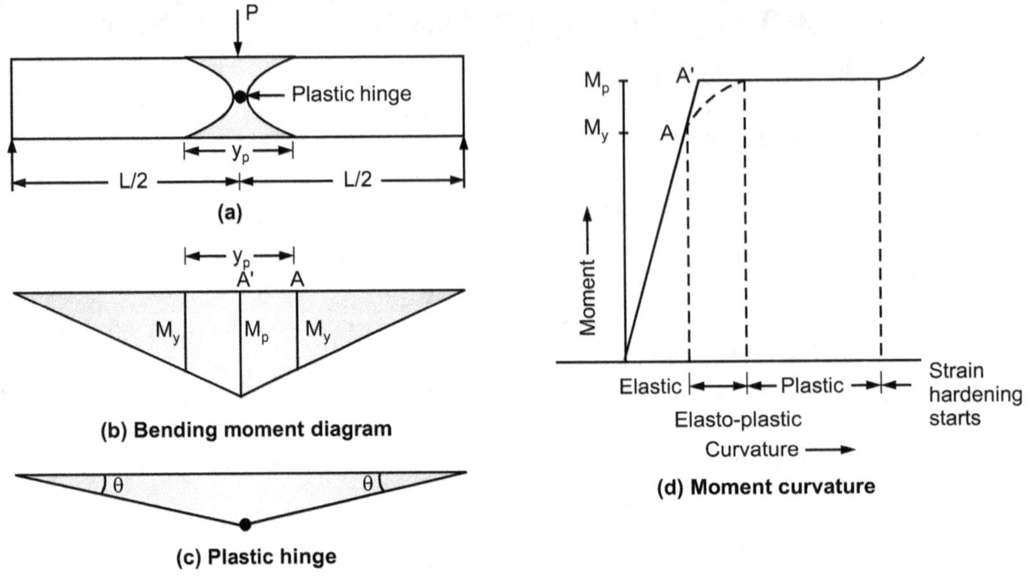

Fig. 9.3: Plastic Hinge

9.7 PLASTIC MOMENT OF RESISTANCE

- An idealized stress-strain curve for mild steel is shown in Fig. 9.4.

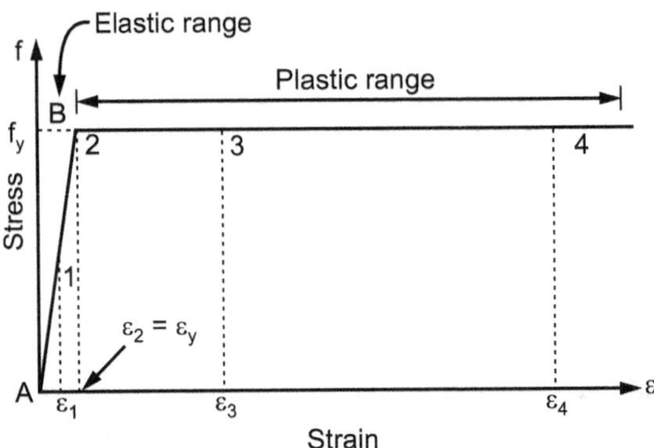

Fig. 9.4: Idealised Elasto-Plastic Stress-Strain Curve

- When an I-shape beam is subjected to external load its behaviour under gradually increasing moment is as shown in Fig. 9.5. Bending stress diagrams corresponding to the points 1234 on the curve (Fig. 9.4) are shown in Fig. 9.5.

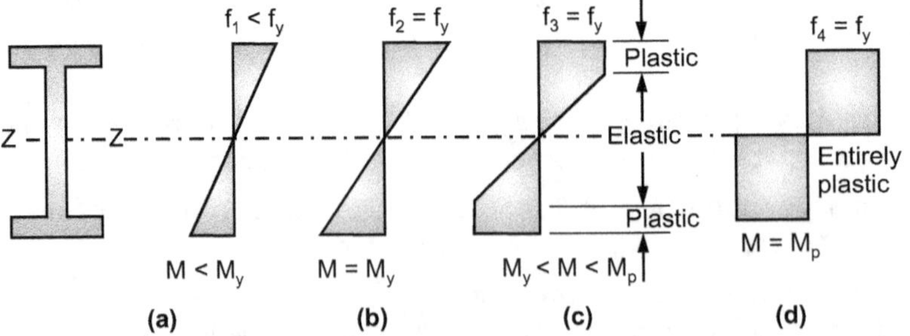

Fig. 9.5: Bending Stress Distribution at Different Stages of Loading

- In the elastic range AB stress is linearly proportional to strain

  i.e. $M \propto f \propto e$ ($\because M = \dfrac{f}{y} \cdot I$ and $f = E \cdot e$). Hence the moment of resistance at say point 1 on AB will be;

  $$M_1 = f_1 \cdot Z_e \quad \text{(as per flexural formula)}$$

  where, $f_1 =$ Stress at extreme fibre

                            $Z_e =$ Elastic section modulus

- At point 2, $f_2 = f_y$. Hence, the moment corresponding to this point is known as first yield moment and is given by

  $$M_2 = M_y = f_2 \cdot Z_e = f_y \cdot Z_e \qquad \qquad \text{... (9.1)}$$

- The extreme fibres after attaining yield stress do not take any more stress. On increasing the load further, the yield progresses inwards, i.e. the stresses are redistributed inwardly towards the neutral axis.

- For example, say for point 3 of Fig. 9.4. Thus the outer fibers are plastified while the inner fibers near the neutral axis are elastic as shown in Fig. 9.6. The moment capacity at this point is sum of contributions from plastic and elastic portions.

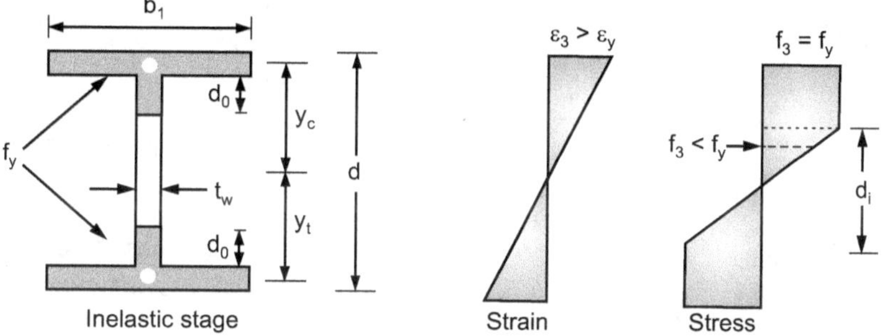

Fig. 9.6: Strain and Stress Distribution in Section Corresponding to Point 3 of Curve of Fig. 9.4

$$M_3 = f_y \, (b_f \, t_f + d_o \, t_w) \, (y_c + y_t) + f_y \, \frac{t_w \, d_1^2}{6}, \qquad \qquad \text{... (9.2)}$$

where, $d_o =$ Depth up to which the web is plastified,

                          $d_1 =$ Depth of the web which is not plastified, and

          $y_c, \, y_t =$ Lever arm distance from neutral axis as shown in Fig. 9.6.

- When load is increased further, the plasticity spreads towards N.A. The whole beam gets plastified and a plastic hinge is formed. The moment at which a plastic hinge is formed is known as plastic moment M_p and is given by

$$M_p = f_y \cdot Z_p = f_y \left[b_f \, t_f \, (d - t_f) + t_w \left(\frac{d}{2} - t_f \right)^2 \right] \qquad \qquad \text{... (9.3)}$$

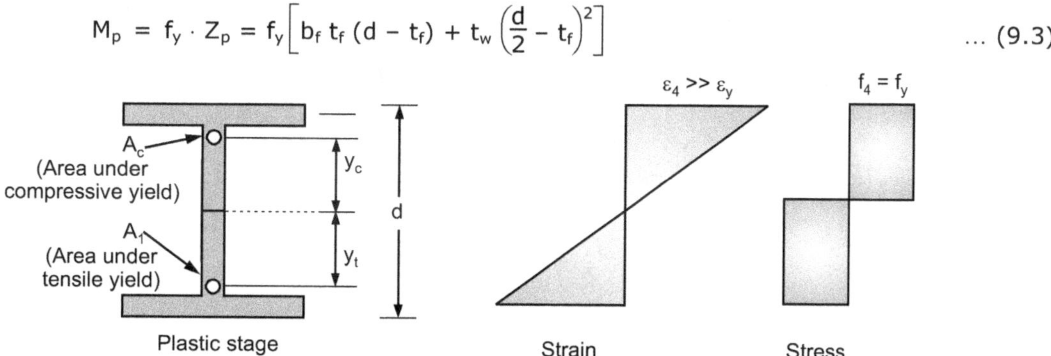

Fig. 9.7: Strain and Stress Distribution in Section Corresponding to Point 4 of Curve of Fig. 9.4

- The M_p value at this stage is generally about 10-20% more than M_y. In reality the beam can carry a load higher than its plastic moment capacity due to strain hardening but this is neglected for design purposes. Furthermore, the residual stresses cause yielding to initiate at a lower load, but however, M_p values are unaffected. It is because residual stresses are self-equilibrating.

- Fig. 9.8 shows the moment-rotation characteristics of the four classes of cross-sections. The shaded diagram depicts the bending stress patterns. The plastic sections exhibit sufficient ductility ($\theta_2 > 6\theta_1$) where θ_1 is rotation at the onset of plasticity and θ_2 is the lower limit of rotation for treatment as plastic section. The compact sections have relatively lower limit for rotation treatment as a plastic sections, but are capable of reaching their full plastic moment values. For semi-compact sections, the bending resistance is limited to the yield moment only and for slender members local or lateral buckling occurs in the elastic range.

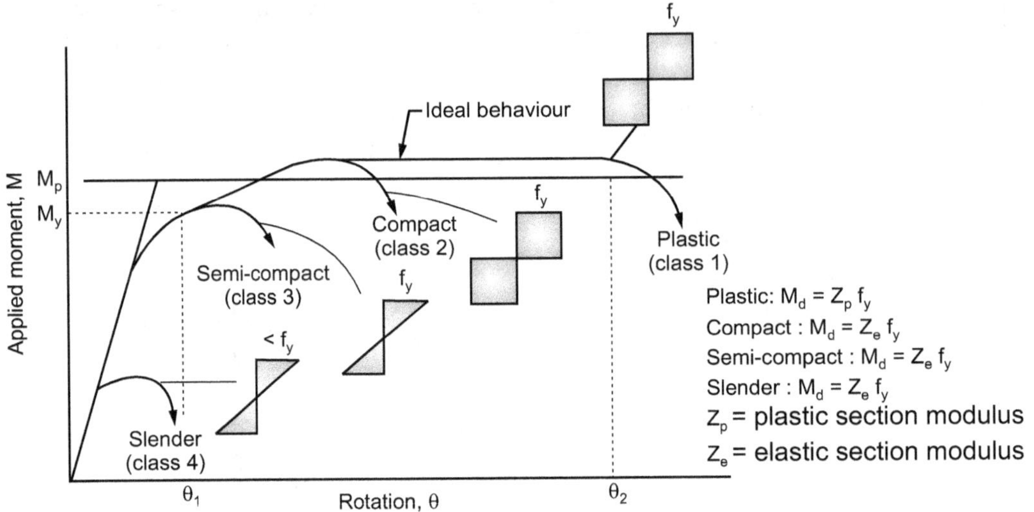

Fig. 9.8: Moment-rotation behaviour of the four classes of cross-sections

- For lateral buckling occurs in the elastic range. The design moment capacity M_d for the classes of sections defined above is indicated on the plot itself (of course partial factor of safety for materials should be applied to the design moment capacity).

9.8 PLASTIC MODULUS OF SECTION

- On studying behaviour of bending in Section the plastic moment is given by equation (9.3) as
$M_p = f_y \cdot z_p = f_y \sum y \cdot dA$.
where, $Z_p = \sum y \cdot dA$ and is called plastic modulus.
The plastic moment resisted by internal couple composed of total compression and total tension. The neutral axis is situated at such a position where equilibrium condition, Total tension = Total compression is satisfied.
Consider a T-section with plastic stress distribution as shown in Fig. 9.9.

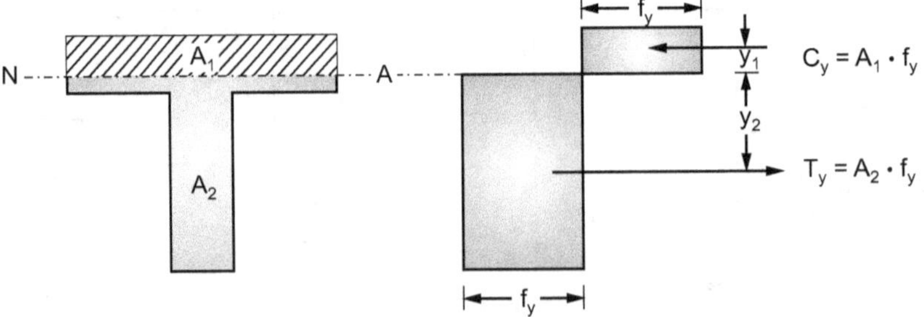

Fig. 9.9: T-Beam

- Assuming yield stress in compression is equal to yield stress in tension, then mathematically

$$f_y \cdot A_1 = f_y \cdot A_2$$

i.e. $\qquad A_1 = A_2$

i.e. $\qquad$ Area above N.A. $=$ Area below N.A.

In other words, the neutral axis divides the cross-section into two equal parts and hence this neutral axis may be called as plastic neutral axis or equal area axis

$$\therefore \qquad A_1 = A_2 = \frac{A}{2}$$

Thus, plastic moment of resistance,

$$M_p = f_y A_1 \cdot y_1 + f_y A_2 y_2$$

where, y_1 and y_2 are the perpendicular distance between centroids of A_1 and A_2 from N.A.

$$= f_y \left[\frac{A}{2} \cdot (y_1 + y_2) \right] \qquad \left(\because A_1 = A_2 = \frac{A}{2} \right)$$

$$= f_y \cdot Z_p$$

$$\therefore \qquad Z_p = \frac{A}{2}(y_1 + y_2)$$

Z_p is called as *plastic modulus* for section. For symmetric sections, the plastic neutral axis passes through c.g. of the section and equal plastic zones are formed simultaneously.

9.9 SHAPE FACTOR (S)

- The ratio of plastic moment to elastic moment is called shape factor

Thus, $\qquad S = \dfrac{M_p}{M_y} = \dfrac{f_y \cdot Z_p}{f_y Z_e} = \dfrac{Z_p}{Z_e}$

The shape factor nearly equal to unity (1) shows the section is efficient in resisting bending.

9.10 LOAD FACTOR

- Ultimate loads which cause the collapse are determined by multiplying the working loads by a load factor.

Load factor, $\qquad \lambda = \dfrac{\text{Plastic moment}}{\text{Allowable moment}} = \dfrac{M_p}{M_a} = \dfrac{f_y \cdot Z_p}{\dfrac{f_y}{\text{F.S.}} \cdot Z_e}$

$$= \text{F.S.} \left[\frac{Z_p}{Z_e} \right] = \text{F.S.} \times \text{Shape factor}$$

$\therefore \qquad$ Load factor $=$ Factor of safety $\times$ Shape factor

9.11 COLLAPSE FACTOR

- Collapse load is the load causing failure of structure in plastic range. It can be evaluated by equating collapse moment to plastic moment.

$$\text{Collapse moment, } M_e = \frac{P_c \cdot L}{4} \text{ for simply supported beam with central point load}$$

$$= \frac{w_c \cdot L^2}{8} \text{ for S.S. beam with u.d.}l. \text{ over entire span}$$

Solved Examples

Ex. 9.1: *Find the plastic modulus and shape factor for a rectangular section of width b and depth d.*

Sol.:

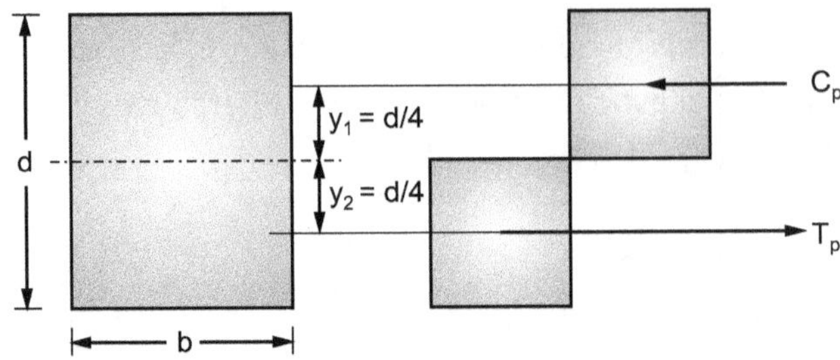

Fig. 9.10

Plastic modulus, $\qquad Z_p = \dfrac{A}{2}(y_1 + y_2) = \dfrac{b \cdot d}{2}\left(\dfrac{d}{4} + \dfrac{d}{4}\right) = \mathbf{\dfrac{bd^2}{4}}$

Elastic modulus, $\qquad Z_e = \dfrac{I_{xx}}{y_{max}} = \dfrac{\dfrac{bd^3}{12}}{\dfrac{d}{2}} = \dfrac{bd^2}{6}$

$\therefore$ Shape factor, $\qquad S = \dfrac{Z_p}{Z_e} = \dfrac{\dfrac{bd^2}{4}}{\dfrac{bd^2}{6}} = \dfrac{6}{4} = \mathbf{1.5}$

Ex. 9.2: *Determine the shape factor about xx axis of the ISMB 400 with root radius omitted. Take $f_y = 250$ MPa.*

Sol.:

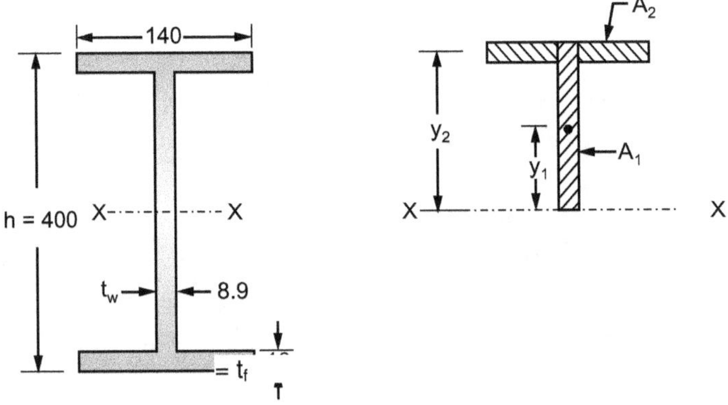

Fig. 9.11

Dividing the section into areas A_1 and A_2 above N.A.

$$A_1 = \frac{h}{2} \cdot t_w = \frac{400}{2} \times 8.9 = 1780 \text{ mm}^2$$

$$A_2 = (b - t_w) \cdot t_f = (140 - 8.9) \times 16 = 2097.6 \text{ mm}^2$$

$$y_1 = \frac{1}{2} \cdot \left(\frac{h}{2}\right) = \frac{h}{4} = \frac{400}{4} = 100 \text{ mm}$$

$$y_2 = \left(\frac{h}{2} - \frac{t_f}{2}\right) = \left(\frac{400}{2} - \frac{16}{2}\right) = 192 \text{ mm}$$

$\therefore$ Plastic modulus $= 2\,(A_1\,y_1 + A_2\,y_2)$

$$Z_p = 2\,(1780 \times 100 + 2097.6 \times 192) = 1.1615 \times 10^6 \ mm^3$$

$$I_{xx} = \frac{140 \times (400)^3 - (140 - 8.9) \times (400 - 2 \times 16)^3}{12} = 202208017 \ mm^4$$

$$y_{max} = \frac{h}{2} = \frac{400}{2} = 200 \ mm$$

$\therefore$ Elastic modulus, $Z_e = \dfrac{I_{xx}}{y_{max}} = \dfrac{202208017}{200} = 1.011 \times 10^6 \ mm^3$

$\therefore$ Shape factor, $s = \dfrac{Z_p}{Z_e} = \dfrac{1.1615 \times 10^6}{1.011 \times 10^6} = 1.1488$

Ex. 9.3: *Determine the shape factor for a circular section of radius R.*

Sol.:

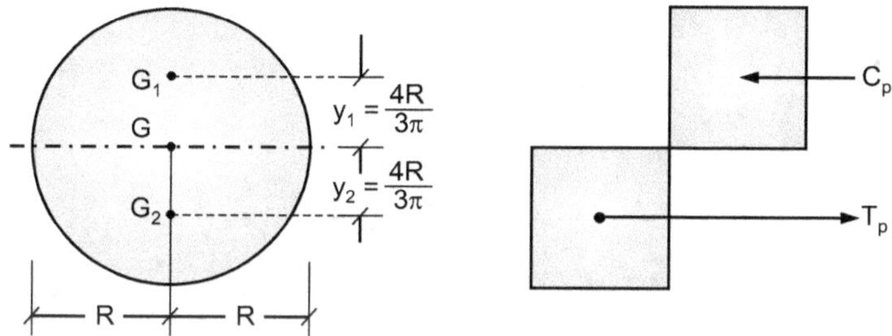

Fig. 9.12

Plastic modulus, $Z_p = \dfrac{A}{2}\,(y_1 + y_2) = \dfrac{\pi R^2}{2}\left(\dfrac{4R}{3\pi} + \dfrac{4R}{3\pi}\right) = \dfrac{4}{3}\,R^3$

Elastic modulus, $Z_e = \dfrac{I_{xx}}{y_{max}} = \dfrac{\dfrac{\pi D^4}{64}}{\dfrac{D}{2}} = \dfrac{\pi D^3}{32} = \dfrac{\pi R^3}{4}$

Shape factor, $S = \dfrac{Z_p}{Z_e} = \dfrac{\dfrac{4}{3}\,R^3}{\dfrac{\pi}{4}\,R^3} = \dfrac{16}{3\pi} = 1.6977$

Ex. 9.4: *Find the shape factor of a thin hollow circular section of external diagram 'D' and internal diameter 'd'.*

Sol.: $d = D - 2t$

Plastic modulus,

$$Z_p = \frac{A}{2}\,(\bar{y}_1 + \bar{y}_2)$$

$$A = \frac{\pi}{4}\,(D^2 - d^2)$$

$\bar{y}_1$ is the centroid of half of the hollow circle.

     $A_1 =$ Area of exterior semicircle

$$= \frac{1}{2} \times \frac{\pi}{4}\,D^2 = \frac{\pi}{8}\,D^2$$

     $A_2 =$ Area of interior semicircle

$$= \frac{1}{2} \times \frac{\pi}{4}\,d^2 = \frac{\pi}{8}\,d^2$$

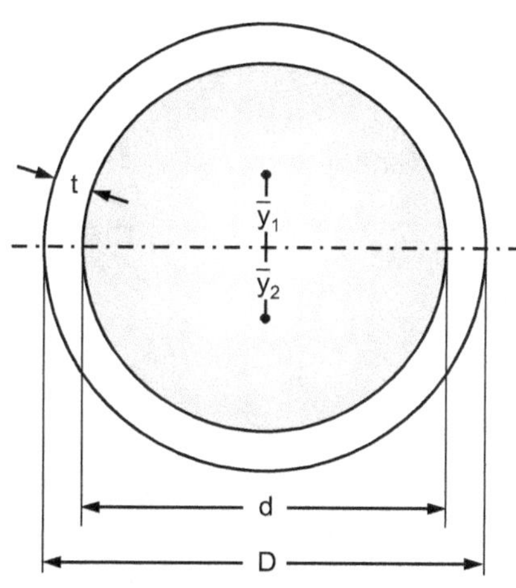

Fig. 9.13

$$y_1 = \text{Centroid of exterior semicircle} = \frac{4R}{3\pi} = \frac{3D}{3\pi}$$

$$y_2 = \text{Centroid of interior semicircle} = \frac{4r}{3\pi} = \frac{2d}{3\pi}$$

$\therefore \quad \bar{y}_1 = \text{Centroid of upper hollow semicircle}$

$$= \frac{A_1 y_1 - A_2 y_2}{A_1 - A_2} = \frac{\frac{\pi}{8} D^2 \times \frac{2D}{3\pi} - \frac{\pi}{8} d^2 \times \frac{2d}{3\pi}}{\frac{\pi}{8} D^2 - \frac{\pi}{8} d^2}$$

$$= \frac{\frac{\pi}{8}\left[\frac{2D^3}{3\pi} - \frac{2d^3}{3\pi}\right]}{\frac{\pi}{8}(D^2 - d^2)} = \frac{\frac{2}{3\pi}(D^3 - d^3)}{D^2 - d^2}$$

$$\bar{y}_1 + \bar{y}_2 = 2 \times \frac{\frac{2}{3\pi}(D^3 - d^3)}{D^2 - d^2} = \frac{4}{3\pi}\left[\frac{D^3 - d^3}{D^2 - d^2}\right]$$

$$\therefore \quad Z_p = \frac{A}{2}\left(\bar{y}_1 + \bar{y}_2\right) = \frac{\frac{\pi}{4}(D^2 - d^2)}{2} \times \frac{4}{3\pi}\left[\frac{D^3 - d^3}{D^2 - d^2}\right] = \frac{D^3 - d^3}{6}$$

Substituting $d = D - 2t$

$$Z_p = \frac{1}{6}[D^3 - (D - 2t)^3]$$

$$= \frac{1}{6}[D^3 - (D^3 - 6D^2 t + 12\,Dt^2 - 8t^3)]$$

$$= D^2 t \ (\text{Ignoring terms involving } t^2 \text{ and } t^3)$$

Elastic modulus

$$Z_e = \frac{I_{xx}}{y_{max}} = \frac{\frac{\pi}{4} D^4 - \frac{\pi}{64} d^4}{\frac{D}{2}} = \frac{\pi}{2}\left[\frac{D^4 - d^4}{D}\right]$$

$$= \frac{\pi}{32D}[D^4 - (D - 2t)^4] = \frac{\pi}{32}[D^4 - (D^4 - 8D^3\,t + 24\,D^2 t^2 - 32Dt^3 - 16t^4)]$$

$$= \frac{\pi}{32D}[D^4 - D^4 + 8D^3 t + 0] \qquad (\text{Ignoring terms involving } t^2, t^3 \text{ and } t^4)$$

$$= \frac{\pi}{32D} \times 8D^3\,t = \frac{\pi D^2 t}{4}$$

$\therefore \quad$ Shape factor,

$$S = \frac{Z_p}{Z_e} = \frac{D^2 t}{\frac{\pi D^2 t}{4}} = \frac{4}{\pi} = 1.273$$

Ex. 9.5: *Determine the plastic modulus and the shape factor for the T-section shown in Fig. 9.14.*

Sol.: Total area of section,

$$A = 150 \times 10 + 200 \times 10 = 3500 \text{ mm}^2$$

As plastic neutral axis is equal area axis, it divides the section into two equal parts.

i.e. $\quad A_1 = A_2 = \dfrac{A}{2} = \dfrac{3500}{2} = 1750 \text{ mm}^2$ (Fig. 9.14)

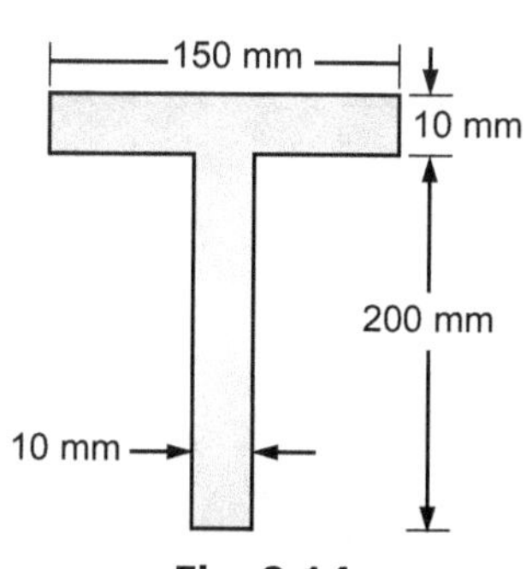

Fig. 9.14

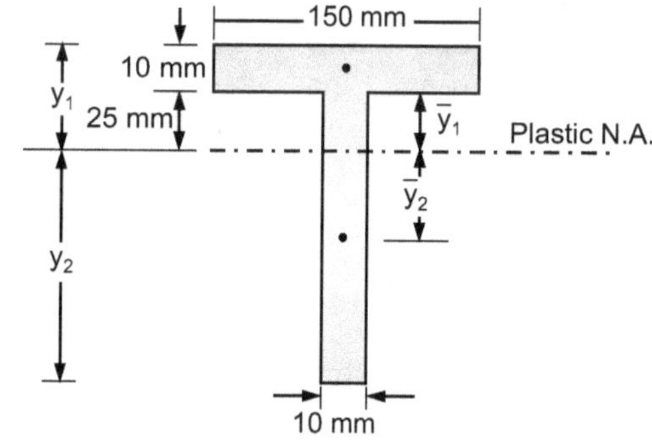

Fig. 9.15

$$A_2 = 10 \times y_2 = 1750$$

$$\therefore \quad y_2 = 175 \text{ mm}$$

$$y_1 = \text{Depth of T section} - y_2 = 210 - 175 = 35 \text{ mm (Fig. 9.16)}$$

Plastic modulus

$$Z_p = \frac{A}{2}\left(\bar{y}_1 + \bar{y}_2\right)$$

$$\bar{y}_1 = \frac{150 \times 10\left(25 + \frac{10}{2}\right) + 10 \times 25 \times 12.5}{150 \times 10 + 10 \times 25}$$

$$= \text{Centroid of upper half area from plastic N.A.} = 27.5 \text{ mm}$$

$$\bar{y}_2 = \frac{175}{2} = 87.5 \text{ mm}$$

$$= \text{Centroid of lower half area}$$

$$\therefore \quad Z_p = \frac{3500}{2}(27.5 + 87.5) = 201250 \text{ mm}^3$$

Depth of centroidal axis from top flange

$$\bar{y} = \frac{150 \times 10 \times 5 + 10 \times 200 \times (100 + 10)}{150 \times 10 + 10 \times 200}$$

$$= 65 \text{ mm}$$

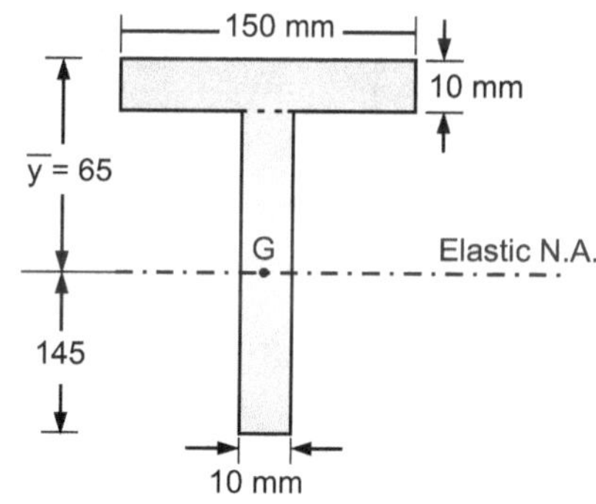

Fig. 9.16

M.I. about centroidal axis

$$I_{xx} = \frac{150 \times 10^3}{12} + 150 \times 10\left(65 - \frac{10}{2}\right)^2 + \frac{10 \times 200^3}{12} + 10 \times 200(110 - 65)^2$$

$$= 16129166.67 \text{ mm}^4$$

Elastic modulus

$$Z_e = \frac{I_{xx}}{y_{max}} = \frac{16129166.67}{145} = 111235.63 \text{ mm}^3$$

$$\therefore \quad \text{Shape factor}$$

$$S = \frac{Z_p}{Z_e} = \frac{201250}{111235.63} = 1.809$$

Ex. 9.6: *An ISNT 150 is simply supported on a span of 3 m and is subjected to a uniformly distributed load on the whole span. Find the shape factor for the section. Find the shape factor for the section. Find also the collapse load per metre run. Take the yield stress = 250 N/mm².*
$Z_{xx} = 54.6 \times 10^3$ mm³, $b_f = 150$ mm, $t_w = t_f = 10$ mm.

Sol.: Plastic neutral axis divides the section into two equal areas. Since area of flange is more than web area, the plastic neutral axis lies within the flange. Let 'n' be the depth of plastic N.A., then

$$A_1 = A_2$$

$$150 \times n = 150 (10 - n) + 140 \times 10$$

$$\therefore \quad n = 9.7 \text{ mm}$$

Taking moments of compression and tension areas about the plastic neutral axis.

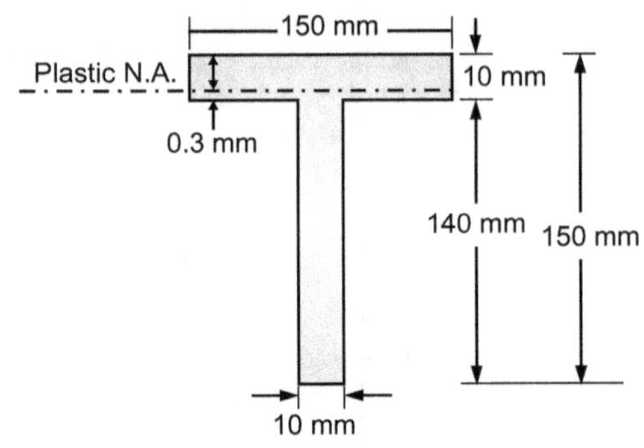

Fig. 9.17

Plastic modulus, $\quad Z_p = 150 \times 9.7 \times \dfrac{9.7}{2} + 150 \times 0.3 \times \dfrac{0.3}{2} + 10 \times 140 \times \left(\dfrac{140}{2} + 0.3\right)$

$$= 105483.5 \text{ mm}^3$$

Shape factor, $\quad S = \dfrac{Z_p}{Z_e} = \dfrac{Z_p}{Z_{xx}} = \dfrac{105483.5}{54.6 \times 10^3} = 1.93$

Let collapse load (u.d.*l.*), be w_c N/mm

Equating collapse moment (M_c) to plastic moment (M_p).

i.e. $\qquad\qquad M_c = M_p$

$$\dfrac{w_c \cdot l^2}{8} = f_y \cdot Z_p$$

$$w_c \times \dfrac{3000^2}{8} = 250 \times 105483.5$$

$$\therefore \qquad\qquad w_c = 23.44 \text{ N/mm} = 23.44 \text{ kN/m}$$

Ex. 9.7: *Determine the shape factor, plastic moment capacity and central collapse point load of the I-section shown in Fig. 9.18. Span of beam is 5 m.*

Sol.: $\qquad\qquad I_{XX} = \dfrac{BD^3 - bd^3}{12}$

$$= \dfrac{150 \times 450^3 - (150 - 9) \times 44^3}{12}$$

$$= 305306658 \text{ mm}^4$$

Section modulus, $\quad Z_e = \dfrac{I_{XX}}{y_{max}} = \dfrac{305306658}{450/2}$

$$= 1356918 \text{ mm}^3$$

Plastic modulus, $\quad Z_p =$ Sum of moments of compression and tension areas about the plastic neutral axis

$$= 2\left[150 \times 18 \left(\dfrac{450}{2} - \dfrac{18}{2}\right) + 9 \times \left(\dfrac{414}{2}\right) \times \left(\dfrac{414/2}{2}\right)\right]$$

$$= 1552041 \text{ mm}^3$$

Shape factor, $\quad S = \dfrac{Z_p}{Z_e} = \dfrac{1552041}{1356918} = 1.143$

Plastic moment capacity,

$$M_p = f_y \cdot Z_p$$

$$= 250 \times 1552041 = 388.01 \times 10^6 \text{ Nmm} = 388.01 \text{ kNm}$$

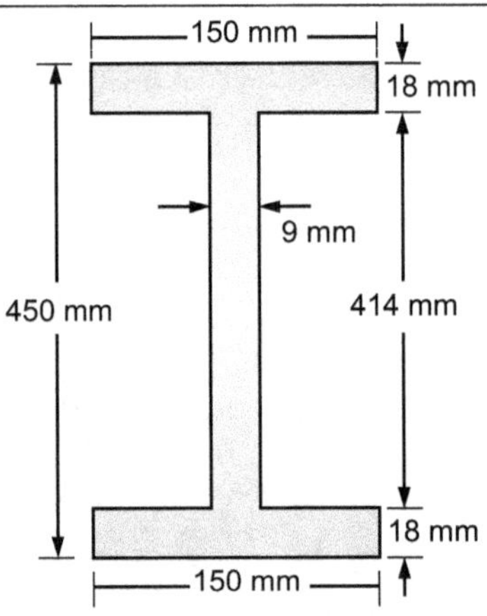

Fig. 9.18

Let central collapse point load be P_c

Collapse moment $M_c = \dfrac{P_c \cdot L}{4}$

Equating collapse moment M_c to plastic moment M_p

$$\frac{P_c \cdot L}{4} = 388.01$$

$$P_c = \frac{4 \times 388.10}{5} = 310.408 \text{ kN}$$

Ex. 9.8: *Determine the shape factor for the steel beam section shown in Fig. 9.19. Find also the plastic moment of resistance and uniformly distributed collapse load if the span of the beam distributed collapse load if the span of the bream is 4 m. Take $f_y = 250$ N/mm^2.*

Sol.:   External dimension $B \times D = 150 \times 300$ mm

Internal dimension $b \times d = 130 \times 280$ mm

∴   M.I. about centroidal xx axis

$$I_{xx} = \frac{BD^3 - bd^3}{12}$$

$$= \frac{150 \times 300^3 - 130 \times 280^3}{12}$$

$$= 99686667 \text{ mm}^4$$

Section modulus, Z_e

$$Z_e = \frac{I_{xx}}{y_{max}} = \frac{99686667}{\dfrac{300}{2}} = 664578 \text{ mm}^3$$

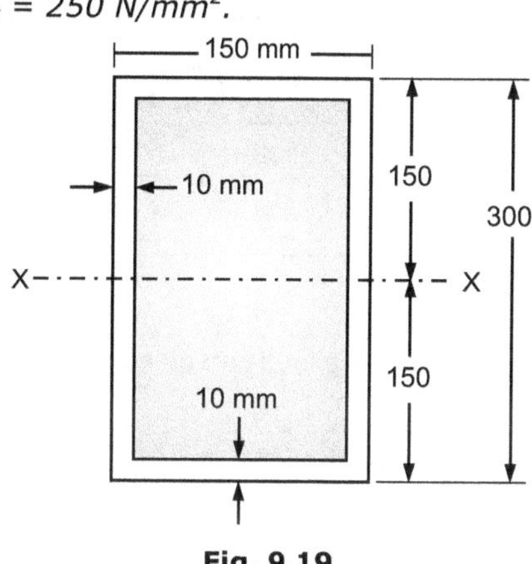

Fig. 9.19

Plastic modulus, Z_p

$$Z_p = \text{Sum of moments of areas about plastic N.A.}$$

$$= 2\left(150 \times 10 \times 145 + 4\left(10 \times 140 \times \frac{140}{2}\right)\right) = 827000 \text{ mm}^3$$

$$\text{Shape factor} = \frac{Z_p}{Z_e} = \frac{827000}{664578} = 1.244$$

Plastic moment of resistance

$$M_p = f_y \cdot Z_p = 250 \times 827000 = 206.75 \times 10^6 \text{ Nmm} = 206.75 \text{ kNm}$$

Let collapse u.d.l. be w_c kN/m

Equating collapse moment to plastic moment

i.e. $M_c = M_p$

$$\frac{w_c \cdot L^2}{8} = M_p$$

$$w_c = \frac{8\,M_p}{L^2} = \frac{8 \times 206.75}{4^2} = 103.375 \text{ kN/m}$$

Important Points

- The methods of analysis and design of steel structures are: Elastic method, plastic method and advanced methods.

- Steel possesses reserved strength beyond yield strength. This reserved strength is utilized in plastic method of design.

- In idealized stress-strain curve used for plastic analysis upper and lower yield points are regarded as one point and effect of strain hardening is ignored.

- When the section enters in plastic range and due to increase in load at certain stage the entire section becomes fully plastic and simply and simply rotates under constant moment, then this maximum moment of resistance is called plastic moment M_p.

- The portion of the member near the critical section, simply rotates at constant moment M_p is called plastic hinge.

- Plastic moment $M_p = f_y \cdot Z_p = f_y \cdot \Sigma y \cdot dA$

- Plastic modulus $Z_p = \Sigma \, y \cdot dA$

- The plastic neutral axis is situated where the cross-sectional area is equally divided into two parts.

- Plastic modulus, $Z_p = \dfrac{A}{2}(y_1 + y_2)$

- The ratio of plastic moment to elastic moment is called shape factor.

$$S = \frac{M_p}{M_y} = \frac{f_y \cdot Z_p}{f_y \cdot Z_e} = \frac{Z_p}{Z_e}$$

Section	Z_p	Z_e	Shape factors
Rectangle	$\dfrac{bd^2}{4}$	$\dfrac{bd^2}{6}$	1.5
Circle	$\dfrac{4}{3}R^3$	$\dfrac{\pi}{4}R^3$	1.6977

- Collapse load can be calculated by equating collapse moment (M_c) to plastic moment (M_p)

Practice Questions

1. Define the following:
   (i) plastic moment, (ii) plastic hinge, (iii) plastic modulus, (iv) shape factor.

2. State how collapse load in found.

3. Determine the plastic modulus, shape factor and plastic moment for a rectangle 20 mm × 300 mm if f_y = 250 Mpa.
   (**Ans.** $Z_p = 4.5 \times 10^6$ mm³, S = 1.5, M_p = 112.5 kNm)

4. A steel rod 40 mm in diameter spans over a length of 2 m. Find the collapse point load to be applied at the centre.
   (**Ans.** P = 5.33 kN)

5. A square rod of steel is 100 mm × 100 mm. Find the point load that has to applied at centre of rod of span 0.80 m.
   (**Ans.** P = 312.50 kN)

6. Determine the plastic modulus, shape factor and plastic moment about the major axis of bending for a channel section of total depth 400 mm, width of flanges 200 mm, thickness of web and flange 8 mm and 15 mm respectively.

7. For an ISMB 300, calculate the shape factor about XX axis with root radius omitted.
   Take f_y = 250 MPa and for ISMB 300, width of flange b = 140 mm, t_f = 13.1 mm, t_w = 7.7 mm.

8. Determine the shape factor for an I section consisting of 8 mm thick web and 12 mm thick flanges. The depth of web excluding flanges is 300 mm. The width of flanges is 120 mm.

9. Determine the shape factor for a channel section having flange width 100 mm, the depth of web excluding flanges 200 mm, t_w = 8 mm and t_f = 12 mm.

10. Determine the shape factors for a Tee-section consisting of 80 mm × 10 mm flange and a web of 8 mm × 80 mm attached to it. The total depth of section is 90 mm.

■■■